Qualitative Mathemati
Social Sciences

In this book Lee Rudolph brings together international contributors who combine psychological and mathematical perspectives to analyse how qualitative mathematics can be used to create models of social and psychological processes. Bridging the gap between the fields with an imaginative and stimulating collection of contributed chapters, the volume updates the current research on the subject, which until now has been rather limited, focussing largely on the use of statistics.

Qualitative Mathematics for the Social Sciences contains a variety of useful illustrative figures, introducing readers from the social sciences to the rich contribution that modern mathematics has made to our knowledge of logic, structures, and dynamic systems. A beguiling array of conceptual systems, topological models and diagrammatic reasoning is discussed, transcending the application of statistics and bringing a fresh perspective to the study of social representations.

The wide selection of qualitative mathematical methodologies discussed in this volume will be hugely valuable to higher-level undergraduate and postgraduate students of psychology, sociology and mathematics. It will also be useful for researchers, academics and professionals from the social sciences who want a firmer grasp on the use of qualitative mathematics.

Lee Rudolph is professor in the Department of Mathematics and Computer Science at Clark University, USA. In his recent work, he has explored applications of topology to robotics and to social and cultural psychology.

Contributors: Ralph Abraham, Jerome R. Busemeyer, Rainer Diriwächter, Damir D. Dzhafarov, Ehtibar N. Dzhafarov, Dan Friedman, Akio Kawauchi, Yair Neuman, Che Tat Ng, Alexander Poddiakov, Lee Rudolph, Jaan Valsiner, Paul Viotti.

The series **Cultural Dynamics of Social Representation** is dedicated to bringing the scholarly reader new ways of representing human lives in the contemporary social sciences. It is a part of a new direction – cultural psychology – that has emerged at the intersection of developmental, dynamic and social psychologies, anthropology, education, and sociology. It aims to provide cutting-edge examinations of global social processes, which for every country are becoming increasingly multi-cultural; the world is becoming one 'global village', with the corresponding need to know how different parts of that 'village' function. Therefore, social sciences need new ways of considering how to study human lives in their globalizing contexts. The focus of this series is the social representation of people, communities, and – last but not least – the social sciences themselves.

In this series

Symbolic Transformation: The Mind in Movement through Culture and Society
Edited by Brady Wagoner

Trust and Conflict: Representation, Culture and Dialogue
Edited by Ivana Marková and Alex Gillespie

Social Representations in the 'Social Arena'
Edited by Annamaria Silvana de Rosa

Qualitative Mathematics for the Social Sciences: Mathematical Models for Research on Cultural Dynamics
Edited by Lee Rudolph

Qualitative Mathematics for the Social Sciences

Mathematical models for research on cultural dynamics

Edited by Lee Rudolph

Routledge
Taylor & Francis Group

LONDON AND NEW YORK

First published 2013
by Routledge
2 Park Square, Milton Park, Abingdon, Oxfordshire OX14 4RN

Simultaneously published in the USA and Canada
by Routledge
711 Third Avenue, New York, NY 10017

First issued in paperback 2014
Routledge is an imprint of the Taylor and Francis Group,
an informa business

Publisher's Note
This book has been prepared from camera-ready copy provided
by the editor.

Trademark notice: Product or corporate names may be trademarks or
registered trademarks, and are used only for identification and
explanation without intent to infringe.

British Library Cataloguing in Publication Data
A catalogue record for this book is available from the British Library

Library of Congress Cataloging in Publication Data
Qualitative mathematics for the social sciences: Mathematical models
for research in cultural dynamics / edited by Lee Rudolph.
    xi, 465 p.: ill.; 25 cm
    1. Psychology—Qualitative research.   2. Research—Mathematical
models.   I. Rudolph, Lee, 1948–.
    BF76.5.M3748 2012
    150.1/5118

ISBN 978-0-415-44482-8 (hbk)
ISBN 978-1-138-80852-2 (pbk)
ISBN 978-0-203-10080-6 (ebk)

Typeset with LATEX2e in Palatino by Lee Rudolph

Contents

Series Editor's Preface

Jaan Valsiner

This book focuses on complex formal mathematical models that are—as the reader will see—far beyond the comfort level of most psychologists who are well versed in statistics. It lets the reader struggle through various systems of qualitative mathematics and wrestle with formal models. Without doubt—this book is for a mathematically savvy reader.

Mathematics is important for psychology. Breakthroughs in basic research often happen at times when scientists from one field enter another and address its relevant questions in ways that the "insiders" have not managed to master. As an example, the influx of physicists into genetics after World War II has led to revolutionary breakthroughs in the field. This is well captured by the story of James Watson and Francis Crick discovering the structure of DNA in the early 1950s—both were outsiders to the field, yet their perspective allowed them to solve the problem in tough competition with established research laboratories. In a similar vein, psychology in the 21st century needs new "outsiders" to enter the field. It needs new formal approaches. The time of reliance upon the pathways of inductive generalization—supported by statistical techniques—is over. It is the deductive primacy of mathematical modeling—especially that by contemporary qualitative mathematics—that needs to find its way into psychology.

The study of social representation is the first beneficiary of any introduction of qualitative mathematics into psychology. It is the complexity—both structural and dynamic—of such representation that cannot be reduced to its constituent elements. The mathematical systems included in this book—quantum probability, knot theory, dynamical systems, topology, etc.—are relatively little known among psychologists. I hope this book makes them more familiar with the readership in the social sciences.

The idea of putting together this kind of a volume emerged in the intellectual discussions in the "K-Group" at Clark University over the past decade. It was in this intellectually vibrant environment where the goal of bringing qualitative mathematics to psychology became formulated. Its leading force was—and is— the creative and poetic Editor of this volume, Lee Rudolph. Without his wisdom and readiness to tell psychologists that from the mathematical standpoint reliance upon statistics is not sufficient for science, this book would have been impossible. Having spent the last decade of his active academic life in Clark University Psychology Department has informed him well about the problem psychologists have—fear to think with the help of creative formal tools. For Lee's mind of a poet—when he does not do mathematics he writes poetry—psychologists' ways of thinking about tools for thought seemed incredibly narrow. Such tools are all around —yet psychology's mathematical preparation remains limited to high-school calculus and normatively taught 'cookbook' statistics. In the latter, mathematical thinking becomes substituted by loyalty to "the scientific method" that is accepted in terms of loyalty rather than rational analysis. A science that follows an orthodoxy is at best a "normal science" in Thomas Kuhn's terms— and a sect of true believers, at the worst. Hopefully the present volume leads its readers to inquire further into the complex realm of modeling of complicated social and psychological processes.

Worcester, February 2012   Jaan Valsiner

Preface

Lee Rudolph

This book brings together chapters by mathematicians—Ralph Abraham (in collaboration with Dan Friedman, an economist, and Paul Viotti, a political scientist), Damir Dzhafarov (in collaboration with Ehtibar Dzhafarov, a psychologist), Akio Kawauchi, and Che Tat Ng —alongside chapters by social scientists: Jerome Busemeyer, Rainer Diriwächter, Ehtibar Dzhafarov, Yair Neuman, and Alexander Poddiakov and Jaan Valsiner. I am grateful to all these authors for their cooperation in this project, and for the opportunities they have afforded me to continue thinking hard—and, I hope, fruitfully—about the nature and role of mathematical modeling in the social sciences. I am especially grateful to Jaan Valsiner, for instigating this project and helping determine its general parameters, as laid out in our introductory chapter.

What is logic? What is structure? What is dynamics? And how do logic, structure, and dynamics interact and impinge upon one another in the course of the human process of studying human processes, that is, of social science broadly construed? These questions are all open-ended, they are all obviously (to me) important— and to none of them does mathematics supply a single, neat, cut-and-dried, off-the-shelf answer. On the contrary: there are *many* logics (more or less mathematical), *many* notions of structure (ditto), and at least a few different (though related) mathematical theories of "dynamical systems". This book is an introduction for social scientists to some mathematics, likely new to them, that they may find useful and inspiring of new perspectives on their lifework.

Although the contributed chapters have been parceled out into three parts (one each with *Logic*, *Structure*, and *Dynamics* in its title), I found while editing that the chapters delivered by the contributing authors mostly turned out *a posteriori* to be deeply involved in at least two, sometimes all three, of those *a priori* divisions. Confronted by this minor chaos, I decided to embrace it, with several consequences.

- In editorial footnotes, I have drawn attention to odd angles and unexpected interconnections both among the contributed chapters and with other parts of the literature.
- For each part, I have written a prefatory essay surveying that part's nominal topic, from the viewpoints of mathematics and of the human sciences.
- In lieu of general conclusions, I have included an addendum on diagrammatic reasoning inspired by the variety of ways in which such reasoning was (or wasn't) made use of in the contributed chapters.
- An extensive Index of Notations and Concepts provides further scaffolding to assist readers with different backgrounds in explorations of chapters on subjects less familiar to them.

As I view things now, the process of reading and editing the authors' contributions has been only the first part—the setting of initial conditions—of a continuing, dynamic process that I turn over, hopefully, to the readers for them to steer (and be steered by) as they can.

Conventions, notations, etc.

Most of the following conventions will be familiar to mathematicians, but perhaps not to social scientists. A few are idiosyncratic.

- The general rule for *slanted roman type* is to use it only to emphasize the first use in the text of a word or phrase that is being, or is about to be, formally defined (implicitly or explicitly), or is the English translation of a standard term (or abbreviation) from another language. If I have taken sufficient care, the preceding sentence will be the sole exception to the rule it states.
- *Italics*, **boldface**, SMALL CAPITALS, and underlined type indicate various kinds of emphasis. Quoted matter in a foreign language is also generally set in italics. Sans-serif type is used to set the names of various sorts of abstractions. A typewriter font is used to set off `program code`, URLs, and the like.
- Like Raymond (2003, p. 14), this book uses "logical or 'British'-style quoting in accordance with established hacker custom," a custom that hackers inherited from mathematicians in general and mathematical logicians in particular; in particular,

whenever "attributed quoted material" (Rudolph, 2008, p. 145) from a cited source is indicated as such by being placed inside a pair of quotation marks, any terminal punctuation mark (like the comma terminating the quotation from Raymond, 2003) belongs to the quoted matter.

- Besides the use just mentioned, double quotation marks are also used in places to set off an English translation of a mathematical formula (see examples below).
- Single quotation marks variously indicate
  - a phrase that 'everybody knows',
  - an informal 'user-friendly' colloquialism or nonce definition like 'noncinition',
  - the rhetorical device often dismissed as 'mere scare quotes' (context should make it clear which case applies), and also
    - when <u>mentioning</u> a mathematical symbol rather than <u>using</u> it (*cf.* Quine, 1940, p. 23 *ff.*; see examples below).

  Raymond (*ibid.*) calls this last usage "philosophers' quotes".
- Displayed equations (and similar mathematical content) are labeled with numbers if (and only if) they are referred to elsewhere in the text; *e.g.*, see equation (0.1), below.
- The symbol '□' indicates either the end of an explicitly indicated mathematical proof, or (at the end of the statement of a mathematical theorem) the omission of a proof.

Throughout, '$\mathbb{N}$', '$\mathbb{Z}$', '$\mathbb{R}$', and '$\mathbb{C}$' stand for the sets of counting numbers (positive integers), integers, real numbers, and complex numbers, respectively. The empty set is denoted by '$\varnothing$'. The symbol '$\in$' denotes set membership; *e.g.*, $0 \in \mathbb{Z}$ means "0 is a member (or element) of the set of integers". The symbol '$\subset$' denotes set inclusion; *e.g.*, $\mathbb{N} \subset \mathbb{Z}$ means "the set of counting numbers is a subset of the set of integers". 'Set-building' notation of the form $\{x : P(x)\}$ is, as usual, read as "the set of all x such that $P(x)$ is true", where "$P(x)$" is a proposition about x; variants of this notation can be used, *e.g.*, $\{x \in \mathbb{R} : P(x)\}$. The *assignment symbols* ':=' and '=:' are often used, as in equation (0.1), to state definitions symbolically: the expression on the same side of '=' as ':' is defined to *mean* (*i.e.*, to be an abbreviation for) the expression on the other side.) Bourbaki notation is used for bounded real intervals, *i.e.*,

$$[a,b] := \{x \in \mathbb{R} : a \leq x \leq b\}, \quad]a,b] := \{x \in \mathbb{R} : a < x \leq b\},$$

$$[a,b[:= \{x \in \mathbb{R} : a \leq x < b\}, \quad]a,b[:= \{x \in \mathbb{R} : a < x < b\}$$

for any real numbers a and b (the last of these leaves (a,b), sometimes used for $]a,b[$, free to denote unambiguously the *ordered pair* that has a as its first member and b as its second). The unbounded interval $\{x \in \mathbb{R}: a \leq x\}$ may be denoted $[a, \infty[$ or $\mathbb{R}_{\geq a}$.

The symbols '$\vee$', '$\wedge$', and '$\sim$' stand, respectively, for the operations of *disjunction* ("exclusive or"), *conjunction* ("and"), and *negation* ("not") in Boolean algebra. Disjunction and conjunction are closely related to the operations *union* (denoted by '$\cup$') and *intersection* (denoted by '$\cap$') on sets: in fact, for any sets X and Y,

$$X \cup Y := \{z: z \in X \vee z \in Y\}, \quad \{z: z \in X \wedge z \in Y\} =: X \cap Y. \qquad (0.1)$$

The *Cartesian product* of X and Y is the set

$$X \times Y := \{(x,y): x \in X \wedge y \in Y\}$$

of all ordered pairs (x,y) in which x is an element of X and y an element of Y. The symbols '$\Rightarrow$' and '$\Longleftrightarrow$' indicate, respectively, *logical implication*[1] and logical equivalence. Notations used in only one chapter are defined where introduced.

Tables, figures, numbered equations, and footnotes are numbered in (separate) sequence throughout each chapter; editorial footnotes are distinguished from authorial footnotes by setting their numerical markers in italics.[2]

Acknowledgments

During the course of some of the preparatory work for this book, Lee Rudolph was Principal Investigator of National Science Foundation award DMS-0308894, an Interdisciplinary Grant in the Mathematical Sciences. The objective of the IGMS program (now discontinued) was to "enable mathematical scientists to undertake research and study in another discipline", with the "expected outcome" being "sufficient familiarity with another discipline so as to open opportunities for effective collaboration by the mathematical scientist with researchers in another discipline" (National Science Foundation, 2001). Twelve months' full support from that grant (and later partial support from NSF Award IIS-0713335) is gratefully acknowledged.

---

[1]Specifically, 'material implication': $(P \Rightarrow Q) \Longleftrightarrow \sim(P \wedge \sim Q)$.

[2][By editorial fiat, note 1 is authorial and this note is editorial. (Ed.)]

References

National Science Foundation. (2001). Interdisciplinary Grants in the Mathematical Sciences (IGMS): Program Solicitation NSF 01-115. Retrieved September 13, 2011, from `http://www.nsf.gov/pubs/2001/nsf01115/nsf01115.htm`

Quine, W. V. O. (1940). *Mathematical Logic*. New York, NY: W. W. Norton.

Raymond, E. S. (2003). *The Art of Unix Programming*. Boston, MA: Addison Wesley.

Rudolph, L. (2008). The finite-dimensional Freeman Thesis. *Integrative Psychological and Behavioral Science*, 42(2), 144–152.

1

Introduction: Mathematical Models and Social Representation

Lee Rudolph and Jaan Valsiner

> Data are the natural enemy of hypotheses.
> (attributed to Goethe)

The research program of social representations has been based on holistic phenomena (Moscovici, 1961, 1976, 1988, 2001) that capture a blueprint of complexity and present it in an abstracted form. Yet, given the complexity of the phenomena, the abstracted forms of social representing—activity—rather than representation—its product—cannot be reductionist. It is impossible to reduce a complex object of social representing—such as patriotism, love, trust—to a combination of three, five, 160, or any other number of various 'factors' that are conceptualized as 'independent' from one another. Or, likewise —the creative construction of ever-new symbolic forms (Cassirer, 1929/1957; Wagoner, 2009) or iconic tactics of representing (de Rosa, 2007, 2012) make the process of social representing highly volatile, yet deeply human. While living in the particular moment that is never to return we create images of stability that we present as eternal! This is probably the greatest fiction one can create: yet other human beings willingly accept them, and use them for organizing their own here-and-now lives.

Keywords: complexity, development, dynamic phenomena, evolutionary epistemology, Gestalt, hermeneutic circle, mathematical modeling, meaning-making, measurement, mesogenesis, microgenesis, numbers, ontogenesis, patterns, qualitative mathematics, qualitative psychology, social representation, statistics, structures.

How is such sharing of illusions possible? The static images—frames for organizing the dynamic flow of living experience—are encoded into abstractions that by way of generalization can be transposed to new circumstances. Thus (to remind us of the first object of the study of social representation, Moscovici, 1961) the abstracted framework of confessing that had been put into place in France over centuries could be transposed from religious to therapeutic contexts, and set up the framework for psychoanalysis in the 20th century. Yet such abstracting, generalizing, and transpositioning was not the result of a single mind, but a result of a collective effort by people of a large number of generations. The continuity of the act—confessing—over generations leads to abstraction and generalization of the form of conduct in ways encoded in extra-individual cultural spaces: in texts, catechisms, forms of furniture (confessionals in churches), parents' demands of the children, public execution of persons convicted as 'witches' or as 'spies', etc.

The act of social representation starts from individual affective minds, becomes encoded in redundant forms in the social material world, and is constantly re-constructed by other individuals who build their symbolic resources (Zittoun, 2007) on the basis of such social worlds. Every act of our individual conduct in a social world participates in the construction of social representation of our strivings, our ways of living. Of course such multiplicity—social representing being present everywhere, and therefore nowhere—creates a special difficulty for both the theory of social representation and its empirical applications. From the viewpoint of contemporary 'mainstream' psychology, which is naïvely dedicated to the 'measurement' of characteristics that are postulated to exist in the person or in the environment, social representations are ephemeral and vague phenomena that can be accessed only through some tangible ('measurable') medium.

Much of the social representation perspective has found that medium in the meanings of language terms. However, meanings in language can be studied in different ways: in relation to other meanings (*e.g.*, structure of meaning system, elaborated by techniques of cluster analysis, topological analysis of nomological networks, etc.), or in relation to the process of making sense. The latter entails modeling of the phenomena through semiotic mediating systems, primary (language) and secondary (iconic, indexical signs). Modeling entails abstraction and generalization.

Modeling as meaning-making, meaning-making as modeling

Mathematical modeling of social representation phenomena brings the topic of this book to a wider issue relevant for the social sciences as a whole: **what kind of mathematics is appropriate to use in modeling complex dynamic wholes?** Aside from social representation almost all other human activities are filled with such constructed wholes (Diriwächter & Valsiner, 2008). Psychology's appropriation of statistics as a tool for arrival at general knowledge through inductive generalization has been demonstrably a social artifact (Gigerenzer et al., 1989). The turning of statistics from a simple tool into a criterion of "scientific purity"(Porter, 2003) is an act of symbolic construction of a powerful myth that continues to organize the social practices in the discipline. Yet popularity of tools and consensual validation of their value does not guarantee their productivity. We claim in this book that psychology's present knowledge base of mathematics is limited and outdated. Psychology needs to widen —in an experientially experimental fashion, open to adaptation and qualified adoption (or rejection) of new tools and ways of using them —its consideration of mathematical systems for its efforts to arrive at generalizable knowledge. The era of post-modernist empiricism, be it quantitative or qualitative in kind, is over (Valsiner, 2009). It is time for psychology to widen its intellectual horizons.

This Introduction is an attempt to make sense of which particular versions of mathematical modeling might fit in the social sciences, both as it has been and as it might be. Obviously we cannot cover the whole area of contemporary mathematics in a book like this; thus, admittedly, our selection of perspectives remains limited. Mathematicians, given their full dedication to the deductive style of reasoning, have no easy time in meeting psychologists, whose almost completely inductive knowledge construction credo creates deep divides. Yet we give a try at overcoming that difference.

Our perspective brings two histories—those of psychology and mathematics—to bear upon the issue of how to model complex psychological phenomena. This merging has an interesting difficulty: while psychology has largely denied, or partially presented, features of its history, mathematics has preserved its history in the present. While a psychologist may easily dismiss a theoretical idea from the 1950s as 'old' or 'outdated', a mathematician cannot do the same when referring to Pythagoras (dated to 500 B.C.E.) or Ancient Indian mathematicians. All the knowledge base of mathematics—once established—stays as such (Dauben, 2003; Zeeman, 1966). Of course there are fads and fashions in mathematics that dominate in one or another historical period, but the systematic models stay. Or, if they change, they may incorporate previous models into themselves as a special case (*e.g.*, Riemannian geometry simultaneously incorporating both its Euclidean predecessor and the non-Euclidean, actually anti-Euclidean, Bolyai–Lobachevsky geometry).

Conceptual perils of complexity

The problem of psychology from its early days of autonomy has been to try to make sense of phenomena from the wrong side —that of elements which might accumulate into wholes—rather than from the side of wholes (which may differentiate to show the presence of their parts). The urge to reduce all complexity to artificially constructed simplicity stays until our times and proceeds further, for instance:

> A measure of pleasures [...] is unattainable. A sum of pleasures [...] is not itself a pleasure [...], a difference of pleasures is not a pleasure, and a "balance of pleasures over pains" is unmeaning. All these are properties of intensive quantities generally, and reveal the fundamental impossibility of a Calculus of intensive quantities. For a strict Hedonist, the problem of weighing two small pleasures against a big one ought to be meaningless, since an aggregate of two pleasures does not form a single quantity of pleasure. (Russell, 1897, p. 335)

Psychology's modus operandi has been to turn any kind of phenomenon, simple or complex, into a sign designated by real numbers: a practice that has been noted as at least an "unresolved conceptual problem" in psychology (Luce, 1997) and perhaps an impasse for arriving at general psychological knowledge (Rudolph, 2006a; see also Chap. 10). Instead of granting 'objectivity' to psychology by the socially legitimized practice of 'assigning numbers', it may be useful to look into what contemporary mathematics can offer to model precisely the kinds of complex phenomena—

feelings, social representations, decision making under uncertainty, etc.—that our everyday world brings to psychologists' laboratory doorsteps and abandons there, hoping psychology can take care of them. Yet it cannot: as it has not considered the philosophical underpinnings of what quality and quantity are, and how they are related. It would be easy to see quantifiable phenomena in psychology as examples of quality, but not vice versa (Bozzi, 1969).

At times, psychology has tried to handle complexity (see Diriwächter, Chap. 7, this volume). Of course there were efforts to start from the whole in psychology's history. The 'Gestalt perspectives' of the late 19th century[1] did insist on the wholeness of the psychological qualities. Their starting point was an

> attempt to answer a question: what is melody? The most
> obvious answer: the sum of the individual tones which
> make up the melody. But opposed to this is the fact
> that the same melody may be made up of quite different
> groups of tones, as happens when the same melody
> is transposed into different keys. If the melody were
> nothing other than the sum of the tones, then we would
> have to have here different melodies, since different
> groups of tones are involved. (Ehrenfels, 1988c, p. 121)

Thus, we are living through and experiencing integrated wholes—music, speech, activity contexts, ornaments that surround us, etc.—rather than within a flow of non-integrated, unexperienced elementary 'stimuli'. Yet what is (usually) available to our analytic schemes are tools to break down the wholes into their elementary constituents (of great salience, if not individuated meaning or significance, to the psychologist undertaking the analysis)—thus losing precisely the qualities of the phenomena that are central and significant to the human subject (who may be that same psychologist) for whom those phenomena are, exactly, experiences. Psychology has been hindered by that ambivalence—analyzing into elements while knowing that it is not the elements, but the wholes to which they belong, that matter—since its independent status in the 1870s.

---

[1] Aside from Ehrenfels, the participants in the discourse on the holistic nature of psychological phenomena included Carl Stumpf, Alexius Meinong, Hans Cornelius, Felix Krueger, and others. The theme of the whole not being reducible to its parts was central to scholarly discussions at the turn of the 20th century.

As patterns of generalized kind, Gestalt qualities are the basis for innovation. The process of completion of the Gestalt is always open-ended (as the person faces the uncertainty of the impending future) and hence calls for free generation by the creative activity of imagination:

> The mind that organizes psychical elements into new combinations does more than merely displace the component elements amongst themselves: he creates something new. [*Der Geist, welcher psychische Elemente in neue Verbindungen bringt, ändert hierdurch mehr als Kombinationen; er schafft Neues.*] (Ehrenfels, 1988b, p. 109; Ehrenfels, 1988d, p. 149)

Thus Gestalt qualities are synthesized by the person—the *"Gestaltmacher"* (Gestalt-maker; Stern, 1935, p. 153): they are not ontological givens. They are in a constant process of being reconstructed anew through the relation of the psyche with the world.[2] Imagination —fantasy—plays the role of synthesizer (*"schöpferische Tätigkeit der Phantasie"*, Ehrenfels, 1988d, p. 149).[3]

The result of such creativity was the recognition of emergence of Gestalt qualities of "higher order" (Ehrenfels, 1916, 1988a): new qualities that may defy verbal description, yet operate precisely in our relations with our environments. Thus, we may recognize the composer of a melody heard for the first time—obviously by way of some generalized 'similarity' between the new tune and others heard before; yet we cannot explain how we do it. Or we construct (*e.g.*, by building or simulating a 'neural network' machine, and appropriately 'training' it) an explanation that explains nothing: to observe that two complex systems, given the same 'input', produce 'outputs' that are very similar (by some measure), can be of interest and quite suggestive, but is very weak evidence that their 'throughputs' (which are precisely their 'hows') are at all alike.

---

[2]Over a century later, within the socio-cultural perspective, a similar focus is brought into science by the concept of *actuations* (Rosa, 2007).

[3]We might, for instance, imagine that other *kinds* of "numbers"—complex numbers, *p*-adic numbers, what have you—could be as useful as 'real' numbers. It is notable that later in his work, Ehrenfels (1922) made a contribution to number theory: a book explicating the Sieve of Eratosthenes and the Prime Number Theorem from a Gestalt perspective. Complex numbers are a necessary component of quantum probability (see Chap. 3, particularly p. 98 *ff*); *p*-adic numbers may provide new insights into dynamical systems (Kennison, 2002; Khrennikov, 2000).

Gestalt level and Gestalt purity

Together with the emergence of qualitatively higher forms of *Gestalten* comes the question of their maintenance and dissipation. The hierarchy of Gestalt qualities could be tested by how they preserve interventions that might eliminate them—how enduring are the particular level of Gestalt qualities:

> A rose is a Gestalt of higher level than a heap of sand: this we recognize just as immediately as that red is a fuller, more lively color than grey [...]. *For a fixed degree of multiplicity of parts, those Gestalten are the higher which embrace a greater multiplicity of parts.* [...] One imagines the given Gestalten (a rose, a heap of sand) to be subject to gradual, accidental and irregular interventions. Whichever of the two Gestalten thereby survives the wider spectrum of changes is of the higher level. (Ehrenfels, 1988a, p. 118; emphasis added)

The resistance to dissipation is thus the proof of the higher order nature of Gestalt qualities. Yet Ehrenfels used a quantitative criterion, "greater multiplicity of parts", to determine which of the Gestalts is of higher level. The general idea of Gestalts being of different levels is in line with the notion of flexible nature of forms: all organismic forms exist as inherently transforming themselves, or as adaptable to external demands. The Gestalt-maker leads the re-construction of the multitude of forms. The crucial feature of the focus on forms is their fluidity—flexibility for transformations into new Gestalten.

Aside from transformation of the Gestalt level, there exists a 'pure form' of the Gestalt, an ideal case. As Ehrenfels described,

> [...] ideal forms of the mathematically exact sphere and of the regular polyhedra are Gestalten of maximal purity, *i.e.*, it is not even logically possible for the purity to be surpassed, but they are of relatively low Gestalt level (Ehrenfels, 1988a, p. 119)

Thus, holistic simplicity (of an ideal sphere) guarantees "purity" but reduces the "level". Writing of development, Ehrenfels saw the opposite—increase in level and decrease in "purity":

> The process of ontogenetic development of the organism from seed to full maturity reveals—at least insofar as it is visible for us—an ascent in level, bound up with *a decline of purity of formedness, the latter brought about by the relatively chaotic effects of the environment.* (*ibid.*, emphasis added)

Gestalt level and purity are values in themselves. The ideals of form are the extreme—infinite—conditions towards which we strive, but which we cannot reach as the circumstances diminish our efforts. Here Ehrenfels brings in value criteria, where "purity" becomes central.[4] In contrast, in the domain of practical uses of the Gestalt ideas (in various versions of Gestalt therapies) it is the Gestalt level that is expected to vary and transform the person.

Mathematics and modeling of wholes

Mathematics has been described as "the study of pattern" (Whitehead, 1941, pp. 674, 680) and (not necessarily more ambitiously) as "the science of patterns" (Devlin, 1997; Resnik, 1997; Steen, 1988). It is usual and natural for humans to perceive patterns, even patterns that are not 'really there'; our perceptual systems create "perceptual illusions" of wholes from configurations of points, corners (Kanizsa, 1969) or sounds (Benussi, 1913).[5] Furthermore, the human mind can contemplate objects that do not exist—a "round triangle" is an example that has fascinated thinkers since the 1880s when Alexius Meinong attempted to understand the nature of such objects in his *Gegenstandstheorie* (Meinong, 1907, *passim*; 1915, p. 14).

In this connection, the traditional use of the word "illusion" is tendentious; it can be disputed along the following lines (see also Carini, 2007). Start with the axiom that a 'whole' that is perceived is *ipso facto* 'correctly' perceived. Then, for the person who perceives a whole, what is—or may be—'illusory' is not the whole: it is the

---

[4]Not surprisingly, the "purity" focus led Ehrenfels to the acceptance of basic eugenic notions of 'purification of the society' that of course became grossly socially abused in the politics of the 20th century.

[5]The phrases "perceptual illusion" and "illusion of perception" burst into English psychological literature *circa* 1880 (Sully, 1880, 1881; Montgomery, 1882); "optical illusion"'s had been discussed much earlier (Dunn, 1762). In German, *Täuschung der Wahrnehmung* is attested as early as Tennemann (1798).

felt need or imposed demand to identify the perceptually present and correctly **perceived whole** as something else, namely, a certain **unperceived whole** that is perceptually and physically absent from the present situation of the perceiver (and might even be physically absent from the entire universe, past, present, and future— if, say, it is a "round triangle"). Contrariwise, for a(nother, or the same) person (perhaps a psychologist) who is observing the situation, what is illusory is the conviction that the 'whole' known to the perceiver is in some manner or degree less (or more) 'real' than the 'unwhole' known to the observer, which the perceiver somehow *should and would* be perceiving—were not the universe (or the observer) somehow setting successful snares. On this view, the ascription of 'illusion' is a category error, a failure of the ascriber's (formal or informal) ontology and epistemology to adequately fit the phenomena of construction by the human mind (starting with the human perceptual system).

We don't even informally attempt to spell out here what might be required of an ontology (or epistemology) that it be "adequate" for such a purpose.[6] We do suggest that one very desirable component of "adequacy" would be the provision of a common language for all constructions of the human mind—or at least, those that arise in psychology and those that arise in mathematics. To that end, we suggest a bit of formal terminology. Take *pattern* to be defined however you like—formally, informally ("I know it when I see it"[7]), or not at all; then define a *structure* to be either a pattern, or a structure of structures (so, *e.g.*, a pattern of patterns is a structure; if *relations* are posited to be certain kinds of patterns, then relations, patterns of relations, etc., are structures; and so on)—in somewhat standard mathematical jargon, "structure" so defined is the *recursive closure* of "pattern". Making use of that definition, we propose that mathematics is not the study (or science) of patterns only, it is **the study (or science) of *structures*.**[8] Meanwhile, and conveniently, psychology (like other sciences, and other kinds of constructions of the human mind) is a richly generative **source of structures**. At the minimum, mathematics can provide 'bookkeeping services'—

---

[6] For ontology, a start on such an attempt appears in Chap. 6, based on ideas of Rudolph (2009, section 2, "Four Ranges in an Organism's Environment").

[7] As in *Jacobellis v. Ohio*, 378 U.S. 184, 197 (1964) (Stewart, J., concurring).

[8] This motto is, of course, not original with us; see Chap. 6 for some further discussion.

descriptive models—for such exogenously generated structures, independently of the bookkeepers' stance(s) on the extent to which the structures are (or have been, or will be) 'really there'.

Keeping structures in mind

Some bookkeeping (not necessarily of mathematical type) is surely necessary for undertakings such as science, given the empirical certainty that the capacity to generate structures exercised by even one human (let alone by human groups extended over space and time) vastly outstrips the ability of human groups (let alone a single human) to keep entire structures 'in mind' (*viz.*, in attention). This may be a matter not only—or not primarily—of 'too much information', but of how the 'information' is structured. It has been hypothesized (Rudolph, 2006c) that there is a quite small upper bound on a certain topologically inspired numerical characteristic, the *intrinsic dimensionality*, of any structure that can humanly be 'kept in mind'.

> *Low-Dimensional Hypothesis (weak form).* In Psychology, few if any *interesting* phenomena need more than two or three independent quantitative dimensions to describe them adequately. [...] *(strong form).* Humans do not —and perhaps cannot—apprehend a phenomenon *as* a phenomenon (unitary, and of psychological interest) unless it has an adequate quantitative description of low intrinsic dimension. (Rudolph, 2006c, p. 77)

Note that, although the "weak form" of the Low Dimensional Hypothesis is "a claim about Psychology, the practice of psychologists", the "strong form" is "a claim about psychology: how humans behave and what they find interesting" with both formal consequences for mathematical models and empirically testable consequences. Formal consequences, for two specific kinds of models, are discussed by Rudolph (2006b, 2006c, 2008). Empirical consequences (for a third kind) are discussed by Rudolph, Han, and Charles (2009) and Rudolph (2012). The large and growing recent literature on "dimensional reduction" (*e.g.*, Ek, Jaeckel, Campbell, Lawrence, & Melhuish, 2009; Roy, Gordon, & Thrun, 2005; Yoo, Zhou, & Zomaya, 2008) may also be relevant—not so much for its discussions of consequences, significance, or meaningfulness of dimensional reduction as for its being evidence that many scientists and engineers find dimensional reduction usefully applicable to a wide variety of phenomena with assuredly high intrinsic dimensionality (*e.g.*, conformations of protein molecules, configurations of many-jointed robotic appendages, or facial expressions). Presumably they find it useful because it helps their human minds better apprehend/simulate/guide/teach [to robot learners!]/... such phenomena.

Whatever the merits of any particular hypothesis about the profoundly human processes—be they schematization, Gestaltmaking, or whatever—by which structures are transformed, it is certain that structures *are* transformed/transforming continually. A problem faced by all living beings (and their social counterparts), in making sense of these continual transformations, is that structures detected/constructed here-and-now are, here and now, already in the process of transforming into new ones. Further, structures have their histories of previous transformations—which may no longer be visible. Fig. 1.1 is an impressionistic sketch of that paradoxical state of modeling—we work with materials in the transition state (no longer *A*—not yet *B*) trying to project into that zone either a model of state *A* or that of state *B*.

Thus all meaning-making efforts, including mathematical models, that are addressed to the situations of living beings (and their social counterparts) must address not what patterns and structures are 'really there' but rather the ones that are not yet there (yet might come to be): this is true *even for* 'merely' *descriptive* (often either synchronic or time-independent) models, and all the more

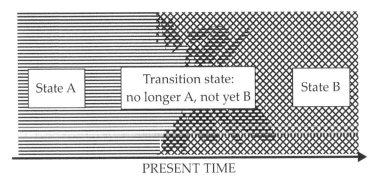

PRESENT TIME

Figure 1.1: Where is 'really there' in dynamic phenomena?

true for *predictive* and *prescriptive* (in particular, most dynamic or merely diachronic) models. As a practice or discipline of engagement with structure, mathematical modeling can enrich and liberate meaning-making—it need not (be used to) impoverish or enslave it.

Multi-level structured phenomena of development

In the history of psychology this paradox has been discussed within the tradition of *Aktualgenese* or microgenesis (Abbey & Diriwächter, 2008; Valsiner, 2005). Human immediate living experience is primarily *microgenetic*, occurring as the person faces the ever-new next time moment in the infinite sequence of irreversible time. In order to create stability of psychological kind, the person creates semiotic devices—meaning fields—that temporarily stabilize the "lurking chaos" (Boesch, 2005) of experiencing ever-new moments. Such semiotic construction is constant and overabundant—the creativity of human psyche in generating new meanings while living one's life is hyper-productive. Most of the semiotic devices created are abandoned—some before their use (intermediate Gestalts of the process of *Aktualgenese*; Valsiner & van der Veer, 2000, chapter 7), others after their use and under the conditions of no need for further use. Of course many semiotic devices—meanings—are retained over ontogenesis, and some have been retained throughout the course of human cultural history.[9]

Thus, personal cultures are tools for creating subjective stability on the background of inevitable uncertainties of experiencing. They are assisted by the collective cultural canalization of these experiences into culturally structured activity settings. Such settings operate as a *mesogenetic* organizational level of human cultural ways of being (Saada-Robert, 1994). The mesogenetic level consists of relatively repetitive situated activity frames, or settings. Thus, a situated activity context such as praying, or going to school or to a bar, or taking a shower/bath, are all recurrent frames for human action that canalize the subjective experiencing by setting up a range of possible forms for such experiencing.

Although psychologists and cognitive scientists have proposed various models of microgenesis (Catán, 1986; Flavell & Draguns,

---

[9]Architectural encoding of cultural values—through temples, public squares, military fortifications, mausolea and marketplaces—is an example of maintenance of ontogenetic formations across generations.

1957; Werner, 1956; Werner & Kaplan, 1962, etc.), including micro-genesis of mathematical development (Bickhard, 1988, 1991; Piaget, 1950; Radford, Bardini, & Sabena, 2005; Valsiner, 1987, etc.), *mathematical models* of microgenesis seem to be rare, though not entirely non-existent.[10] If this void is not a necessary one, is there some mathematics that could be useful for understanding microgenesis?

Our characterization above of microgenesis itself suggests the quite minimal mathematical system $\mathbb{N}$ of counting numbers 1, 2, 3, ..., a suggestion made explicitly if informally by von Uexküll.

> In the course of apperception, it always happens that certain moment-signs are noted especially. We call this power of noting certain moment-signs more than others, "attention." And since we describe as rhythm any regularly recurring change whatsoever, we may in the last instance refer the power to form numbers to a rhythm of the attention. [...] To our attention it is a matter of indifference towards what kind of content it is directed— whether on objects, or sensations, or feelings. As soon as there appears a regular change in the attention, it can be subjected to the rule of the simplest rhythm, *i.e.* it can be counted. (Uexküll, 1920, p. 57)

Besides formalized versions of the statements (in our characterization) about "moments"—that they are generated in "infinite sequence" by "ever-new next"ness—the Peano Postulates that mathematically characterize $\mathbb{N}$ contain two other axioms: that *counting begins* (there is a first counting number, namely 1, and that $\mathbb{N}$ *contains no more and no less than what is needed to count* (the Axiom of Mathematical Induction, which essentially says that no natural number is superfluous in the sense that it could never be reached by any conceivable idealized act of counting).

The first of these two axioms is at least implicit in microgenesis as applied to the experience of any one person as observed by a person (Self or Other). The second is less obviously true in the context of microgenesis—in fact, the sad, familiar case is that Self often observes an Other's sequence of moments come to an end

---

[10]Bickhard and Campbell (1996) give a very interesting discussion of how such a model might be constructed along topological lines, but do not give an explicit model themselves.

(after which there are no more to count), even though Self's own experience of Self seems always (or nearly always) to anticipate a "next time moment". In any case, Mathematical Induction (MI) is a nearly indispensable tool in all parts of mathematical practice,[11] and is nearly always adopted.[12]

That the counting number system apparently arises out of microgenesis does not, of course, imply that microgenesis is necessarily in the domain of the "almost unlimited applicability" of counting (Uexküll, 1920, p. 57). It does, however, at least suggest that mathematical tools are not *a priori* alien to an analysis of microgenesis— perhaps even a microgenetic analysis of microgenesis.

Finally, the most enduring aspect of human cultural life—the development of the person through the whole life course—is *ontogenetic*. In the process of ontogenesis, selected experiences (some directly from the microgenetic domain, others through recurrent mesogenetic events) are transformed into relatively stable meaning structures that guide one within one's life course. As Fig. 1.2 suggests, there are no isomorphisms (structure-preserving one-to-one correspondences) between the three organizational levels of human living. On the contrary, some microgenetic events—

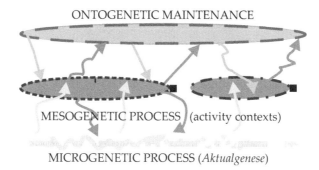

ONTOGENETIC MAINTENANCE

MESOGENETIC PROCESS (activity contexts)

MICROGENETIC PROCESS (*Aktualgenese*)

Figure 1.2: Relations between ontogenesis, mesogenesis, and microgenesis (*Aktualgenese*).

---

[11] Henri Poincaré (1902/1905, p. 13), a founder of modern dynamical systems theory (see Chap. 10) and topology, wrote that MI is "necessarily imposed upon us, because it is only the affirmation of a property of the mind itself".

[12] Edward Nelson, a contemporary dynamicist, probabilist, and logician, is a proponent of the "predicative mathematics" championed by Poincaré who has argued (Nelson, 1986) that MI is impredicative (and therefore to be avoided where possible) and may perhaps even lead to inconsistencies in arithmetic (Nelson, 2011). Ultrafinitists (*e.g.*, Ésénine-Volpine, 1961; Isles, 1992) reject MI outright.

unexpected, one-time events in a person's life that are not guided by the mesogenetic collective-cultural framing—may have a major impact on the ontogenetic level. Thus, surviving a near-accident scenario, or a special feeling of unity with a particular partner in the act of full mutuality, may become relevant for the construction of one's ontogenetic life trajectory. Other one-time deeply affective moments within personal lives, such as the death of a mother, father, friend, etc., may be assisted by mesogenetic events that become integrated into the ontogenetic structure of developing subjectivity.

It is obvious that *lack* of one-to-one correspondences between these levels is adaptive for the successful survival of the person within his or her life course. The encounter at the microgenetic level with dramatic—and traumatic—life events is inevitable (it is only in their details that they are unexpectable) during the course of living. The people who live encounter death of others around them, personal losses, changes between peace- and war-times, and so on. Yet it is important that such highly affective experiences do not hinder their basic progression through their life courses.

How do mathematicians model?

Though mathematics as a subject may be 'about' structures of diverse kinds, mathematics as a *practice* appears more to be 'about' certain ways of clear thinking (about structures)—or rather certain ways to clarify thinking. Of course when a common person, or common social scientist, meets a mathematician she (or he) might not understand that clarity—as it is customarily formulated within (though only partly based upon and enacted using) the framework of deductive inference.

Psychology's history shows ambivalence in its relations to mathematics. While Immanuel Kant doubted that psychological phenomena can be abstracted to the level of mathematical precision, his follower on his *Lehrstuhl* in Königsberg, Johann Friedrich Herbart, charted out ways in which it could happen (see Chap. 7). The relationships of quantity and quality as ideal properties were central for Hegel's dialectics in the early 1800s—yet with implications for philosophy rather than mathematics of psychology.

William James had a "rooted dislike for mathematics" (Knight, 1954, p. 18), and found Russell's "abstract world of mathematics and pure logic" to be "vicious abstractionism" (James, 1910, p. 105); yet, although "too ignorant of the development of mathematics to feel very confident of [his] own view" (*ibid.*, p. 84), James had proposed —though no one "seems to have noticed"—what today at least is immediately recognizable as a (proto-) *mathematical* "humanistic" theory of identity of "improvised human 'artefacts'" via "relations of comparison" (*ibid.*, p. 83).[13] In somewhat of a mirror image, Kurt Lewin—despite obvious admiration for mathematics, and an intuition (felt "already in 1912 as a student") that "the young mathematical discipline 'topology' might be of some help in making psychology a real science" (Lewin, 1936, p. vii)—created, in his conjoined theories of "topological and vector psychology", a body of work that London (1944) justly derided as "mathematically spurious" (p. 290), "an interminable use of a few *definitions* ripped out of their proper context" (p. 287; italics in the original), in short a "misuse of the auxiliary concepts of physics and mathematics" (the article's title).[14]

In Chap. 2, we discuss the internal logic(s) of mathematics and mathematical modeling, and its/their similarities and dissimilarities to some of the various other logics proper to other domains. In the remainder of this introduction, we will concentrate on external features of the mathematical modeling process.

Styles and goals of mathematical modeling

We focus our discussion by quoting three statements, by mathematicians, on mathematical modeling.

---

[13]James's proposed theory might, for instance, have fit into the psychophysically inspired "behavioral approach" to comparison using the mathematics of "V-spaces" (applied to the Paradox of the Heap in Chap. 4), or into category theory (applied to the structure of "family resemblances" in Chap. 9).

[14]The topologist and dynamicist René Thom had a more charitable view: *"Lewin a fondé une psychologie topologique dans laquelle il avait également introduit la notion de champ : le problème c'est qu'alors la topologie n'avait pas été créée !"* (Lewin founded a topological psychology in which he also introduced the notion of [vector] field: the problem is that the topology hadn't yet been created!) (Thom, Giorello, & Morini, 1983, p. 87). Abraham (1995) develops Lewin's ideas, rephrased in terms of continuous dynamical systems and chaos theory. Rudolph (2009) gives a first step in an alternative mathematical rehabilitation of "topological psychology" using *finite topological spaces*. Lewin's "vector psychology" is discussed briefly in Chap. 10. Leeper (1943) gave a consolidated contemporaneous review and (sympathetic) critique of topological and vector psychology.

(A) The symbiosis of mathematics and the sciences has been described as an hermeneutical circle. That is, experimental data determines a model, the model suggests new experiments, new data refines the model, and so on. This infinite loop is the motor for the advance of science.

(B) [T]he criteria of credibility evaluation [of a model] depend both on the extramathematical scientific status of the model, *i.e.* how structured the theoretical universe is to which the model belongs, and on the analytical transparency of the mathematical treatment: Theoretically based and analytically treated models are not necessarily "better" than ad-hoc models treated by procedures of dubious convergence; however, they can be checked theoretically, whereas the other models depend solely on empirical control. (Bohle-Carbonnell et al., 1984, p. 62; bracketed text added)

(C) Now what are the goals of applied mathematics [read: mathematical modeling]? One can distinguish at least three. They are description, prediction and prescription; that is to say, tell me what is, tell me what will be, tell me what to do about it. (Davis, 1991, p. 2; bracketed text added)

Statement (A) (the first paragraph of Chap. 11 in this volume, p. 321), particularly the second sentence, provides a baseline. It is practically a standard definition of idealized mathematical modeling, as propounded by mathematicians, engineers, and scientists (including social scientists) for at least the last 85 years: originally in scholarly or technical papers and monographs (Jourdain, 1925; Tinbergen, 1935, etc.); later in handbooks and textbooks for practitioners (Harary & Norman, 1953; Bellman, 1961; Harder, 1966/1969, etc.) and proposals for curricular innovation or reform (Begle, 1955; Burghes, 1980; Freudenthal, 1961, etc.); eventually in lecture notes and schoolbooks from graduate down to primary level (*e.g.*, Burghes & Borrie, 1981; Bell et al., 2007). Two typical depictions of mathematical modeling as "hermeneutical circle"—one adapted from a figure in a review article on sociological methodology (Cook & Weisberg, 1982), the other from a best-selling calculus text explicitly addressed to students of both "physical" and "social" sciences (Stewart, 2009) —appear in Fig. 1.3 (in both adaptations, the labels on boxes and arrows are quoted exactly from the original figures).

Often this definition is stated restrictively, in that limits (implicit or explicit) are placed on those kinds of "mathematics", "sciences", or both, that are expected or permitted to participate in an hermeneutic circle. For instance, as Maki (1980, p. 767) points out, "many individuals who are interested in model building [...] view mathematical models as sets of equations", even (perhaps especially?) in the "social" sciences: thus, "A mathematical model of a process is nothing more than a set of equations which one would like to generate trends similar to those observed for the real world" (Cole, 1973, p. 24), and "The term *mathematical model*, when applied to the social sciences, refers to the use of mathematical equations to depict the behavior of persons, groups, communities, states, or nations" (Miller & Salkind, 2002, p. 47).

On the other hand, it is sometimes made clear (and occasionally explicit) that any kind of mathematics whatever, and any kind of science, might potentially find themselves participating in an hermeneutic circle with each other: "there are also many individuals who consider the notion of a mathematical model to be considerably more general" (Maki, 1980, p. 767). As "an example of an area of study in which there are models which seem to be mathematical in nature, but which are not sets of equations," Maki offers "the area of preference theory and voting", where models "continue to evolve (much like more traditional models in the physical sciences), yet [...] are not easily characterized by a system of equations." Preference

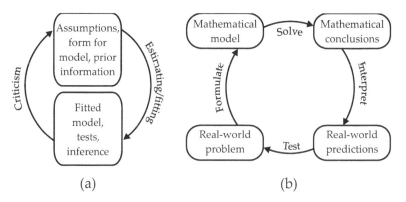

Figure 1.3: Two variants on the "hermeneutical circle", adapted from (a) a review paper on regression in sociology (Cook & Weisberg, 1982, p. 314), (b) a college calculus text (Stewart, 2009, p. 25).

theory is discussed in Chap. 12. Other mathematical models of non-equational type appear in Chaps. 4, 7, 8, and 9.

From such a point of view, mathematical modeling is both a <u>formalization</u> and a <u>form</u> of making meaning by living reflectively: we conceive of the world in a certain way; as we experience life, we fit the world to our concepts but also change our concepts of the world; and so on and so forth. (This notion of "living reflectively" is broader than, but clearly related to, the process that has been described—in a primarily pedagogical context— as "the development of reflective knowledge" through "dialogical epistemology" by Skovsmose, 1990, pp. 769 and 779.)

Among those philosophers and historians of science and mathematics who have written on mathematical modeling, there is less consensus that the process described in **(A)** is (or should be) the actual (or ideal) state of affairs. Indeed (see Fig. 1.4), several recent radically alternative descriptions of the hermeneutics of mathematical modeling (Frigg, 2010; Giere, 1990; Godfrey-Smith, 2009) have reduced or entirely removed the element of circularity characteristic of the 'standard' description. Although these ideas seem not yet to have led to any changes in the practice of mathematical modeling, we mention them for completeness and in case they may intrigue some reader and perhaps provoke precisely such a change.

Statement **(B)** is more methodological than foundational or definitional; reading it in the light of statement **(A)**, we can say that it

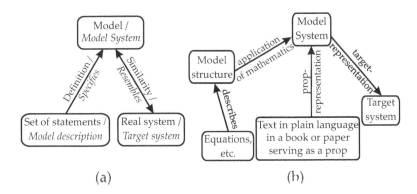

Figure 1.4: Two non-cyclic schemata for modeling, adapted from (a) Giere (1990) (roman labels) and Godfrey-Smith (2009) (italic labels), (b) Frigg (2010).

goes towards explicating the nature of a properly functioning hermeneutic circle (of mathematical type). One significant difference between mathematical models and models proffered otherwise is that, although all kinds of models use analogies and hermeneutic circles, in the mathematical hermeneutic circle what counts as proof in the model (Bohle-Carbonell et al.'s "mathematical treatment") is (much) less negotiable; that is, in the 'fitting' process, what has to give, has to give somewhere else (Bohle-Carbonell et al.'s "extramathematical [...] theoretical universe").

What is notably lacking from the standard definition of mathematical modeling (as typically presented), and from many explications of its methodology, is the role of "prescription". Davis's statement **(C)** explicitly adds prescription to modeling's "goals" alongside "description" and "prediction", glossing it as "Tell me how to change things along certain lines that are perceived to be desirable" (*ibid.*)—an imperative that we quote precisely because we believe it to be desirable to keep always in mind that, however value-free a given instance of mathematical modeling may be *qua* mathematics, nonetheless every modeler has (and applies, and possibly modifies) values while undertaking the act of modeling. A good reason—though not the only one—to keep this in mind is Booß-Bavnbek's warning (1991, pp. 77–78): "society has been transformed into a laboratory for scientific experiments, where mathematical models play the role of an opiate, tranquilising the feeling of uncertainty."[15]

Not only as regards values, but quite generally, inattention to the (human) role of the (human) modeler in the human activity of mathematical modeling is likely to be a mistake, and must surely be so for mathematical modeling in the social sciences.

Mathematical modeling as a human activity

In the beginning there is an act—to do or not to do. Activities are Gestalts, "higher-level" constructs (cognitive or perceptual, collaborative or individual) of acts—done and not done, actual and potential. A useful understanding of any human activity could involve some understanding of any or all of the following:
   – who undertakes that activity (and who refrains from undertaking it—or—how the same person undertakes it in setting X but

---

[15]N.B.: The most common side-effect of opiate tranquilization is constipation.

does not do so in setting Y);
- why they do so (insofar as that can be determined: *e.g.*, by considering individual actors' and non-actors' explicit declarations of internal and external motivation, or implicit and explicit norms of implicitly and explicitly declared groups of actors);
- whom—or what—the undertaking of that activity (or the refraining from that activity) benefits (and harms, and affects neutrally or leaves unaffected) and how it does so; and only then,
- what those who undertake that activity (and those who deliberately refrain from it) do.

In general, the last item is the hardest to understand, and a useful understanding of it can depend strongly on useful understandings of (at least) the preceding items.

An example from political science[16] illustrates one way in which modelers' self-alienation from their model can undercut its value.

> Most voters would probably be alienated and outraged upon hearing the hypothetical (but theoretically possible) election night report: "Mr. O'Grady did not obtain a seat in today's election, but if 5,000 of his supporters had voted for him in second place instead of first place, he would have won!" Any evaluation of the desirability of using the Single Transferable Vote as a social choice rule should take this possibility into consideration. (Doron & Kronick, 1977, p. 310)

Here the modelers' determination of the 'fit' of their model has been made, not by any empirical process of (for instance) actually surveying "most voters" (or even the statisticians' sacred proxy for that unattainable plenum, the 'representative sample'), but rather by the modelers' *imagining* the results of such a survey—while seemingly omitting (or repressing any desire) also to imagine how such results might be modified by, *e.g.*: providing the surveyed voters with educational (or fictional, or frankly propagandistic) materials explaining Single Transferable Vote (and the modelers' model and analysis of it); quantifying the relative likelihoods of the "hypothetical (but theoretically possible)" outcome and other (equally hypothetical)

---

[16]Similar examples from other social sciences are easy to find.

outcomes; or turning the lens of "social choice" theory not on elections alone but on the multiply self-referential system incorporating the institutions that run elections, make "election night report"s, formulate "social choice rules", and foster studies of studies of "social choice". As it is, whether by omission or repression, the hermeneutical circle has been broken. Alienation of mathematical modelers from the models that they work with may spring from a desire to avoid contamination by 'subjectivity', coupled with a belief that 'purity' and 'objectivity' imbue mathematics with great prophylactic powers. It seems certain that the converse phenomenon —many social scientists (particularly of a 'qualitative' bent) avoid mathematics entirely—springs from the opposite beliefs: that mathematical 'objectivity' is sterile and incompatible with the 'subjective' experiences those scientists embrace. In the two cases mathematics is given opposite valences, but in both the nature of actual mathematical practice is misapprehended as being *necessarily* reductive— a form of the "vicious abstractionism" James discerned in Russell's logicism and decried as

> a way of using concepts which may be thus described: We conceive a concrete situation by singling out some salient or important feature in it, and classing it under that; then, instead of adding to its previous characters all the positive consequences which the new way of conceiving it may bring, we proceed to use our concept privatively; reducing the originally rich phenomenon to the naked suggestions of that name abstractly taken, treating it as a case of 'nothing but' that concept, and acting as if all the other characters from out of which the concept is abstracted were expunged. (James, 1910, pp. 249–250)

Although mathematics of course can be (and often is) used reductively, to treat it as "nothing but" a reductionist tool is both mistaken and ... reductive!

One source of such misapprehensions may be confusion about the nature of 'axioms' in modern mathematics. It may be understood that "axioms are taken as bases for knowing—following them proceeds without doubt" (Valsiner, 2009, p. 46), but not also that "choice of one's axioms is always deeply filled with doubt—this distinguishes the act of axiom construction from religious conversion" (*ibid.*) and from mysticism of all sorts.

James famously proposed three axioms for mystical (religious) experience.

(1) Mystical states, when well developed, usually are, and have the right to be, absolutely authoritative over the individuals to whom they come.

(2) No authority emanates from them which should make it a duty for those who stand outside of them to accept their revelations uncritically.

(3) They break down the authority of the non-mystical or rationalistic consciousness, based on the understanding and the senses alone. They show it to be only one kind of consciousness. They open out the possibility of other orders of truth, in which, so far as anything in us vitally responds to them, we may freely continue to have faith. (James, 1902, p. 414)

A somewhat parallel set of axioms for the axiomatic method (in mathematics or elsewhere) might run as follows.

(1) Axioms are, and have the right to be, absolutely authoritative over the reasoners who adopt them *in the context of their present reasoning*.

(2) No authority emanates from axioms which should make it a duty for anyone, including those who reason from them, to accept their consequences uncritically *either inside or (especially) outside the context of their present reasoning*.

(3) Axioms break down the authority of non-axiomatic reasoning, based on what is 'understandable' and 'sensible'. *They also cast themselves in doubt.* They show themselves to be only particular axioms. They open out the possibility of other axioms, to which, so far as anything in us vitally responds to them, we may freely turn our attention and our capacity to generate new reasoning.

Naturally, we do not propose these as the last word on axioms. But we do offer them, particularly the recursive axiom (3), for serious consideration.

Mathematical modeling and evolutionary epistemology

Over fifty years ago, when Eugene Wigner published "The Unreasonable Effectiveness of Mathematics in the Physical Sciences" (Wigner, 1960), he conspicuously avoided asserting any especial "effectiveness of mathematics" in the non-physical sciences. In fact, he speculated at the end of his paper that

> if we could, some day, establish a theory of the phenomena of consciousness, or of biology, which would be as coherent and convincing as our present theories of the inanimate world [...] it is quite possible that an abstract argument [could] be found which shows that there is a conflict between such a theory and the accepted principles of physics. The argument could be of such abstract nature that it might not be possible to resolve the conflict, in favor of one or of the other theory, by an experiment

and concluded, rather mystically,

> The miracle of the appropriateness of the language of mathematics for the formulation of the laws of physics is a wonderful gift which we neither understand nor deserve. We should be grateful for it and hope that it will remain valid in future research and that it will extend, for better or for worse, to our pleasure, even though perhaps also to our bafflement, to wide branches of learning. (Wigner, 1960, p. 14)

Wigner's article has provoked many responses,[17] though not from particularly "wide branches of learning": they seem to be primarily from physicists, mathematicians, philosophers of mathematics and of science, and theologians. Nanjundiah (2003) gives a rare response from biology, and Luce (1997) one from psychology (drawing particular attention to the tension between 'continuous' and 'discrete' mathematics).

For the purpose of understanding mathematical modeling as a human activity, the most interesting responses to Wigner may be

---

[17] At least, his article's title has done so. The assessment of Longo and Viarouge (2010, p. 17) is that "everybody quotes" the article "due to its so very effective and memorable title", but "very few read" it; and that Wigner in fact "presents examples that are not astounding".

from the perspective of "Evolutionary Epistemology". Campbell (1974, p. 413) introduced that name for the study of the proposition that "epistemic activities, such as learning, thought, and science" have evolved alongside and by the same processes as human biology and sociality. Rav (1989, p. 73) argued explicitly for the claim that we should "add mathematics" to Campbell's list of evolved epistemic activities. A few years earlier Kreisel (1978, pp. 85–86) had already made that claim (without argument) in a rhetorical question: "Would it be *obviously* more 'reasonable' if we were not effective in thinking about the external world in which we have evolved?"[18]

In fact, Konrad Lorenz's foundational essay on evolutionary epistemology (Lorenz, 1941, before it was so named) can profitably be read as a 'presponse' to Wigner.

> Is not human reason with all its categories and forms of intuition something that has organically evolved in a continuous cause-effect relationship with the laws of the immediate nature, just as has the human brain? Would not the laws of reason necessary for a priori thought be entirely different if they had undergone an entirely different historical mode of origin, and if consequently we had been equipped with an entirely different kind of central nervous system? Is it at all probable that the laws of our cognitive apparatus should be disconnected with those of real external world? Can an organ that has evolved in the process of a continuous coping with the laws of nature have remained so uninfluenced that the theory of appearances can be pursued independently of the existence of the thing-in-itself, as if the two were totally independent of each other? (Lorenz, 1962, p. 23)[19]

About mathematics specifically, Lorenz writes:

> Nothing that our brain can think has absolute a priori validity in the true sense of the word, not even mathematics with all its laws. The laws of mathematics are but

---

[18]Kreisel's conflation of "effective [...] thinking about the external world" with thinking mathematically about it—*i.e.*, modeling it mathematically—is deliberate.

[19]Here and below, we quote at length because Lorenz (1941, 1962) and Lorenz and Wuketits (1983) are all very difficult to lay hands on.

an organ for the quantification of external things, and what is more, an organ exceedingly important for man's life, without which he never could play his role in dominating the earth, and which thus has amply proved itself biologically, as have all the other "necessary" structures of thought. Of course, "pure" mathematics is not only possible, it is, as a theory of the internal laws of this miraculous organ of quantification, of an importance that can hardly be overestimated. But this does not justify us in making it absolute. [...] [T]he number one applied to a real object will never find its equal in the whole universe. It is true that two plus two equals four, but two apples, rams or atoms plus two more never equal four others because no equal apples, rams or atoms exist! (Lorenz, 1962, p. 27)

It would be entirely conceivable to imagine a rational being that does not quantify by means of the mathematical number (that does not use 1, 2, 3, 4, 5, the number of individuals approximately equal among themselves, such as rams, atoms or milestones, to mark the quantity at hand) but grasps these immediately in some other way. [...] It can very well be purely coincidental, in other words brought about by purely historical causes, that our brain happens to be able to quantify extensive quantities more readily than intensive ones. It is by no means a necessity of thought and it would be entirely conceivable that the ability to quantify intensively [...] could be equally valuable and replace numerical mathematics. [...] A mind quantifying in a purely intensive manner would carry out some operations more simply and immediately than our mathematics [...] [which] is and remains only an organ, an evolutionarily acquired, "innate working hypothesis" which basically is only approximately adapted to the data of the thing-in-itself. (Lorenz, 1962, p. 29)

The very essence of mathematical modeling as hermeneutical circle is that the mathematics, as well as the data, is subject to change.

Unsere Arbeitshypothese lautet also: Alles ist Arbeitshypothese. [Our working hypothesis should read as follows:

Everything is a working hypothesis.]   (Lorenz, 1941, p. 109; 1962, p. 29)

This slogan is very much in line with our proposed "axioms for the axiomatic method" (p. 23).

If, in fact, mathematics has so far been relatively <u>in</u>effective in the social sciences, perhaps it is because 'we' (not just *Homo sapiens sapiens* or even—say—*Hominidæ* or *Mammalia*, but all life have evolved in the "external world" of physics, chemistry, and life-in-general for so much longer than in the "external world" of human sociality, that is, human life-in-particular. Yet we (humans) need not simply give up on the attempt to make it effective.

Conclusion

It is sometimes said that mathematicians make bad journalists because a mathematician would rather be correct than interesting. Neither of us is a journalist, and only one is a mathematician. Nonetheless, we would not mind being correct (and certainly we wish to avoid knowingly stating falsehoods), and we would like to be interesting; above all, however, our purpose here has been to be provocative.

References

Abbey, E., & Diriwächter, R. (Eds.). (2008). *Innovating Genesis: Microgenesis and the Constructive Mind in Action.* Charlotte, NC: Information Age Publishers.

Abraham, R. (1995). Erodynamics and the dischaotic personality. In F. D. Abraham & A. R. Gilgen (Eds.), *Chaos Theory in Psychology* (pp. 157–167). Westport, CT: Praeger Publishers/Greenwood Publishing Group.

Begle, E. G. (1955). *Lectures on experimental programs in collegiate mathematics.* (Mimeographed notes from NSF Summer Mathematics Institute, Oklahoma Agricultural and Mechanical College)

Bell, M., Bell, J., Bretzlauf, J., Dillard, A., Flanders, J., & Hartfield, R. (2007). *Everyday Mathematics: Teacher's Reference Manual (Gr. 4–6).* Columbus, OH: McGraw-Hill Wright Group.

Bellman, R. E. (1961). *Mathematical Model-Making as an Adaptive Process*. Santa Monica, CA: RAND Corporation.

Benussi, V. (1913). *Psychologie der Zeitauffassung [Psychology of Time Comprehension]*. Heidelberg, DE: Winter.

Bickhard, M. H. (1988). Piaget on variation and selection models: Structuralism, logical necessity, and interactivism. *Human Development, 31*, 274–312.

Bickhard, M. H. (1991). A pre-logical model of rationality. In L. P. Steffe (Ed.), *Epistemological Foundations of Mathematical Experience* (pp. 68–77). New York, NY: Springer.

Bickhard, M. H., & Campbell, R. L. (1996). Topologies of learning and development. *New Ideas in Psychology, 14*(2), 111–156.

Boesch, E. E. (2005). *Von Kunst bis Terror: über den Zwiespalt in der Kultur [From Art to Terror: on Conflict in Culture]*. Göttingen, DE: Vanderhoeck & Ruprecht.

Bohle-Carbonell, M., Booß, B., & Jensen, J. H. (1984). Innermathematical vs. extramathematical obstructions to model credibility. In E. Y. Rodin & X. J. R. Avula (Eds.), *Mathematical Modelling in Science and Technology (Zürich, 1983)* (pp. 62–65). Oxford, GB: Pergamon.

Booß-Bavnbek, B. (1991). Against ill-founded, irresponsible modelling. In M. Niss, W. Blum, & I. Huntley (Eds.), *Teaching of Mathematical Modelling and Applications* (pp. 70–82). Chichester, GB: Ellis Horwood.

Bozzi, P. (1969). *Unità, Identità, Causalità—Una Introduzione Allo Studio Della Percezione [Unity, Identity, Causality—an Introduction to the Study of Perception]*. Bologna, IT: Capelli Editore.

Burghes, D. N. (1980). Mathematical modelling: A positive direction for the teaching of applications of mathematics at school. *Educational Studies in Mathematics, 11*(1), 113–131.

Burghes, D. N., & Borrie, M. S. (1981). *Modelling with Differential Equations*. Chichester, GB: Ellis Horwood.

Campbell, D. T. (1974). Evolutionary epistemology. In P. A. Schilpp (Ed.), *The Philosophy of Karl Popper, Part 1* (pp. 413–463). La Salle, IL: Open Court.

Carini, L. (2007). *Perception, Cognition, Metaphysics, and Our Cultural Natures*. (Preprint)

Cassirer, E. (1957). *The Philosophy of Symbolic Forms. Vol. 3: The Phenomenology of Knowledge*. New Haven, CT: Yale University Press. (Original work published 1929)

Catán, L. (1986). The dynamic display of process: Historical development and contemporary uses of the microgenetic method. *Human Development*, 29, 252–263.

Cole, H. S. D. (1973). *Models of Doom: A Critique of* The Limits to Growth. New York, NY: Universe Books.

Cook, R. D., & Weisberg, S. (1982). Criticism and influence analysis in regression. *Sociological Methodology*, 13, 313–361.

Dauben, J. (2003). Mathematics. In D. Cahan (Ed.), *From Natural Philosophy to the Social Sciences* (pp. 129–162). Chicago, IL: University of Chicago Press.

Davis, P. J. (1991). Applied mathematics as a social instrument. In M. Niss, W. Blum, & I. Huntley (Eds.), *Teaching of Mathematical Modelling and Applications* (pp. 1–9). Chichester, GB: Ellis Horwood.

de Rosa, A. S. (2007). From September 11 to the Iraq War. In S. K. Gertz, J. Valsiner, & J.-P. Breaux (Eds.), *Semiotic Rotations: Modes of Meanings in Cultural Worlds* (pp. 135–177). Charlotte, NC: Information Age Publishing Inc.

de Rosa, A. S. (Ed.). (2012). *Social Representations in the "Social Arena"*. Abingdon, GB: Routledge.

Devlin, K. (1997). *Mathematics: The Science of Patterns: The Search for Order in Life, Mind and the Universe*. New York, NY: Macmillan.

Diriwächter, R., & Valsiner, J. (2008). *Striving for the Whole: Creating Theoretical Syntheses*. New Brunswick, NJ: Transaction Publishers.

Doron, G., & Kronick, R. (1977). Single transferable vote: An example of a perverse social choice function. *American Journal of Political Science*, 21(2), 303–311.

Dunn, S. (1762). An attempt to assign the cause, why the sun and moon appear to the naked eye larger when they are near the horizon. *Philosophical Transactions of the Royal Society of London*, 52, 462.

Ehrenfels, C. von. (1916). Höhe und Reinheit der Gestalt [Gestalt level and Gestalt purity]. In *Koskmogonie* (pp. 93–96). Jena, DE: E. Diederichs.

Ehrenfels, C. von. (1922). *Das Primzahlengesetz entwickelt und dargestellt auf Grund der Gestalttheorie [The Prime Number Theorem Developed and Presented on the Basis of Gestalt Theory]*. Leipzig, DE: O. B. Reisland.

Ehrenfels, C. von. (1937). über gestaltqualitäten. *Philosophia (Belgrad)*,

2, 139-141.

Ehrenfels, C. von.   (1988a).   Gestalt level and Gestalt purity. In B. Smith (Ed.), *Foundations of Gestalt Theory* (pp. 118–120). München, DE: Philosophia Verlag.   Retrieved June 5, 2012, from `http://ontology.buffalo.edu/smith/book/FoGT/Ehrenfels_Purity.pdf` (translated by B. Smith)

Ehrenfels, C. von.   (1988b).   On Gestalt qualities.   In B. Smith (Ed.), *Foundations of Gestalt Theory* (pp. 82–117). München, DE: Philosophia Verlag.   Retrieved June 5, 2012, from `http://ontology.buffalo.edu/smith/book/FoGT/Ehrenfels_Gestalt.pdf` (translated by B. Smith)

Ehrenfels, C. von.   (1988c).   On Gestalt qualities (1932).   In B. Smith (Ed.), *Foundations of Gestalt Theory* (pp. 121–123). München, DE: Philosophia Verlag.   Retrieved June 5, 2012, from `http://ontology.buffalo.edu/smith/book/FoGT/Ehrenfels_Gestalt_1932.pdf` ("a heavily revised and corrected version [by B. Smith] of the translation by Mildred Focht" of Ehrenfels (1937))

Ehrenfels, C. von. (1988d). Über "Gestaltqualitäten [On Gestalt qualities]. In R. Fabian (Ed.), *Psychologie, Ethik, Erkenntnistheorie* (pp. 128–167). München, DE: Philosophia Verlag.

Ek, C. H., Jaeckel, P., Campbell, N., Lawrence, N., & Melhuish, C. (2009). Shared Gaussian process latent variable models for handling ambiguous facial expressions. In L. Beji (Ed.), *Intelligent Systems and Automation, AIP Conference Proceedings, Vol. 1107* (pp. 147–153). Melville, NY: American Institute of Physics.

Ésénine-Volpine, A. S. Y. (1961). Le programme ultra-intuitionniste des fondements des mathématiques [The ultra-intuitionist program for foundations of mathematics]. In *Infinitistic Methods (Proceedings of the Symposium on the Foundations of Mathematics, Warsaw, 1959)* (pp. 201–223). Oxford, GB: Pergamon.

Flavell, J. H., & Draguns, J. (1957). A microgenetic approach to perception and thought. *Psychological Bulletin, 64,* 197–217.

Freudenthal, H. (Ed.). (1961). *The Concept and the Role of the Model in Mathematics and Natural and Social Sciences.* New York, NY: Gordon and Breach.

Frigg, R. (2010). Models and fiction. *Synthese, 172,* 251–268.

Giere, R. N. (1990). *Explaining Science: A Cognitive Approach.* Chicago, IL: University of Chicago Press.

Gigerenzer, G., Swijtink, Z., Porter, T. M., Daston, L., Kruger, L., & Beatty, J. (1989). *The Empire of Chance: How Probability Changed Science and Everyday Life*. Cambridge, GB: Cambridge University Press. (Ideas in Context [G. Gigerenzer and L. Daston, Series Eds.], Vol. 12)

Godfrey-Smith, P. (2009). Models and fictions in science. *Philosophical Studies*, *143*, 101–116.

Harary, F., & Norman, R. Z. (1953). *Graph Theory as a Mathematical Model in Social Science*. Ann Arbor, MI. (Issue 2 of University of Michigan Research Center for Group Dynamics Publications)

Harder, T. (1969). *Introduction to Mathematical Models in Market and Opinion Research: With Practical Applications, Computing Procedures, and Estimates of Computing Requirements* (P. H. Friedlander & E. H. Friedlander, Trans.). New York, NY: Gordon and Breach. (Original work published 1966; originally published as *Elementare mathematische Modelle der Markt- und Meinungsforschung*, München, DE & Wien, AT: R. Oldenbourg)

Hersh, R. (Ed.). (2006). *18 Unconventional Essays on the Nature of Mathematics*. New York, NY: Springer. Retrieved from `http://dx.doi.org/10.1007/0-387-29831-2`

Isles, D. (1992). What evidence is there that 2^{65536} is a natural number? *Notre Dame Journal of Formal Logic*, *33*(4), 465–480.

Jacobellis v. Ohio, 378 U.S. 184, 197. (1964).

James, W. (1902). *The Varieties of Religious Experience: A Study in Human Nature*. New York, NY: The Modern Library.

James, W. (1910). *The Meaning of Truth: A Sequel to 'Pragmatism'*. London, GB: Longmans, Green, and Company.

Jourdain, P. E. B. (1925). The purely ordinal conceptions of mathematics and their significance for mathematical physics. *The Monist*, *25*, 140–144.

Kanizsa, G. (1969). Perception, past experience and the "impossible experiment". *Acta Psychologica*, *31*, 66–96.

Kennison, J. (2002). The cyclic spectrum of a Boolean flow. *Theory and Applications of Categories*, *10*, 392–409.

Khrennikov, A. Yu. (2000). *p*-adic discrete dynamical systems and collective behaviour of information states in cognitive models. *Discrete Dynamics in Nature and Society*, *5*, 59–69.

Knight, M. (1954). Introduction. In M. Knight (Ed.), *William James: A Selection from his Writing on Psychology* (pp. 11–41). Middlesex, GB: Penguin Books.

Kreisel, G. (1978). Untitled [Review of the book *Wittgenstein's Lectures on the Foundations of Mathematics, Cambridge, 1939*]. *Bulletin of the American Mathematical Society*, *84*, 79–90.

Leeper, R. W. (1943). *Lewin's Topological and Vector Psychology: A Digest and a Critique*. Eugene, OR: University of Oregon Press.

Lewin, K. (1936). *Principles of Topological Psychology* (F. Heider & G. M. Heider, Trans.). New York, NY, and London, GB: McGraw-Hill.

London, I. (1944). Psychologists' misuse of the auxiliary concepts of physics and mathematics. *Psychological Review*, *51*, 266–291.

Longo, G., & Viarouge, A. (2010). Mathematical intuition and the cognitive roots of mathematical concepts. *Topoi*, *29*, 15–27.

Lorenz, K. (1941). Kants Lehre vom Apriorischen in Lichte gegenwärtiger Biologie [Kant's doctrine of the *a priori* in the light of contemporary biology]. *Blätter für Deutsche Philosophie*, *15*, 94–125. (Reprinted in Lorenz & Wuketits, 1983; translated as Lorenz, 1962)

Lorenz, K. (1962). Kant's doctrine of the *a priori* in the light of contemporary biology (C. Ghurye, Trans.). In L. V. Bertalanffy & A. Rapoport (Eds.), *General Systems: Yearbook of the Society for General Systems Research* (Vol. VII, pp. 23–35). Ann Arbor, MI: Society for General Systems Research.

Lorenz, K., & Wuketits, F. M. (1983). *Die Evolution des Denkens [The Evolution of Thought]* (2nd ed.). München, DE: R. Piper.

Luce, R. D. (1997). Several unresolved conceptual problems of mathematical psychology. *Journal of Mathematical Psychology*, *41*, 79–87.

Maki, D. P. (1980). Untitled [Review of the book *Mathematical Modelling Techniques*]. *Bulletin of the American Mathematical Society*, *3*(1), 766–770.

Meinong, A. (1907). *Über die Stellung der Gegenstandstheorie im System der Wissenschaften [On the Place of Object Theory in the System of Sciences]*. Leipzig, DE: R. Voigtländer.

Meinong, A. (1915). *Über Möglichkeit und Wahrscheinlichkeit. Beiträge zur Gegenstandstheorie und Erkenntnistheorie [On Possibility and Probability. Contributions to Object Theory and Epistemology]*. Leipzig, DE: J. A. Barth. (Reprinted in Alexius Meinong Gesamtausgabe [Complete Edition], R. Haller and R. Kindinger, Eds., in collaboration with R. M. Chisholm, Vol. VI: XIII–XXII, Graz, AT: Akademische Druck- u. Verlagsanstalt, pp. 1–728)

Miller, D. C., & Salkind, N. J. (2002). *Handbook of Research Design and Social Measurement*. Thousand Oaks, CA: Sage.

Montgomery, E. (1882). Causation and its organic conditions. Part III. *Mind: A Quarterly Review of Psychology and Philosophy, 7,* 209–230, 381–397, 514–532.

Moscovici, S. (1961). *La psychanalyse, son image et son public [Psychoanalysis, Its Image and Public]*. Paris, FR: Presses Universitaires de France.

Moscovici, S. (1976). *Society Against Nature: The Emergence of Human Societies*. Atlantic Highlands, NJ: Humanities Press.

Moscovici, S. (1988). Notes towards a description of social representations. *European Journal of Social Psychology, 18,* 211–250.

Moscovici, S. (2001). The phenomenon of social representations. In G. Duveen (Ed.), *Social Representations* (pp. 18–77). New York, NY: New York University Press.

Nanjundiah, V. (2003). Role of mathematics in biology. *Economic and Political Weekly, 38*(35), 3671– 3673 and 3676–3677.

Nelson, E. (1986). *Predicative Arithmetic*. Princeton, NJ: Princeton University Press. Retrieved June 15, 2011, from http://www.math.princeton.edu/~nelson/as/pa.pdf

Nelson, E. (2011). Warning signs of a possible collapse of contemporary mathematics. In M. Heller & W. H. Woodin (Eds.), *Infinity: New Research Frontiers* (pp. 75–85). Cambridge, GB: Cambridge University Press.

Piaget, J. (1950). *Introduction à l'épistémologie génétique. Tome I: La pensée mathématique [Introduction to Genetic Epistemology. Volume I: Mathematical Thinking]*. Paris, FR: Presses Universitaires de France.

Poincaré, H. (1905). (W. J. Greenstreet, Trans.). London, GB: Walter Scott. (Original work published 1902)

Porter, T. M. (2003). The social sciences. In D. Cahan (Ed.), *From Natural Philosophy to the Social Sciences* (pp. 254–290). Chicago, IL: University of Chicago Press.

Radford, L., Bardini, C., & Sabena, C. (2005). Perceptual semiosis and the microgenesis of algebraic generalizations. In M. Bosch (Ed.), *Fourth Congress of the European Society for Research in Mathematics Education* (CERME 4) , *17–21 February 2005, Sant Feliu de Guíxols, Spain* (pp. 684–695). Barcelona, ES: Fundemi IQS.

Rav, Y. (1989). Philosophical problems of mathematics in the light of evolutionary epistemology. *Philosophica*, 43(1), 49–78. (Recent issues in the philosophy of mathematics, II; reprinted in Hersh 2006, pp. 71–96)

Resnik, M. D. (1997). *Mathematics as a Science of Patterns*. Oxford, GB: Clarendon Press.

Rosa, A. (2007). Acts of psyche: Actuations as synthesis of semiosis and action. In J. Valsiner & A. Rosa (Eds.), *The Cambridge Handbook of Sociocultural Psychology* (pp. 205–237). Cambridge, GB: Cambridge University Press.

Roy, N., Gordon, G., & Thrun, S. (2005). Finding approximate POMPD solutions through belief compression. *Journal of Artificial Intelligence Research*, 22, 1–40.

Rudolph, L. (2006a). The fullness of time. *Culture & Psychology*, 12, 157–186.

Rudolph, L. (2006b). Mathematics, models and metaphors. *Culture & Psychology*, 12, 245–265.

Rudolph, L. (2006c). Spaces of ambivalence: Qualitative mathematics in the modeling of complex, fluid phenomena. *Estudios Psicologías*, 27, 67–83.

Rudolph, L. (2008). The finite-dimensional Freeman Thesis. *Integrative Psychological and Behavioral Science*, 42(2), 144–152.

Rudolph, L. (2009). A unified topological approach to *Umwelt*s and life spaces, Part I: *Umwelt*s and finite topological spaces. In R. I. S. Chang (Ed.), *Relating to Environments: A New Look at Umwelt* (pp. 185–206). Charlotte, NC: Information Age Publishing.

Rudolph, L. (2012). *The Hole in Emotion Space: Topological Consequences for Circumplex Models of Affect*. (In preparation)

Rudolph, L., Han, L., & Charles, E. P. (2009). *Modeling Emotional Development via Finite Topological Spaces and Stratified Manifolds*. (Draft manuscript, Clark University)

Russell, B. (1897). On the relations of number and quantity. *Mind (N.S.)*, 6(23), 326–341.

Saada-Robert, M. (1994). Microgenesis and situated cognitive representations. In N. Mercer & C. Coll (Eds.), *Explorations in Socio-Cultural Studies, Vol. 3: Teaching, Learning, and Interaction* (pp. 55–64). Madrid, ES: Fundación Infancia y Aprendizaje.

Skovsmose, O. (1990). Reflective knowledge: Its relation to the mathematical modelling process. *International Journal of Mathematical Education in Science and Technology*, 21(5), 765–779.

Steen, L. A. (1988). The science of patterns. *Science, 240,* 611–616.

Stern, W. (1935). *Allgemeine Psychologie auf personalistischer Grundlage [General Psychology from the Personalistic Standpoint].* The Hague, NL: Martinus Nijhoff.

Stewart, J. (2009). *Calculus: Concepts and Contexts.* Florence, KY: Cengage Learning.

Sully, J. (1880). Illusions of memory. *The Cornhill Magazine, 41,* 416–433.

Sully, J. (1881). Illusions of introspection. *Mind: A Quarterly Review of Psychology and Philosophy, VI,* 1–18.

Tennemann, W. G. (1798). *Geschichte der Philosophie [History of Philosophy]* (Vol. 1). Leipzig, DE: J. A. Barth.

Thom, R., Giorello, G., & Morini, S. (1983). *Paraboles et catastrophes: entretiens sur les mathématiques, la science et la philosophie [Parables and Catastrophes: Interviews on Mathematics, Science and Philosophy]* (Vol. 118). Paris, FR: Flammarion.

Tinbergen, J. (1935). Quantitative Fragen der Konjunkturpolitik [Quantitative questions of economic policy]. *Weltwirtschaftliches Archiv, 42,* 366–399.

Uexküll, J. von. (1920). *Theoretische Biologie [Theoretical Biology].* Frankfurt, DE: Suhrkamp.

Valsiner, J. (1987). *Culture and the Development of Children's Action: A Cultural-Historical Theory of Developmental Psychology.* New York, NY: Wiley.

Valsiner, J. (2005). Scaffolding within the structure of dialogical self: Hierarchical dynamics of semiotic mediation. *New Ideas in Psychology, 23*(3), 197–206.

Valsiner, J. (2009). Integrating psychology within the globalizing world: A *requiem* to the post-modernist experiment with *Wissenschaft. Integrative Psychological and Behavioral Science, 43*(1), 1–21.

Valsiner, J., & van der Veer, R. (2000). *The Social Mind: Construction of the Idea.* Cambridge, GB: Cambridge University Press.

Wagoner, B. (2009). The experimental methodology of constructive microgenesis. In J. Valsiner, P. C. M. Molenaar, M. C. D. P. Lyra, & N. Chaudhary (Eds.), *Dynamic Process Methodology in the Social and Developmental Sciences* (pp. 99–121). New York, NY: Springer.

Werner, H. (1956). Microgenesis and aphasia. *Journal of Abnormal Social Psychology, 52,* 347–353.

Werner, H., & Kaplan, B. (1962). *Symbol Formation*. New York, NY: Wiley.

Whitehead, A. N. (1941). Mathematics and the good. In P. A. Schilpp (Ed.), *The Philosophy of Alfred North Whitehead* (pp. 666–681). Evanston and Chicago, IL: Northwestern University Press.

Wigner, E. (1960). The unreasonable effectiveness of mathematics in the natural sciences. *Communications in Pure and Applied Mathematics*, *13*(1), 1–14. (Richard Courant Lecture in Mathematical Sciences, delivered at New York University, May 11, 1959)

Yoo, P. D., Zhou, B. B., & Zomaya, A. Y. (2008). Machine learning techniques for protein secondary structure prediction: An overview and evaluation. *Current Bioinformatics*, *3*(2), 74–86.

Zeeman, E. C. (1966). Mathematics and creative thinking. *The Psychiatric Quarterly*, *40*(2), 348–354.

Zittoun, T. (2007). Review symposium: Symbolic resources in dialogue, dialogical symbolic resources. *Culture & Psychology*, *13*(3), 365–376.

Logics of Modeling

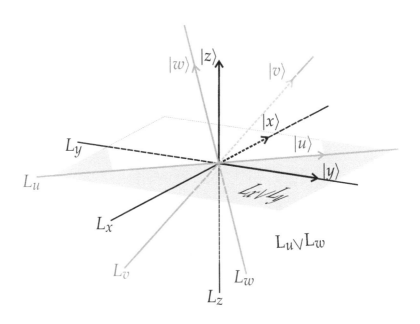

2

Logic in Modeling, Logics as Models

Lee Rudolph

> Logic is often informally described as the study of
> sound reasoning. [...] In an enormous
> development beginning in the late 19th century, it
> has been found that a wide variety of different
> principles are needed for sound reasoning in
> different domains, and "a logic" has come to
> mean a set of principles for some form of sound
> reasoning. (Mossakowski, Goguen, Diaconescu, &
> Tarlecki, 2005, p. 113)

My impulse on first reading the passage quoted above was to pro-
claim a slogan: **Logic is a model of the notion of soundly reasoned
persuasion.** Then I was moved to propose two definitions: for a
given domain, _a logic_ is a model of soundly reasoned persuasion
within that domain, and a _formal_ logic is a mathematical model of
soundly reasoned persuasion within that domain.

In my slogan and definitions I acknowledge the classical notion
that persuasion can have both _dialectical_ (reasoned) and _rhetori-
cal_ (unreasoned) components. I do not claim these components
can always be clearly distinguished or, if distinguished, entirely
isolated from each other. In accord with my editorial attention
to social process aspects of modeling, I have brought "persuasion"
forward explicitly to accompany the "reasoning" that stands unaccompanied in the epigraph. I acknowledge that self-persuasion

Keywords: axiomatic method, computation, dialectic, eisegesis, excluded middle,
exoterica, formal logic, informal logic, persuasion, proof, reason, rhetoric.

39

is as important as other-persuasion[1] and I neither exclude it from logic nor claim it is entirely *non*-social. I further acknowledge that any given purported instance of "reasoning" can be more or less "sound", according to a more or less public and explicit standard or standards (what Mossakowski et al. call "a set of principles"), and that to whatever extent a particular attempt at persuasion is purported to be "reasoning", whoever attempts that persuasion also purports to value "soundness" positively, and its absence or inadequate enactment negatively.

The limits of formality

Although my definitions of 'a logic', and especially of 'a formal logic', are not standard, they may be useful, especially in the context of modeling. In this section I describe some alternative definitions of 'formal logic' (without the indefinite article) in particular, and list some 'formal logics' in my sense, before turning in the rest of the chapter to the *informal* logic of mathematical modeling.

Meanings of 'formal logic'

John Burgess, a mathematical and philosophical logician as well as a philosopher of logic and mathematics, writes

> Logic, whether classical or extra- or anti-classical, is concerned with form. (On this traditional view of the subject, the phrase "formal logic" is pleonasm and "informal logic" oxymoron.) (Burgess, 2009, p. 2)

Thereafter neither phrase appears in his text. Whether or not Burgess's parenthetical assessment of the terminological entailments of the "traditional view of the subject" is correct (as a matter of fact, I trust his judgment without reservation), I have been surprised by the relative difficulty of finding definitions of the phrase "formal logic" that have been committed to print (or electronic media) by authoritative contemporary writers. For instance, the Stanford Encyclopedia of Philosophy has no entry for "Formal

---

[1] An account of self-persuasion by a group of mathematical logicians (of the truth of a conjecture that was, in fact, eventually proved), constituted entirely of *informal* (inductive and diagrammatic) reasoning, is quoted in Chap. 13, p. 416.

Logic", and though about a dozen entries use the phrase, none—not even "Informal Logic" (Groarke, 1996/2008)—defines it.

Among the definitions (or characterizations) that I have found, perhaps the broadest is given by Richard Jeffrey in the first sentence of his *Formal Logic: Its Scope and Limits*, though (depending on, *e.g.*, what one expects or demands of a "science") his second sentence may narrow the definition quite a bit.

> Formal logic is the science of deduction. It aims to provide systematic means for telling whether or not given conclusions follow from given premises, i.e., whether arguments are valid or invalid. (Jeffrey, 1967/2006, p. 1)[2]

A more precise statement (which is, however, meant to characterize more or less the same collection of logicians' practices; this is clear when both are read in context) is given by W. V. Quine:

> [...] the business of formal logic is describable as that of finding statement forms which are *logical*, in the sense of containing no constants beyond the logical vocabulary, and (extensionally) *valid*, in the sense that all statements exemplifying the form in question are true. (Quine, 1953, p. 436; definitional italics in the original)

A much narrower definition is given by I. M. Bocheński in his "source book in the history of logic, covering the whole period from the pre-Socratics to Gödel" (Mates, 1960, p. 57), again at the very beginning of the book (Bocheński, 1956/1961, §1, pp. 1–3). That definition takes up a bit more than two full pages, and is both too detailed and too nuanced to cut down to much less. I believe that the following extracts do not, at least, badly misrepresent it.

Bocheński first "discard"s from "all that has been termed 'logic' in the course of western thought"

> whatever most authors either expressly ascribe to some other discipline, or call 'logic' with the addition of an adjective, as for example epistemology, transcendental logic, ontology etc.

---

[2]Burgess edited, and contributed a supplement to, the posthumous fourth edition of Jeffrey's text; I draw the conclusion that pleonasm is not necessarily an obstacle to using and even defining "formal logic".

He is left with the meaning of "logic" reduced to (essentially) syllo-gistic and its descendants. To cut that down to "formal logic" he discards even more.

> [S]emiotic and methodological problems are closely con-nected with logic; in practice they are always based on semiotics and completed in methodology. What remains over and above these two disciplines we shall call *formal logic*.[3]

The use of 'formal logic'

'Formal logics' as variously defined above is concerned with 'truth', but despite the observation of McKinsey and Suppes (1954, p. 54) that

> It should not be forgotten that about nineteen hundred years elapsed between the time when Pilate asked his famous question "What is truth?" [...] and Tarski [1933] made a satisfactory reply

there remain—nearly 80 years further on—plenty of working scien-tists, mathematicians, and philosophers who are not persuaded that the reply of Tarski (1933) is in fact "satisfactory" for their several purposes.[4] Meanwhile, to all appearances, working scientists in

---

[3]Regrettably, the Wikipedia article *History of Logic* (n.d.)—and other resources that copy from it, both off- and on-line (*e.g.*, the Google eBook *Logic and Metalogic*, 2011, which incorporates the October 28, 2011, version of *History of Logic* as a chapter with 251 author/editors, partly pseudonymous, mostly anonymous)—quotes nearly word for word (with a citing footnote but no quotation marks) the description (Bocheński, 1956/1961, p. 63) of

> what later came to be called Aristotle's assertoric syllogistic [...] in which no less than three great discoveries are applied for the first time in history: variables, purely formal treatment, and an axiomatic. It constitutes the beginning of formal logic.

so as to give the impression that, to Bocheński, necessary and sufficient conditions for something to be "formal logic" are that it applies those "three great discoveries".

[4]It seems that in fact any such claim on Tarski's part was evanescent: Tarski's original (1933) Polish title applies his *"pojęcie prawdy"* (concept of truth) to *"ję-zykach nauk dedukcyjnych"* (the languages of the deductive sciences); but in his German translation (Tarski, 1936), it applies only to *"formalisierten Sprachen"*, that is, formal(ized) languages. This may, of course, merely acknowledge that most science is *inductive* as much as it is *deductive*.

general (with a few exceptions) neither claim nor have any interest in claiming that discourse in the sciences (at least, in *their* science) is or should be conducted or communicated in a "formal language".

Worries (or complacencies) about the meaning of 'truth' aside, purely practically the 'formality' of 'formal logic' has found little to recommend it to working scientists (including mathematicians in general). As Georg Kreisel points out (about—in effect—the purging of all "constants beyond the logical vocabulary," in the words of Quine, 1953, or the discarding of "semiotic and methodological problems" by Bocheński, 1956/1961),

> Certainly, *if* one is operating on masses of meaningless symbols as one says, then the possibility of formalization or mechanization is a pretty obvious prerequisite for precision. But it would be odd, so to speak illogical, to conclude from this that the reliability of an argument is enhanced by ignoring everything one knows about its subject matter, and by treating it as a manipulation of meaningless symbols! (Kreisel, 1985, p. 145)

Some years earlier, Kreisel observed that his teacher Wittgenstein's

> obviously offhand comment [to Turing] [...], to the effect that [formalizing mathematical proofs] [...] is 'easy', overlooks [...] the distinction between (absolute) effort and the ratio: effort/reward, familiar from economics (utility and marginal utility). (Kreisel, 1978, p. 88)

Whether or not because of this combination of dubiously founded 'reliability',[5] great effort, and practical *unreliability* (at least, a great

---

[5]Imre Lakatos (1962), after quoting (formal) *Logic for Mathematicians* (Rosser, 1953, p. 11) to the effect that a mathematician "should not forget that his intuition is the final authority", asked

> But why on earth have *"ultimate"* tests, "final authority"?[33] Why foundations, if they are admittedly subjective? Why not honestly admit mathematical fallibility, and try to defend the dignity of *fallible* knowledge from cynical scepticism, rather than delude ourselves that we can invisibly mend the latest tear in the fabric of our "ultimate" intuitions? (Lakatos, 1962, p. 184; original italics. Note 33 cites Bourbaki, 1949, p. 8.)

Mutatis mutandis, this seems to me to apply (and with greater force) to any claim that formal logic should, allegedly for foundational reasons, be granted an especially high credibility when used *outside* mathematics.

potential for it), 'formal logic' in the sense(s) of Jeffrey, Quine, Bocheński, and so on, is rarely if ever *used* by scientists (or most mathematicians), either in their daily practice or in the presentation of their work (the results of that practice) to others.

The uses of 'formal logics'

Table 2.1 lists some 'formal logics' in my sense that have been proposed for, *inter alia*, use in the human sciences. Only two are much used there today, namely, probability and statistics (which, it should be noted, were not joined—to whatever extent they are— until quite late: their two "logics" were and are quite different in form and content, *cf.* Aldrich, 2009). Those two, especially statistics, are now used nearly everywhere: "Indeed there is no logic so irresistible as the logic of statistics" (Newton, 1863).

Rather than describe the virtues or failings of any of these formal logics in particular, I propose instead to spend the rest of this chapter discussing what seems to me a more basic question: how does 'a formal logic' in my sense, *as a mathematical model*, derive credibility from mathematics? I argue that it derives credibility in the same way that any mathematical model—or mathematics as a whole— does; and that that is via the *informal logic* of mathematical practice.

The informal logic of mathematical practice

Gian-Carlo Rota, like Burgess both a mathematician and a philosopher, wrote

> philosophy, like mathematics, relies on a method of argumentation that seems to follow the rules of some logic. But the method of philosophical reasoning, unlike the method of mathematical reasoning, has never been clearly agreed upon by philosophers, and much philosophical discussion since its Greek beginnings has been spent on method. (Rota, 2006, p. 221)

As a mathematician who is an occasional outside observer of philosophy and philosophers, I do see (what appears to me to be) ample evidence that "the method of philosophical reasoning" has indeed not yet "been clearly agreed upon by philosophers". I am less sure than

Table 2.1: Formal (mathematized) logics used or proposed for use in the human sciences or related areas.[*]

INTRODUCED OR ELABORATED	LOGIC	MATHEMATICS EMPLOYED AND OTHER NOTABLE FEATURES
1654–1718	discrete probability[1]	arithmetic ($\mathbb{N}$); first "logic of uncertainty"
1812	analytical probability[2]	real analysis ($\mathbb{R}$)
1847	algebraicized logic[3]	algebraic notation; "laws of thought"; Boolean algebra
1840s	statistical logic[4]	samples and populations
1860s–1900s	semiotic logic[5]	triadic system
1873	predicate logic[6]	isolation of intuition by axiomatization
1879	concept logic[7]	two-dimensional formulæ
1908–1920	modal logics[8]	necessity, possibility, etc.
	intuitionist logic[9] multi-valued logics[10] }	Aristotelian "Principle of the Excluded Middle" *denied*
1910	*Principia Mathematica*[11]	"logicism" (reduction of mathematics to pure logic)
1920–1930	deontic logic[12]	logic of obligation
1937	quantum logic[13]	Hilbert spaces; $\mathbb{C}$; non-Boolean algebra; non-commutativity
	vague logic[14]	vague predicates
1965	fuzzy logic[15]	combination of probability theory and set theory
1960s–2000	belief logic[16]	theory of evidence
	temporal logics[17]	time-dependent truth values
	co-genetic logic[18]	formal parallel to Baldwin's "genetic logic"[*]
2006–	steps toward finitistic *ambivalence logic*[19]	finite *non-discrete* topological spaces; use of $\mathbb{R}$ deprecated

[*]Valsiner (2008) focuses on Baldwin's work (*c.*1895–*c.*1915) but surveys other logics, formal and informal. [1]Pascal, Fermat, et al. (*cf.* Hald, 2003). [2]Laplace (1814). [3]Boole (1847), De Morgan (1847). [4]Quetelet (1846). [5,6]Peirce (1873, 1893, 1903/1998). [7]Frege (1879). [8]Lewis (1910,1932). [9]Brouwer (1908). [10]Łukasiewicz (1920). [11]Whitehead and Russell (1912). [12]Mally (1926). [13]Birkhoff and von Neumann (1936). [14]Black (1937). [15]Zadeh (1965). [16]Dempster (1967); Shafer (1976). [17]Burgess (1979, 1980, 1984); Anisov (2002). [18]Herbst (1995). [19]Rudolph (2006a,b,c); see also this volume, pp. 180–181.

Rota that "the method of mathematical reasoning" is in fact "clearly agreed upon", either by philosophers or by mathematicians; but surely (for nearly all mathematicians) it is not *only* formal logic, even if it incorporates *some* formal logic at times (and makes talismanic gestures towards formal logic at others). In this section I identify what seems to me to be the informal logic of mathematical practice, and urge mathematicians—mathematical modelers in particular— to bring it into their awareness and embrace it.

Mathematics, computation, and persuasion

Mathematics is concerned with much more than calculation/computation alone.[6] Nonetheless, performing computations and reasoning about computation (and the performance thereof) have always been —and continue to be—activities of the first importance among mathematicians. Leibniz, for one, had great hopes for computation.

> It was Leibnitz's dream that man should discover: (a) a precisely definable universal symbolism for making statements of science (Leibnitz called this a *characteristica universalis*); and (b) an algorithm which, when applied to the symbols of any formula of the characteristica universalis, would determine whether or not that formula were true as a statement of science[.] Leibnitz called this algorithm a *calculus ratiocinator*. (Rogers, 1963, p. 934; definitional emphasis in the original)

In fact, Leibniz appears to have believed that truths, calculated by such methods, would necessarily be persuasive:

> *quando orientur controversiæ, non magis disputatione opus erit inter duos philosophos, quam inter duos Computistas. Sufficiet enim, calamos in manus sumere, sedereque ad abacos, et sibi mutuo (accito si placet amico) dicere: calculemus.* ["If controversies were to arise," he says, "there would be no more need of disputation between two philosophers than between two accountants. For it would suffice to take their pens in their hands, to sit down to their desks, and to say to each other (with a friend as witness, if they

---

[6]For many readers, "calculation" has an overwhelming connotation of purely *numerical* computation; I will try to use the word only when that is what I mean.

liked), 'Let us calculate.'"] (Leibniz, 1684?/1840, p. 82; Russell, 1900, p. 200)

Less ambitiously, the paradigmatic 'mathematical model' (for users of models, especially in the human sciences) ends with—if it does not entirely reduce to—computations. So: what *is* computation (or a computation), and why (if it is) is it persuasive?

Formalizing computation[7]

Most mathematicians working today would probably assent to the following semi-formal definition.

> A *computation* is a process whereby we proceed from initially given objects, called *inputs*, according to a fixed set of rules, called a *program*, *procedure*, or *algorithm*, through a series of *steps* and arrive at the end of these steps with a final *result*, called the output. The algorithm, as a set of rules proceeding from inputs to output, must be precise and definite, with each successive step clearly determined. (Soare, 1996, p. 286; definitional emphases in the original)

This and similar modern definitions,[8] presented as (semi-)formalizations of the activities of "an idealized *human* calculating agent" (Soare, 1996, p. 291; emphasis in the original), all go back to the work of Alan Turing.

Turing's definition of his entirely non-physical "computing machines" (known as *Turing machines* since so called by Church, 1937) was quite explicitly framed to capture "the fact that the human memory is necessarily limited" (Turing, 1937, p. 231), and to mimic "the behaviour of the [human] computer" (Leibniz's *Computista*) which "is determined by [...] his 'state of mind'" during the act of computing (p. 250). Turing goes so far as to observe that

---

[7]Much of the matter in this and the two following subsections is recapitulated from the section titled "Actual and idealized computation in mathematics" in Rudolph (2011). My ideas, there and here, owe a great deal to (my understanding of) Azzouni (2006). The historical survey by Soare (1996) was a valuable resource.

[8]Pre-modern mathematicians and philosophers did without such definitions, which are very much in the spirit of *modern* mathematics and mathematical logic.

> It is always possible for the computer to break off from
> his work, to go away and forget all about it, and later
> to come back and go on with it. If he does this he must
> leave a note of instructions (written in some standard
> form) explaining how the work is to be continued. This
> note is the counterpart of the "state of mind". (p. 253)

Among various aspects of "the behaviour of the computer" that
Turing idealizes away when defining his "computing machines", the
most obvious (and most important, for his purposes) is that, in the
human realm, there are many other possibilities "for the computer"
besides breaking off from work, going away, forgetting about it,
and then coming back and going on with it. The brute fact of human
mortality, in particular, is a condition of finitude no less salient
than "the fact that the human memory is necessarily limited", but
is (reasonably—for Turing's purposes) not mirrored in his model.

Here I concentrate on two other important aspects of Turing's,
Soare's, and similar idealizations of human computation: the suppression of 'false steps' (including, but not limited to, outright
errors in execution) and of 'flashes of insight' (veridical or delusive).
These suppressions are seen as right and unproblematic by all but a
handful of mathematicians, philosophers of mathematics, and other
knowledgeable commentators on computation.[9] Since both 'false
steps' and 'flashes of insight' have probably been noticed by every
non-ideal "human calculating agent" in at least some private computations, readers who are not mathematicians (or philosophers,
etc.) may want to see some justification for their suppression from
the idealization of computation.

---

[9] Among this handful, two of the more provocative "knowledgeable commentators" are Roger Penrose and Edward Nelson. Penrose (1990, 1994) has proposed
that humans (in particular, humans acting as mathematicians) are not limited to deterministic, algorithmic calculations because quantum effects in human brains can
lead directly to a (possibly mathematical) "flash of understanding" or insight (1994,
p. 141). Nelson has argued that mathematical practice that incorporates algorithmic calculations, insofar as those allow arbitrary (as contrasted with "bounded")
recursion, with its "impredicativity" and concomitant "actual" (or "completed")
infinities, goes beyond its remit (Nelson, 1995), and may yet lead to a "collapse of
contemporary mathematics" (Nelson, 2011).

At a less foundational but very practical level, of course there is a considerable
literature in computer science/engineering devoted precisely to dealing with 'false
steps' engendered by necessarily fallible hardware, buggy (or subverted) software,
operator error, etc.

Suppressing 'false steps' from ideal computation

The suppression of 'false steps' might be justified by the following sequence of claims.

(A) Reliability, replicability, and (to some extent) efficiency are qualities that (nearly universally) are highly desired of actual human computations, just as they are of many other human activities (*e.g.*, legal proceedings, medical diagnoses and surgical procedures, mass production of objects, etc.), though less so (or not at all) of others.

(B) Much actual, non-idealized mathematical practice of computation is reliable and replicable, and (if not uniquely so, so at least) to a markedly higher degree, than actual, non-idealized practice in law, medicine and surgery, mass production, etc.

(C) Actual, non-idealized mathematical practice (by no means limited to the practice of computation) is

    (i) publicly correctable (and often publicly corrected), even though this practice is (often)

    (ii) privately performed (and prone to private error).

(D) Mathematical practices of actual "human calculating agent"s, to the extent that they are of interest to a public (of other human agents), become public and are eventually corrected as necessary to make them reliable, replicable, and efficient.

Of these claims, (A) and (B) seem to me to be purely empirical and entirely unexceptionable; I won't explicate them further here, nor marshal explicit empirical evidence in their favor. Claim (C) is also empirical, but for it to be seen clearly as such I must place it in the appropriate theoretical framework, borrowed from Jody Azzouni.

What I mean by (i) is that *public* mathematical practice depends very strongly on 'reason' manifested *interpersonally* as a communications protocol. Indeed, the (purported) 'objectivity' of interpersonal reason springs (purportedly) from its <u>im</u>personality; and certainly public mathematical practice, though (apparently, at least beyond mere counting) always embodied in persons, appears substantially impersonal. Thus Edward Nelson can answer what he calls "the central question of the sociology of mathematics", namely, "why is it that mathematicians are such nice people [...] quick to say 'I was wrong'", by writing (emphasis in the original)

the worth of a mathematical work is judged largely by

whether the *proof* is correct (Nelson, 2007, p. 843)

and Azzouni (2006, p. 202; emphasis in the original) can note:

> It's widely observed that, unlike other cases of confor-
> mity, and where social factors *really are* the source of that
> conformity, one finds in mathematical practice *nothing
> like* the variability found in cuisine, clothing, or meta-
> physical doctrine.

What I mean by (ii) is that the *intrapersonal* manifestation of 'reason', upon which *private* mathematical practice depends so strongly
(on essentially all accounts, and in my own experience), certainly
appears (on the evidence of many persons' accounts of their personal experiences—mine included) to be (much?) more error-prone
than interpersonal 'reason' in matters of mathematics.[10] Azzouni
states the general position very clearly.

> Let's turn to the second (*unnoticed*) way that mathemat-
> ics *shockingly* differs from other group practices. *Mistakes
> are ubiquitous in mathematics.* [...] What makes math-
> ematics difficult is (1) that it's *so easy* to blunder in;
> and (2) that it's *so easy* for others (or oneself) to see
> —when they're pointed out—that blunders have been
> made. (2006, pp. 204 and 205; italics in the original)

Claim (D), again empirical, echoes similar claims of Nelson
(2007, p. 843; italics in the original):

> whether the *proof* is correct [...] is something on which
> we all agree (eventually), despite the fact that we may
> have divergent views on the nature of mathematics

and Azzouni (2006, p. 203 and p. 208; italics in the original):

---

[10] If 'interpersonal reason' is, in fact, less "error-prone" than 'intrapersonal
reason', it might simply be because 'several minds are better than one' in that,
being several, they have several perspectives, several arrays of emotions, access
to several (not necessarily identical) bodies of fact, etc.; or it might be that what
are taken to be several manifestations of some single 'faculty of reason', common
(in various degrees) to all persons, actually are (in some cases) manifestations of
several different 'faculties' with (only) certain (possibly superficial) similarities.

Very likely these questions have been asked, and answered (possibly in incompatible ways), many times before. In any case, I won't consider them here.

almost every other group practice (diet, language, cosmetics, and so on) [...]—in contrast to mathematical practice—show[s] great deviation across groups. That is, even when systematic algorithmic rules (such as the ones of languages or games) govern a practice, that practice still drifts over time—unlike, as it seems, the algorithmic rules of mathematics.

[...] What seems odd about mathematics as a social practice is the presence of substantial conformity on the one hand, and yet, on the other, the absence of (sometimes brutal) social tools to induce conformity that routinely appear among us *whenever* behavior really is socially constrained. Let's call this "the benign fixation of mathematical practice."

Although readers standing outside the "substantial conformity" of the mathematical community may find claim (D) exceptionable,[11] as with (A) and (B) I won't try to marshal empirical evidence for it here; some can be found in parts IV–VI of Azzouni (2006), in several other essays collected by Hersh (2006), and in Bonsall (1982).

In short, the exclusion of 'false steps' from an idealized account of human computation might be justified by observing "the benign fixation of mathematical practice" and the quite remarkable (eventual) equifinality of mathematicians' judgments of the correctness of the results of that practice.

Suppressing 'flashes of insight' from ideal computation

The suppression of 'flashes of insight' from an idealized account of human computation might also be justified by claims (A)–(D). Indeed, computation by 'flash of insight' may (or may not) be "efficient", but (on the common understanding of the phrase) surely it can be neither "reliable" nor "replicable" *on a societal level*. By

---

[11] The "unconventional" essays on the nature of mathematics collected by Hersh (2006) earn their epithet in various ways, but not from the support they collectively lend (D): as far as I know, that claim (or something very like it) is accepted by virtually all mathematicians, on what is taken (by virtually all mathematicians) to be compelling empirical evidence (bolstered in the case of some mathematicians —hardly all—by explicit beliefs about the metaphysics of mathematics, and not much diminished even in the cases of mathematicians who express more or less skepticism about formally similar claims sometimes made for "science" in general).

the last phrase, I intend both to acknowledge the (apparent, and apparently undisputed) fact that no known mechanism of interpersonal 'gnosis' (as contrasted with interpersonal reason) applies to the communication of mathematics, and to restrict the domain of the idealization by ignoring both the calculations of 'mathematical savants' and the computations of 'mathematical geniuses'.

This restriction has two aspects. (1) If a computation produced by a 'flash of insight' (occurring to anyone—savant, genius, or otherwise) remains forever private, then from the point of view of mathematics as a social endeavor it might as well never have happened. (2) If, on the contrary, a private computation (however produced) somehow becomes public, then "the benign fixation of mathematical practice" will (ideally) lead to its being redone (corrected if need be) or ignored (if uncorrectable or forever irrelevant).

Of these aspects, (1) is tautologous, and (2) abbreviates a more or less standard idealized account, like the following, of "the science of mathematics" (more specifically, of the "scientific method" as appropriately applied to mathematics):

> mathematicians publishing their work [...] each author accepting full responsibility for the correctness of the whole of his publication including the results that he quotes from other authors. [...] Everything in a publication must be based on the individual understanding of the author, nothing being accepted on authority, no matter how distinguished. [...] A second mathematician using the work of another may himself fall into error in following the proof, but at least he has a strong emotional drive to avoid such error [...] It is true that this scientific method will not detect the errors in theorems that are never used. But that is unimportant; such a theorem is a dead branch anyway. (Bonsall, 1982, pp. 9–10)

In short, the suppression of 'flashes of insight' from an idealized account of human computation might be justified by noting that it appears nothing important *about computation on a societal level* is lost by suppressing them.[12]

---

[12]Of course many mathematicians, *have* (or have believed themselves to have) experienced a 'flash of insight' (intrapersonal gnosis) while 'doing mathematics', just as all sorts of persons doing all sorts of other things have done (or have

Why computation is persuasive (when it is)

Having finished my account of idealized human computation, I can propose answers to the questions "what *is* computation (or a computation), and why (if it is) is it persuasive?" I will argue for my answers in the following section.

As to what "computation" in the general sense "*is*", I do not think that in human terms it should be taken to be either some semi-formalization of idealized human computation (like that quoted from Soare, 1996), or some full formalization (like that given, in parts of his paper that I did not quote, by Turing, 1937). I would say, rather, that "computation" in the general sense should be taken to be some *informal* idealization (like the one discussed briefly in the last several pages) of *actual human calculation(s) and other mathematical activities*, with 'false steps' and 'flashes of insight' fully suppressed *as part of, and for the sake of, the idealization.*[13]

The persuasiveness (such as it is—which is remarkably great, to many people, of highly varied degrees of mathematical training and sophistication) of "computation" in the general sense, so understood, would then follow from its being an idealization of a human activity —actual mathematical practice—that is "unique as a social practice" because of its "benign fixation" (Azzouni, 2006, title and p. 208)[14] and robust reliability.[15]

To base a description (or prescription) of what "a computation" is (or should be taken to be) on the preceding description

---

believed themselves to have done), and have then made this experience public. A famous instance from the history of modern mathematics is Poincaré's account of his discovery of "Fuchsian functions"(Poincaré, 1908, pp. 362–364; see also Gruber, 1981, pp. 51–57, and 1996, pp. 412–420). I am not asserting that nothing important *about mathematics as a social behavior* would be lost if such 'flashes' (real or apparent) were invariably kept private, or if their very possibility were systematically denied.

[13]As already mentioned (note 9, p. 48), there are formalizations of computation (mathematical models built upon the Turing formalization) that incorporate formalized 'false steps'. Likewise, a thriving specialty within mathematical logic and theoretical computer science—initiated by Turing (1939) himself—incorporates formalized (veridical) 'flashes of insight' (now often called "oracles"; *cf.* Chang et al., 1994) into formal models of computation. Thus, formalization per se does not require that 'false steps' and 'flashes of insight' be suppressed.

[14]See also pp. 3–4, above: "mathematics has preserved its history in the present. ... All the knowledge base of mathematics—once established—stays as such".

[15]In the section "Mathematical modeling and evolutionary epistemology" of our Introduction (p. 24 *ff.*), we discuss possible origins of this "robust reliability".

ot what "computation in general" is (or can be taken to be) might be irremediably and viciously circular, though I don't believe it need be; but to use that preceding description to explain why and how "a computation" persuades (when it does) is, I think, both legitimate and easily done, as follows. I propose that "a computation" is persuasive *because of, and to the extent that it is felt and/ or understood to be associated to*, the informal idealization of actual mathematical practice that I claim constitutes "computation".

That is, I propose that precisely because (as Azzouni, 2006, says) mathematics is "*so easy* to blunder in", and yet it is also "*so easy* for others (or oneself) to see—when they're pointed out—that blunders have been made", the persuasiveness of any individual instance of actual mathematical practice does not have to rest on a positive assurance of reliability (*i.e.*, freedom from blunders) of that instance: it can rest instead on the (putative) ease of seeing "that blunders have been made", and on the "benign fixation" and robust reliability *of the entire culturally embedded, and culturally maintained, network* of actual mathematical practice in which that instance subsists, and to/ from which it gives/takes meaning.

Mathematical practice as eisegesis of the exoteric

Table 2.2 collects four definitions from the universally acknowledged "definitive record of the English language" (*OED Online*, 2011). The definienda on the main diagonal of the 2 × 2 array, "esoteric" and "exegesis", are used much more often their off-diagonal counterparts, "exoteric" and "eisegesis". Only one of the four, "esoteric", is regularly used of mathematics, and that use (even by mathematicians and mathematical modelers) is typically uncomplimentary if not downright deprecatory.[16] In this section I argue that, on the contrary, mathematics is an essentially *exoteric* activity. More precisely, I make a claim that extends (A)–(D) on pp. 49–49.

---

[16]Word use in written academic English can be assessed by searching various corpora of journal articles, *e.g.*, JSTOR, MUSE, and Google Scholar. At various dates in 2010 and 2011, I searched JSTOR's "archival and current issues of more than 1,400 scholarly journals across more than 50 academic disciplines" (JSTOR, 2011) for the four words in various forms and combinations; these searches confirmed to my satisfaction that "much more often" is correct. Similarly, I satisfied myself that "typically" is justified by searches of both JSTOR and the larger, more broadly based Google Scholar corpus. I omit the details.

Table 2.2: Four definitions from the Oxford English Dictionary.

Esoteric [...] **A.** *adj.* **1. a.** [...] Of philosophical doctrines, treatises, modes of speech, etc.: Designed for, or appropriate to, an inner circle of advanced or privileged disciples; communicated to, or intelligible by, the initiated exclusively. Hence of disciples: Belonging to the inner circle, admitted to the esoteric teaching. [...] **2.** *transf*[erred sense] [...] **b.** Pertaining to a select circle; private, confidential. (Esoteric, 1891/1971)	**Exoteric** [...] **A.** *adj.* **2.** [...] Of philosophical doctrines, treatises, modes of speech, etc.: Designed for or suitable to the generality of disciples; communicated to outsiders, intelligible to the public. Hence of disciples, etc.: Belonging to the outer circle; not admitted to the esoteric teaching. Of an author: Dealing with ordinary topics; commonplace, simple. [...] **3.** *transf.* **a.** Current among the outside public; popular, ordinary, prevailing. (Exoteric, 1894/1971)
eisegesis [...] The interpretation of a word or passage (of the Scriptures) by reading into it one's own ideas. (eisegesis, 1972/1987)	**Exegesis** [...] **1. a.** Explanation, exposition (of a sentence, word, etc.); *esp.* the interpretation of Scripture or a Scriptural passage. [...] **b.** An explanatory note [...] **c.** An expository discourse. (Exegesis, 1894/1971)

(E) The prevailing _in_formal logic in the domain of mathematical
      practice is, in fact, *eisegesis of the exoteric*.

In support of this claim, I have assembled some usage history of the
four words, in their two antonymous pairs.

Esoteric and exoteric knowledge

In the canonical lexicon of classical Greek (Liddell & Scott, 1852),
ἐσωτερικός (whence "esoteric") and ἐξωτερικός (whence "exoteric") have primary definitions—"inner, intimate" (p. 556) and
"external, belonging to the outside" (p. 483), respectively—close to
the OED's 'transferred senses' **2. b** of Esoteric (1891/1971) and **3** of
Exoteric (1894/1971). Contrariwise, the OED's primary definitions
(incorporating identical range limitations to "philosophical doctrines, treatises, modes of speech, etc.") clearly have been derived,
during the words' post-classical development, by generalization
from the uses described by Liddell and Scott in glosses to their

primary definitions: under ἐσωτερικός, "esp. of those disciples of Pythagoras, Aristotle, etc., who were scientifically taught"; under ἐξωτερικός, "esp. of those disciples of Pythagoras and others *who were not yet initiated into their highest philosophy*" (emphasis in the original).

All these definitions and senses are quite neutral. Although the classical Greek prefixes that became English "es-" in "esoteric" and "ex-" in "exoteric" conveyed the spatial meanings *in, inwards* and *out, outwards*, respectively, I can find no evidence that (in these words) the Greeks took the *dimension* those meanings implicitly span to be a *polarity*, *i.e.*, a dimension equipped with a mapping to the dimension spanned by 'good' (or 'desirable', 'favored', etc.) and 'bad' (or 'undesirable', 'disfavored', etc.). Nor can I see that the OED records any such *denotational* polarity for the English words (unless, maybe, "select" in Esoteric, 1891/1971, **2. b**, and "ordinary" in Exoteric, 1894/1971, **3**, are weak markers of such a polarity).

In present usage, however, there is (outside the many contexts where the words are used historically or technically, *e.g.*, discussions of religious traditions or practices) a highly charged *connotational* polarity, or rather *monopolarity*: "esoteric" is often used to deprecate (and rarely, if ever, to commend), but "exoteric" appears not to have been adopted as either a commendation or a deprecation.

The indirect deprecatory use of "esoteric" (*i.e.*, attributed by a writer to somebody else) appears as early as Shepherd, 1887, p. 236. The earliest direct use quoted by the OED—"precious, morbid, esoteric sensibilities" (atonalism, 1972/1987)—appeared in 1937.

In recent academic English, it appears that most deprecatory uses of "esoteric" are indirect.[17] Here are some examples.

> First of all we must recognize the fact that most students turn up at college badly prepared in mathematics. They believe that even rudimentary arithmetic has an arcane or esoteric quality that makes it incomprehensible except to the elect. (Kusch, 1966, p. 41)

> We need to combat, especially in the schools, the view that mathematics is an esoteric activity entirely revealed to and used by other-worldish people, beyond the ken

---

[17] Not all are: *e.g.*,"[T]he words used are esoteric. Witness, for example, the word 'eisegesis' [...] [an] expression [...] scarcely in common parlance" (Rosenwald, 1977, p. 527) is part of a quite direct attack on the (style of the) author being reviewed.

of ordinary people. (Jones, 1982, p. 797)

> The third criticism of laboratory experiments was that they tended to focus on the study of esoteric and even absurd situations as a means of understanding everyday reality [...] all was not well with social psychology in the summer of 1974 [...]. (Forgas, 2003, p. 254)

> Moreover, the (alleged) complexity and esoteric nature of economic laws sometimes mean that political leaders, not to mention citizens, are deemed too inept or ignorant [...] to make economic decisions. (Swanson, 2008, p. 63)

> The ecology community at that time thought of matrix algebra as an advanced and esoteric mathematical subject. (Briggs et al., 2010, p. 246)

This academic deprecatory use of "esoteric" appears to function as follows. The writer asserts that

(a) **someone else** ("most students"; "ordinary people"; social psychologists criticizing the idea of "laboratory experiments" in the summer of 1974; elite economic theorists, counterpoised to "political leaders, not to mention citizens"; "the ecology community at that time"), or perhaps the writer at an earlier date, has

(b) **attributed** to certain intellectual objects or productions ("rudimentary arithmetic"; mathematics in general; "situations" that are created in "laboratory experiments" instead of arising in "everyday reality"; "economic laws"; "matrix algebra") the property of being "esoteric" (though not necessarily using that word, nor even being aware that such a word exists), *and therefore* has

(c) **refrained** from engaging with or in these objects or productions, and/or **restrained** others from so engaging, *even though*

(d) the writer **knows** (and the reader has now been alerted) that these objects or productions **really aren't** so foreign (so "arcane"; so "entirely revealed" and "other-worldish"; so close to "absurd" in their non-"everyday" un-"reality"; so filled with "complexity", so "advanced") that *We* (the writer and reader) couldn't take them in stride, dismissing them or (easily) assimilating them according to their other (completely obvious) qualities.

That is, the deprecation gets its force from a double denial: the inner circle *They* talk about doesn't exist—and *We* don't want to join it.

This is all very democratic and Enlightened, and suitable to the (surface) rules of contemporary academic culture(s), following which no academic could call another anyone's "disciple" with a neutral or positive intonation and a straight face.[18] It also has— inevitably, I think, given the sheep/goat partition inherent in part (a) —the effect of *creating an inner circle* with **Us** at the center, separated from **Them** by part (d) of the deprecation process, that is, by **Our** special knowledge that there is no special knowledge.

Eisegetic and exegetic interpretation

"Eisegesis" and "exegesis", like "esoteric" and "exoteric", have classical Greek precursors spanning an implicit dimension. As in that case, I find no evidence that this dimension was taken to be a polarity in the classical era: Liddell and Scott (1852) define εἰσήγησις (whence "eisegesis") and ἐξήγησις (whence "exegesis") quite neutrally, as "a bringing in, introduction, proposing, bringing forward" (p. 408) and "a statement, narrative, [...] an explaining, explanation" (p. 478), respectively.[19] Again, eisegesis (1972/1987) and Exegesis (1894/1971), like Esoteric (1891/1971) and Exoteric (1894/1971), contain range limitations, in this case not identical but still close: "(of the Scriptures)" for the former; "*esp.* the interpretation of Scripture or a Scriptural passage" for the primary definition of the latter.

The parallelism between the two pairs of antonyms stops there. The OED's eisegesis (1972/1987) and Exegesis (1894/1971) barely suggest a weak connotational polarity (dependent on a presupposed cultural norm favoring "Scripture" over "one's own ideas"), but even a brief survey of actual uses of the words shows that a highly charged connotational polarity exists. In fact, nearly everyone (with some fairly recent academic exceptions, discussed below) who uses the word "eisegesis" at all, uses it in reference to (Christian) Scriptures, and contrasts it unfavorably (often *very* unfavorably) with "exegesis". On the other hand, with the exceptions already

---

[18]"Contemporary" is a necessary modifier. Russell (1919, p. 79) wrote, with evident commendation, of Peano's "very able school of young Italian disciples."

[19]Their authority for the former may be weak: according to Graves (1908, p. 135), their *locus classicus* (Thucydides 5.30) is also a *hapax legomenon*: "εἰσήγησις appears to be found here only in classical Greek". Graves glosses it with Latin *rogatio*, a (parliamentary) motion, maintaining neutrality while discarding the dimension.

alluded to, very few academic writers writing on subjects unrelated to religion *do* use "eisegesis", whereas "exegesis" is very common throughout academic English—either entirely freed of its "Scriptural" range limitation (*i.e.*, falling under sense **b** or **c** of Exegesis, 1894/1971), or with "Scripture" generalized to 'an important text'.

Not only was εισήγήσις almost absent from classic Greek (note 19), it seems to have left no traces in European writing (unless in later Greek) until it suddenly reappeared (or was freshly coined) as *Eisegese* in mid-19th century German theological treatises (Kurtz, 1846, p. 27; Ströbel, 1855, p. 119). Theology has remained the most common domain of use for the word. Its first sighting in English (in the derived form "eisegetical"; eisegesis, 1972/1987), in 1878, is also in a theological context.[20] Since then it has appeared in documents—variously scholarly, homiletic, and apologetic—produced by and for (mostly evangelical) Christians. The following examples are typical.

> *Vor 1746 gab es eine Exegese, eine Auslegung dessen, was die Schrift enhält; nach 1746 giebt es eine Eisegese, welche das miserable Produkt ihres eigenen Genius taschenspielerisch in die Bibel hineinpracticirt [...]*. [Before 1746 there was exegesis, a laying out of what Scripture contains; after 1746 there is eisegesis, which like a pickpocket slips the miserable product of its own Genius into the Bible [...].] (Ströbel, 1855, pp. 118–119; emphasis in the original)

> How much broader and more wholesome the divine truth than the claptrap eisegesis substituted for it! (Funk & Gregory, 1900, p. 381)

> The cloven hoof of eisegesis tempts us all. (Branton, Brown, Burrows, & Smart, 1955/1956, p. 38)

> 'The Eisegesis Virus' [...] has been found responsible for the 'death' of many church members. (Ham, 2002, p. 16)

That "eisegesis" in these uses is deprecatory (indeed, "derogatory"; *cf.* Grenz, Guretzki, & Nordling, 1999, p. 50) is very clear, and very understandable for authors to whom a soul or souls are at stake.

"Eisegesis" seems to have been introduced to secular academic discourse quite late, by R. H. Dana (1966), specifically in the context

---

[20]Philip Schaff, the OED's source for "eisegetical" (1878, p. 53), was an American theologian and church historian. Schaff's first language was German, and he was demonstrably familiar with Kurtz (1846; see, *e.g.*, Schaff & Jackson, 1887, p. 121).

of personality assessment.[21] With something much less (or much more) than souls at stake, Dana and authors who follow his lead do not use the word as a derogation, but (to my ear) they still do deprecate eisegesis as a process, even as they minimize deprecation of the eisegete by acknowledging that eisegesis *will* occur, and propose counter-processes to minimize its effect.

> Dana (1966) urges that eisegesis be trained out of student psychologists and suggests a way of doing so. (Tallent, 1976, p. 35)

> In using the Rorschach or the Thematic Apperception Test (TAT) *eisegesis* describes interpretive inferences from projective data that are prone to contamination by un-acknowledged assessor fantasy, personalization, and/or bias [...]. (Dana, 1998, p. 167; original italics)

> [C]ase studies can become prime examples of *eisegesis* in which the author injects and creates the issues, problems and principles which are then presented as being dis-covered. The author puts his thumb into the case-study pie, and, *mirabile dictu*, finds exactly the issues and prin-ciples that are required for illustrative or teaching pur-poses. Thus, while purporting to be about the world "out there", case studies may actually more accurately reflect the mind of the author. This is somewhat ironic given the reaction against hypotheticals in favour of "real people" in much of the literature on medical ethics. (Pattison, Dickenson, Parker, & Heller, 1999, p. 43; original italics)

As suggested in the quotation from Pattison et al., and confirmed by Dana (personal communication, September 20, 2011), in this tradition the acknowledgment and management of eisegesis is an ethical requirement, not a moral injunction: "eisegesis gone wild" (Tallent, 1976, p. 55) can be an ethical lapse (*e.g.*, in relationships with power differentials like that between the writer and the subject of a psychological report), but it is not a moral failing.

---

[21]A yet later stream of academic uses of "eisegesis" appears in literary scholar-ship, beginning with Day (1974, p. 134): "Biblical scholarship is said to employ a term, eisegesis, which differs from exegesis in that it designates not so much the attempt to interpret a text as the effort to interpret the interpreter's views or feelings about the text". All uses of "eisegesis" in this stream (that I have tracked down) appear again to be deprecatory (or self-deprecatory, as with Caspel, 1986).

Why should the devil have all the good words?

With usage histories in place, my proof of claim (E), that "The prevailing informal logic in the domain of mathematical practice is, in fact, *eisegesis of the exoteric*", is simple. First I rewrite it as two sub-claims:

(i) within the domain of actual mathematical practice, mathematical knowledge is treated as exoteric, not esoteric;

(ii) within the domain of actual mathematical practice, mathematical interpretation is performed eisegetically, not exegetically.

Then I observe that, denotationally, (i) and (ii) are essentially equivalent to the correspondingly numbered sub-claims of (C) on p. 49, *viz.*, that "Actual, non-idealized mathematical practice … is (i) publicly correctable (and often publicly corrected), even though this practice is (often) (ii) privately performed (and prone to private error)." That —assuming the reader already accepts (C), and agrees that the two sets of sub-claims are "essentially equivalent"—completes the proof of claim (E). … Or, at least, that completes the proof of (E) in its aspect as an empirical claim about what is true "in fact".

But (E) has a *deontic* aspect too: *i.e.*, it "connotes" my "degree of requirement of, desire for," or "commitment to the realization of the proposition" that I mean it to express (SIL, 2004). Specifically:

(1) like essentially all statements in all natural languages, sub-claims (i) and (ii) have connotational as well as denotational content;

(2) as statements in academic English, where (as just seen) "esoteric" and "eisegetic" have strong standard connotations, the default connotational tone of (i) is mildly positive (by default, it denies a deprecation or derogation of "mathematical knowledge", though giving no explicit approbation) and that of (ii) is negative (it avers a deprecation of "mathematical interpretation");

(3) like many statements in any language (and most statements in academic publications), each of (i) and (ii)—by the very fact that I have stated it without disclaimers—is understood to express my "commitment" to its understood connotations.

If this analysis is correct, the simplest completion of the proof of (E) would be to overturn (3) with the relevant disclaimers: minimally, replacing each "is" by "is (or should be)"; more drastically, perhaps hedging the domain specification "within the domain of actual mathematical practice" somehow, to make it clear that, after all, I have not witnessed *all* actual mathematical practice, or even conducted a statistically valid survey of representative exemplars

Dialectically, that would (I suppose) be satisfactory. Rhetorically, I find it more satisfactory to go beyond disclaimers to endorsements, that is, to overturn (2) and (3) simultaneously.[22] So be it.

> **Endorsement I: *pro* exoteric.** It is a positive good that knowledge be available for all persons to avail themselves of it as they will, and that no divisions between "inner" and "outer circles" (as they may dynamically arise) be socially/culturally enforced.

> **Endorsement II: *pro* eisegesis.** It is a positive good that persons interpret knowledge for and among themselves as they will, and that no body of knowledge be socially/culturally afforded a status as uniquely interpretable or uninterpretable (*e.g.*, as "scriptures").

I do not propose **I** and **II** as general axioms (though that might be interesting), nor claim I necessarily would be comfortable applying them entirely generally. I am, however, comfortable applying them to mathematical practice, not least because, relativized to that context and de-deontified, they become *true statements of empirical fact*, as Azzouni (2006) shows to my satisfaction along lines described earlier in this chapter. That is, in actual mathematical practice:

(a) mathematical knowledge *is* (eventually) available for all persons to avail themselves of it as they will;

(b) no divisions between "inner" and "outer circles" are (ultimately) enforceable, nor are they (generally) enforced;

(c) persons engaging in the practice *do* interpret mathematical knowledge for and among themselves as they will, making *and* correcting errors both privately *and* publicly; and

(d) although "the benign fixation of mathematical practice" constantly creates (and recreates) the body of mathematical knowledge, neither it nor (apparently) any other social or cultural process affords that body of knowledge or any part of it a "scriptural" status as uniquely interpretable or uninterpretable.

---

[22]I have no desire to attempt to overturn (1), here or in general. I am sure one reason most "formal logics" fail, in practice, to be even adequate descriptions of—much less replacements for—the "laws of thought" (as Arnauld & Nicole, 1662; Boole, 1851, and others since have proposed them to be), is that *human* languages and logic(s), like *human* thought, subsist in and through structures that are connotational as much as they are denotational.

Now that I have stated and (somewhat) justified my endorsements **I** and **II**, I have concluded my argument for claim (E), and with it my discussion of the informal logic of mathematical practice.

Logic and understanding in mathematics and mathematical modeling

The number theorist G. H. Hardy famously began his memoir, *A Mathematician's Apology*, with the assertion that "The function of a mathematician is [...] to prove new theorems" (1940, p. 1). In this section I discuss the function(s) of mathematicians, particularly when involved in mathematical modeling.

"That astonishingly beautiful complex"

Hardy wrote his *Apology* at the end of his mathematical career, when he was convinced, perhaps correctly, that "his creative powers as a mathematician at last, in his sixties, [had] left him" (Snow, 1967, p. 50). Close to the height of his career, addressing the British Association for the Advancement of Science as president of the Section of Mathematics and Physics, he had described mathematicians' work somewhat differently.

> The function of a mathematician is simply to observe the facts about his own intricate system of reality, that astonishingly beautiful complex of logical relations which forms the subject-matter of his science, as if he were an explorer looking at a distant range of mountains, and to record the results of his observations in a series of maps, each of which is a branch of pure mathematics. Many of these maps have been completed, while in others, and these, naturally, are the most interesting, there are vast uncharted regions. Some, it seems, have some relevance to the structure of the physical world, while others have no such tangible application. (Hardy, 1922, p. 402)

Of his two descriptions of what mathematicians do and how they exist as mathematicians, this earlier one feels more nearly correct to me, particularly when I take Hardy's geographical metaphor seriously (perhaps more seriously than Hardy meant).

As I read the metaphor, Hardy is emphasizing interplay between the "intricate system" of mathematics—that is, the *structure* (actually, structures) collected under that name—and mathematicians, who are defined functionally by their *activities* in relationship with that system. The activities Hardy makes explicit (by using verbal forms) are observing, looking, recording, and completing. But if the metaphor is to cohere at all, these must be complemented by other activities, albeit ones that are present only implicitly, in his nouns, adjectives, and adverbs: judging factuality, feeling of astonishment and beauty and interest, exploring, and mapping. Taken together (especially in the context of mathematics), these activities serve pretty well (it seems to me) as aspects of a single, 'higher-order' activity, Understanding.

On an account like this, *the function of mathematicians is to **understand** mathematics*; a "new theorem" can be significant, but a *proof* (be it of a new theorem or an old one) is truly valuable—not because it certifies that a theorem is true, but because it certifies that a person—functioning as mathematician—understands (some part of) mathematics. As one topologist has observed, the desire to have a 'certificate of understanding' (and not just in mathematics) can be overwhelming.

> [...] I was unable to find flaws in my "proof" for quite a while, even though the error is very obvious. It was a psychological problem, a blindness, an excitement, an inhibition of reasoning by an underlying fear of being wrong. Techniques leading to the abandonment of such inhibitions should be cultivated by every honest mathematician. (Stallings, 1966, p. 88)

An advantage mathematics (and mathematical modeling) has, over other social practices where such desires can arise, is its "benign fixation", which provides a (uniquely?) favorable environment for the development of the "techniques" that Stallings calls for.

The penultimate Proverb of Hell

Hardy also wrote (again, famously)

> that very little of mathematics is useful practically, and that that little is comparatively dull. The 'seriousness' of

a mathematical theorem lies, not in its practical consequences, which are usually negligible, but in the *significance* of the mathematical ideas which it connects. We
may say, roughly, that a mathematical idea is 'significant' if it can be connected, in a natural and illuminating
way, with a large complex of other mathematical ideas.
Thus a serious mathematical theorem, a theorem which
connects significant ideas, is likely to lead to important
advances in mathematics itself and even in other sciences.
(Hardy, 1940, p. 29; italics in the original)

William Blake's coda to his *Proverbs of Hell* (Blake, 1790/1960,
p. 111) is the ambiguous injunction "Enough! or Too much." Immediately before it, Blake forthrightly proclaims an epistemological
axiom: "Truth can never be told so as to be understood, and not
be believ'd." In the context of mathematics, this axiom together
with my preceding account of the function of mathematicians gives
an operational definition of truth in mathematics (and in whatever
mathematics can be used to model) not much like Tarski's: for mathematicians (and mathematical modelers), **truth is what they come
to believe more firmly as they function better** (as mathematicians,
or mathematical modelers).

As such, truth is always conditional and subject to amendment:
but it has, always and unconditionally, a net or web of meaning
that anchors it pretty firmly to many places in the "astonishingly
beautiful complex of logical relations" *and* (in the case of modeling)
to many places in that other "astonishingly beautiful complex",
human experience of the world; and that version of "truth" is good
enough to keep us going, mostly.

References

Aldrich, J. (2009). *Mathematics in the Statistical Society 1883–1933.*
    University of Southampton, School of Social Sciences, Discus-
    sion Papers in Economics and Econometrics No. 0919. Re-
    trieved May 28, 2012, from http://www.southampton.ac.uk/
    socsci/economics/research/papers/documents/2009/
    0919.pdf
Anisov, A. M. (2002). Logika neopredelennosti i neopredelennost'

vo vremeni [The logic of indeterminacy and indeterminacy in time]. *Logical Studies*, *8*, 1–27.

Arnauld, A., & Nicole, P. (1662). *La logique, ou l'art de penser [Logic, or the Art of Thinking]*. Paris, FR: Charles Savreux.

atonalism. (1987). *The Compact Edition of the Oxford English Dictionary* (Vol. III, p. 37). Oxford, GB: Clarendon Press. (First published in *A Supplement to the OED* I, 1972)

Azzouni, J. (2006). How and why mathematics is unique as a social practice. In R. Hersh (Ed.), *18 Unconventional Essays on the Nature of Mathematics* (pp. 201–219). New York, NY: Springer.

Birkhoff, G., & Neumann, J. von. (1936). The logic of quantum mechanics. *Annals of Mathematics*, *37*, 823–843.

Black, M. (1937). Vagueness. *Philosophy of Science*, *4*, 427–455.

Blake, W. (1960). *Blake* (R. Todd, Ed.). New York, NY: Dell Publishing Co. (Original work published 1790; "THE MARRIAGE OF HEAVEN AND HELL is not dated, but the reference to 'thirty-three years' makes it certain that it was written in 1790" [R. Todd, *Notes*, p. 158])

Bocheński, I. M. (1961). *A History of Formal Logic* (I. Thomas, Ed. & Trans.). Notre Dame, IN: University of Notre Dame Press. (Original work published 1956)

Bonsall, F. F. (1982). A down-to-earth view of mathematics. *The American Mathematical Monthly*, *89*(1), 8–15.

Boole, G. (1847). *The Mathematical Analysis of Logic: Being an Essay Towards a Calculus of Deductive Reasoning*. Cambridge, GB: Macmillan, Barclay, & Macmillan.

Boole, G. (1851). *An Investigation of the Laws of Thought*. London, GB: Macmillan.

Bourbaki, N. (1949). Foundations of mathematics for the working mathematician. *Journal of Symbolic Logic*, *14*(1), 1–8.

Branton, J. R., Brown, R. A., Burrows, M., & Smart, J. D. (1955/1956). Our present situation in biblical theology. *Religion in Life*, *XXVI*(1), 5–49.

Briggs, J., Dabbs, K., Holm, M., Lubben, J., Rebarber, R., Tenhumberg, B., et al. (2010). Structured population dynamics: An introduction to integral modeling. *Mathematics Magazine*, *83*(4), 243–257.

Brouwer, L. E. J. (1908). De onbetrouwbaarheid der logische principes [The unreliability of the logical principles]. *Tijdschrift voor Wijsbegeerte*, *2*, 152–158.

Burgess, J. P. (1979). Logic and time. *Journal of Symbolic Logic, 44,* 566–582.

Burgess, J. P. (1980). Decidability for branching time. *Studia Logica, 39,* 203–218.

Burgess, J. P. (1984). Basic tense logic. In D. Gabbay & F. Guenther (Eds.), *Handbook of Philosophical Logic, Vol. II: Extensions of Classical Logic.* (pp. 89–133). Dordrecht, NL: D. Reidel Publishing Company.

Burgess, J. P. (2009). *Philosophical Logic.* Princeton, NJ: Princeton University Press.

Caspel, P. V. (1986). *Bloomers on the Liffey: Eisegetical Readings of Joyce's Ulysses.* Baltimore, MD: Johns Hopkins University Press.

Chang, R., Chor, B., Goldreich, O., Hartmanis, J., Hastad, J., Ranjan, D., et al. (1994). The random oracle hypothesis is false. *Journal of Computer and System Sciences, 49*(1), 24–39.

Church, A. (1937). Untitled [Review of the paper *On computable numbers, with an application to the* Entscheidungsproblem]. *The Journal of Symbolic Logic, 2*(1), 42–43.

Couturat, L. (1900). L'algèbre universelle de M. Whitehead [Review of the book *A Treatise on Universal Algebra, with Applications. Vol. 1*]. *Revue de Métaphysique et de Morale, 8*(3), 323–362.

Dana, R. H. (1966). Eisegesis and assessment. *Journal of Projective Techniques and Personality Assessment, 30,* 215–222.

Dana, R. H. (1998). Projective assessment of Latinos in the United States: Current realities, problems, and prospects. *Cultural Diversity and Mental Health, 4*(3), 165–184.

Day, R. A. (1974). James Joyce à la mode [Review of the book *L'Éxil de James Joyce ou l'art de remplacement*]. *The Sewanee Review, 82*(1), 130–138.

De Morgan, A. (1847). *Formal Logic.* London, GB: Taylor and Walton.

Dempster, A. P. (1967). Upper and lower probabilities induced by a multivalued mapping. *The Annals of Mathematical Statistics, 38*(2), 325–339.

elsegesis. (1987). *The Compact Edition of the Oxford English Dictionary* (Vol. III, p. 230). Oxford, GB: Clarendon Press. (First published in *A Supplement to the OED* I, 1972)

Esoteric. (1971). *The Compact Edition of the Oxford English Dictionary* (Vol. I, p. 894). Oxford, GB: Oxford University Press. (Earlier version first published in *New English Dictionary*, 1891)

Exegesis. (1971). *The Compact Edition of the Oxford English Dictionary* (Vol. I, p. 921). Oxford, GB: Oxford University Press. (Earlier version first published in *New English Dictionary*, 1894)

Exoteric. (1971). *The Compact Edition of the Oxford English Dictionary* (Vol. I, p. 927). Oxford, GB: Oxford University Press. (Earlier version first published in *New English Dictionary*, 1894)

Forgas, J. P. (2003). Why don't we do it in the road … ? Stereotyping and prejudice in mundane situations. *Psychological Inquiry*, 14(3 & 4), 251–257.

Frege, G. (1879). *Begriffsschrift, eine der arithmetischen nachgebildete Formelsprache des reinen Denkens [Concept Notation: A Formula Language of Pure Thought, Modeled on Arithmetic]*. Halle a.S., DE: Louis Nebert.

Funk, I. K., & Gregory, D. S. (1900). Perverse eisegesis in preaching. In I. K. Funk & D. S. Gregory (Eds.), *Homiletic Review* (Vol. XL, p. 381). New York, NY: Funk and Wagnalls.

Graves, C. E. (1908). *The Fifth Book of Thucydides*. London, GB: Macmillan.

Grenz, S. J., Guretzki, D., & Nordling, C. F. (1999). *Pocket Dictionary of Theological Terms*. Downers Grove, IL: InterVarsity Press.

Groarke, L. (2008). Informal logic. In E. N. Zalta (Ed.), *The Stanford Encyclopedia of Philosophy* (Fall 2008 ed.). Stanford, CA: Metaphysics Research Lab, CSLI, Stanford University. Retrieved December 15, 2011, from http://plato.stanford.edu/archives/fall2008/entries/logic-informal/ (Original work published 1996)

Gruber, H. E. (1981). On the relation between 'Aha experiences' and the construction of ideas. *History of Science*, 9, 41–58.

Gruber, H. E. (1996). Insight and affect in the history of science. In R. J. Sternberg & J. E. Davidson (Eds.), *The Nature of Insight* (pp. 398–431). Cambridge, MA: MIT Press.

Hald, A. (2003). *A History of Probability and Statistics and Their Applications Before 1750*. New York, NY: Wiley.

Ham, K. (2002). Eisegesis: A Genesis virus. *Creation*, 24(3), 16–19. Retrieved September 15, 2011, from http://creation.mobi/article/430

Hardy, G. H. (1922). The theory of numbers. *Science (N. S.)*, 56(1450), 401–405.

Hardy, G. H. (1940). *A Mathematician's Apology*. Cambridge, GB:

Cambridge University Press.

Herbst, D. (1995). What happens when we make a distinction: An elementary introduction to co-genetic logic. In T. Kindermann & J. Valsiner (Eds.), *Development of Person-Context Relations* (pp. 67–79). Hillsdale, NJ: Lawrence Erlbaum Associates, Inc.

Hersh, R. (Ed.). (2006). *18 Unconventional Essays on the Nature of Mathematics*. New York, NY: Springer. Retrieved from http://dx.doi.org/10.1007/0-387-29831-2

History of Logic. (n.d.). Wikipedia, The Free Encyclopedia. Retrieved October 28, 2011, from http://en.wikipedia.org/wiki/History_of_Logic)

Jeffrey, R. (2006). *Formal Logic: Its Scope and Limits* (J. P. Burgess, Ed.). Indianapolis, IN: Hackett Publishing Company. (Original work published 1967; fourth edition "Edited, with a New Supplement, by John P. Burgess")

Jones, P. S. (1982). Untitled [Review of the book *Great Moments in Mathematics (Before 1650)*]. *The American Mathematical Monthly, 89*(10), 796–798.

JSTOR. (2011). *Factsheet.* Retrieved June 7, 2012, from http://about.jstor.org/sites/default/files/jstor-factsheet-20120213.pdf.

Kreisel, G. (1978). Untitled [Review of the book *Wittgenstein's Lectures on the Foundations of Mathematics, Cambridge, 1939*]. *Bulletin of the American Mathematical Society, 84,* 79–90.

Kreisel, G. (1985). Mathematical logic: Tool and object lesson for science. *Synthese, 62*(2), 139–151.

Kurtz, J. H. (1846). *Die Einheit der Genesis. Ein Beitrag zur Kritik und Exegese der Genesis [The Unity of Genesis. A Contribution to the Criticism and Exegesis of Genesis].* Berlin, DE: Wohlgemuth.

Kusch, P. (1966). Science and the university. *Proceedings of the Academy of Political Science, 28*(2), 34–45.

Ladd, G. T. (1897). *Philosophy of Knowledge: An Inquiry Into the Nature, Limits, and Validity of Human Cognitive Faculty.* New York, NY: Charles Scribner's Sons.

Lakatos, I. (1962). Infinite regress and foundations of mathematics. *Proceedings of the Aristotelian Society, Supplementary Volumes, 36,* 155–184. (Reprinted with editorial corrections in *Mathematics, Science, and Epistemology: Vol. 2 of Philosophical Papers of Imre Lakatos,* J. Worrall and G. Currie, Eds., Cambridge University Press, 1980, Chap. I, pp. 3–23)

Laplace, P.-S. (1814). *Théorie analytique des probabilités [Analytical Theory of Probability]* (2nd ed.). Paris, FR: Courcier. ("Seconde Édition, revue et aumentée par l'auteur")

Leibniz, G. W. (1684?/1840). De Scientia Universali seu Calculo Philosophico [On universal science, or, The philosophical calculus]. In J. E. Erdmann (Ed.), *Opera Philosophica* (pp. 82–85). Berlin, DE: G. Eichler. (From an unpublished manuscript in the Royal Library of Hanover. A precise dating to 1684 is ascribed to Trendelenburg, 1867, p. 18, note 1, by Windelband, 1893, p. 382, but this is an evidently careless reading of Trendelenburg. The same date is later published without ascription by Ladd, 1897, p. 69 and Couturat, 1900, p. 362, *inter alia*)

Lewis, C. I. (1910). *The Place of Intuition in Knowledge*. Ph. D. thesis, Harvard University.

Lewis, C. I. (1932). Alternative systems of logic. *The Monist, 42*, 481–507.

Liddell, H. G., & Scott, R. (1852). *A Greek-English Lexicon*. New York, NY: Harper and Brothers.

Logic and Metalogic. (2011). eM Publications.

Łukasiewicz, J. (1920). O logice trojwartosciowej [On three-valued logics]. *Ruch Filozoficny, 5*, 170–171.

Mally, E. (1926). *Grundgesetze des Sollens: Elemente der Logik des Willens [Fundamental Laws of Obligation: Elements of Deontic Logic]*. Graz, AT: Leuschner & Lubensky Universitäts-Buchhandlung.

Mates, B. (1960). Untitled [Review of the book *Formale Logik*]. *The Journal of Symbolic Logic, 25*(1), 57–62.

McKinsey, J. C. C., & Suppes, P. (1954). Untitled [Review of the book *La structure des théories physiques*.]. *The Journal of Symbolic Logic, 19*(1), 52–55.

Mossakowski, T., Goguen, J., Diaconescu, R., & Tarlecki, A. (2005). What is a logic? In J.-Y. Beziau (Ed.), *Logica Universalis* (pp. 113–133). Basel, CH: Birkhäuser.

Nelson, E. (1995). Ramified recursion and intuitionism. In A. Fruchard & A. Troesch (Eds.), *Colloque trajectorien à la mémoire de Georges Reeb et Jean-Louis Callot*, Prépublication IRMA (pp. 171–178). Strasbourg, FR: Institut de Recherche Mathématique Avancée.

Nelson, E. (2007). Untitled [Review of the book *18 Unconventional Essays on the Nature of Mathematics*]. *American Mathematical Monthly, 113*(9), 843–848.

Nelson, E. (2011). Warning signs of a possible collapse of contemporary mathematics. In M. Heller & W. H. Woodin (Eds.), *Infinity: New Research Frontiers* (pp. 75–85). Cambridge, GB: Cambridge University Press.

Newton, I. (1863). Report of the commissioner of agriculture for the year 1863. In *Executive Documents Printed by Order of the House of Representatives During the First Session of the Thirty-Eighth Congress, 1836-'64* (Executive Document No. 91, pp. 3–17). Washington, DC: Government Printing Office.

OED Online. (2011, June). New York, NY: Oxford University Press. Retrieved July 31, 2011, from http://www.oed.com/

Pattison, S., Dickenson, D., Parker, M., & Heller, T. (1999). Do case studies mislead about the nature of reality? *Journal of Medical Ethics*, 25(1), 42–46.

Peirce, C. S. (1873). Description of a notation for the logic of relatives, resulting from an amplification of the conceptions of Boole's calculus of logic. *Memoirs of the American Academy of Arts and Sciences, New Series*, 9 Vol. 9, No. 2(2), 317–378.

Peirce, C. S. (1893). Evolutionary love. *The Monist*, 3, 176–200.

Peirce, C. S. (1998). Nomenclature and divisions of triadic relations, as far as they are determined. In N. Houser & C. Kloesel (Eds.), *The Essential Peirce: Selected Philosophical Writings* (Vol. 2, pp. 289–299). Bloomington and Indianapolis, IN: Indiana University Press. (Original work published 1903)

Penrose, R. (1990). *The Emperor's New Mind*. New York, NY: Oxford University Press.

Penrose, R. (1994). *Shadows of the Mind*. New York, NY: Oxford University Press.

Poincaré, H. (1908). L'invention mathématique [Mathematical invention]. *L'Enseignement Mathématique*, 10, 357–371.

Quetelet, A. (1846). *Lettres sur la théorie des probabilités, appliquée aux sciences morales et politiques [Letters on the Theory of Probability, Applied to the Moral and Political Sciences]*. Brussels, BE: Hayez.

Quine, W. V. O. (1953). Mr. Strawson on logical theory [Review of *Introduction to Logical Theory*]. *Mind*, 62, 433–451.

Rogers, H. J. (1963). An example in mathematical logic. *The American Mathematical Monthly*, 70(9), 929–945.

Rosenwald, A. K. (1977). Old World revisited [Review of Tallent, 1976]. *PsycCRITIQUES*, 22(7), 527–528.

Rosser, J. B. (1953). *Logic for Mathematicians*. New York, NY: McGraw-

Hill.

Rota, G.-C. (2006). The pernicious influence of mathematics upon philosophy. In R. Hersh (Ed.), *18 Unconventional Essays on the Nature of Mathematics* (pp. 220–230). New York, NY: Springer.

Rudolph, L. (2006a). The fullness of time. *Culture & Psychology*, 12, 157–186.

Rudolph, L. (2006b). Mathematics, models and metaphors. *Culture & Psychology*, 12, 245–265.

Rudolph, L. (2006c). Spaces of ambivalence: Qualitative mathematics in the modeling of complex, fluid phenomena. *Estudios Psicologías*, 27, 67–83.

Rudolph, L. (2011). *Turtles all the way down: Why on Earth not?* (Under revision for publication)

Russell, B. (1900). *A Critical Exposition of the Philosophy of Leibniz*. Cambridge, GB: Cambridge University Press.

Russell, B. (1919). *Mysticism and Logic and Other Essays*. London, GB: Longmans, Green and Company.

Schaff, P. (1878). *Through Bible Lands*. New York, NY: American Tract Society.

Schaff, P., & Jackson, S. M. (1887). KURTZ, Johann Heinrich. In *Encyclopedia of Living Divines [etc.]* (p. 121). New York, NY: Funk & Wagnalls.

Shafer, G. (1976). *A Mathematical Theory of Evidence*. Princeton, NJ: Princeton University Press.

Shepherd, H. E. (1887). A study of Lord Macaulay's English. *Transactions and Proceedings of the Modern Language Association of America*, 3, *Transactions of the Modern Language Association of America 1887*, 231–237.

Snow, C. P. (1967). *Foreword*. In G. H. Hardy, *A Mathematician's Apology* (reprinted with foreword). Cambridge, GB: Cambridge University Press.

Soare, R. I. (1996). Computability and recursion. *Bulletin of Symbolic Logic*, 2, 284–321.

Stallings, J. (1966). How not to prove the Poincaré Conjecture. In *Topology Seminar (Wisconsin, 1965), Annals of Mathematics Studies, No. 60* (pp. 83–88). Princeton, NJ: Princeton University Press.

Ströbel, K. (1855). Lutherische Antithesen [Lutheran anti-theses] [Review of the book *Unlutherische Thesen. Deutlich fuer Jedermann*]. *Zeitschrift für die gesammte lutherische Theologie und Kirche*, 16(1),

110–133.

Summer Institute in Linguistics. (2004). What is deontic modality? In E. E. Loos (Gen. Ed.), S. Anderson, D. H. Day Jr., P. C. Jordan, & J. D. Wingate (Eds.), *Glossary of Linguistic Terms*. Dallas, TX: SIL International. Retrieved September 15, 2011, from http://www.sil.org/linguistics/GlossaryOflinguisticTerms/WhatIsDeonticModality.htm

Swanson, J. (2008). Economic common sense and the depoliticization of the economic. *Political Research Quarterly, 61*(1), 55–67.

Tallent, N. (1976). *Psychological Report Writing*. Englewood Cliffs, NJ: Prentice Hall.

Tarski, A. (1933). *Pojęcie prawdy w językach nauk dedukcyjnych [The Concept of Truth in the Languages of the Deductive Sciences]*. Warsaw, PL: Towarzystwa Naukowego Warszawskiego.

Tarski, A. (1936). Der Wahrheitsbegriff in den formalisierten Sprachen [The concept of truth in formalized languages]. *Studia Philosophica, 1*, 261–405.

Trendelenburg, F. A. (1867). *Historische Beiträge zur Philosophie [Historical Survey of Philosophy]* (Vol. III). Berlin, DE: G. Bethge.

Turing, A. M. (1937). On computable numbers, with an application to the *Entscheidungsproblem*. *Proceedings of the London Mathematical Society, 42*, 230–265.

Turing, A. M. (1939). Systems of logic based on ordinals. *Proceedings of the London Mathematical Society, 45*, 161–228.

Valsiner, J. (2008). Baldwin's quest: A universal logic of development. In J. Clegg (Ed.), *The Observation of Human Systems: Lessons from the History of Anti-Reductionistic Empirical Psychology* (pp. 45–82). New Brunswick, NJ: Transaction Publishers.

Whitehead, A. N., & Russell, B. (1912). *Principia Mathematica*. Cambridge, GB: Cambridge University Press.

Windelband, W. (1893). *A History of Philosophy* (J. H. Tufts, Trans.). New York, NY: Macmillan. (Translation of *Geschichte der Philosophie*, 1892, Freiburg im Breisgau, DE: J. C. B. Mohr)

Zadeh, L. (1965). Fuzzy sets. *Information and Control, 8*, 338–353.

Zalta, E. N. (Ed.). (2011). *The Stanford Encyclopedia of Philosophy* (Winter 2011 ed.). Stanford, CA: The Metaphysics Research Lab, Center for the Study of Language and Information, Stanford University. Retrieved from http://plato.stanford.edu/

3

Introduction to Quantum Probability for Social and Behavioral Scientists

Jerome R. Busemeyer

This chapter has two related purposes: to generate interest in a new and fascinating approach to understanding behavioral measures based on quantum probability principles, and to introduce and provide a tutorial of the basic ideas in a manner that is interesting and easy for social and behavioral scientists to understand.

It is important to point out from the beginning that in this chapter, quantum probability theory is viewed simply as an alternative mathematical approach for generating probability models. Quantum probability may be viewed as a generalization of classic probability. No assumptions about the biological substrates are made. Instead this is an exploration into new conceptual tools for constructing social and behavioral science theories.

Why should one even consider this idea? The answer is simply this (*cf.* Khrennikov, 2007). Humans as well as groups and societies are extremely complex systems that have a tremendously large number of unobservable states, and we are severely limited in our ability to measure all of these states. Also human and social systems are highly sensitive to context, and are easily disturbed and disrupted by our measurements. Finally, the measurements that we obtain from the human and social systems are very noisy and filled with uncertainty. It turns out that classical logic, classic probability, and classic information processing force highly restrictive assumptions

Keywords: mixed state, quantum event, quantum information processing, quantum logic, quantum measurement, quantum probability, superposition state.

on representations of these complex systems. Quantum information processing theory provides principles that are more general and powerful for representing and analyzing complex systems of this type. Although the field is still in a nascent stage, applications of quantum probability theory have already begun to appear in areas including information retrieval, language, concepts, decision making, economics, and game theory (see Bruza, Busemeyer, & Gabora, 2009; Bruza, Lawless, van Rijsbergen, & Sofge, 2007, 2008).

The chapter is organized as follows. First, we describe a hypothetical yet typical type of behavioral experiment to provide a concrete setting for introducing the basic concepts. Second, we introduce the basic principles of quantum logic and quantum probability theory. Third, we discuss basic quantum concepts including compatible and incompatible measurements, superposition, measurement and collapse of state vectors.

A simple behavioral experiment

Suppose we have a collection of stimuli (*e.g.*, criminal cases) and two measures: a random variable X with possible values x_i, $i = 1, \ldots, n$ (*e.g.*, 7 degrees of guilt); and a random variable Y with possible values y_j, $j = 1, \ldots, m$ (*e.g.*, 7 levels of punishment) under study. A criminal case is randomly selected with replacement from a large set of investigations and presented to the person. Then one of two different conditions is randomly selected for each trial:

Condition Y: Measure Y alone (*e.g.*, rate level of punishment alone).

Condition XY: Measure X then Y (*e.g.*, rate guilt followed by punishment).

Over a long series of trials (say 100 trials per person to be concrete) each criminal case can be paired with each condition several times. We sort these 100 trials into conditions and pool the results within each condition to estimate the relative frequencies of the answers for each condition. (For simplicity, assume that we are working with a stationary process after an initial practice session that occurs before the 100 experimental trials.)

The idea of the experiment is illustrated in Fig. 3.1, where each measure has only two responses, *yes* or *no*. Each trial begins with a presentation of a criminal case. This case places the participant in a state indicated by the little box with the letter z. From this initial state, the individual has to answer questions about guilt and punishment. The large box indicates the first of the two possible measurements about the case. This question appears in a large box because on some trials there is only the second question in which case the question in the large box does not apply. The final stage represents the second (or only) question. The paths indicted by the arrows indicate all possible answers for two *yes/ no* questions.

Classic probability theory

Events. Classic probability theory assigns probabilities to *classic events.* Each event (such as the event $x = X \geq 4$ or the event $y = Y < 3$ or the event $z = X + Y = 3$) is represented algebraically as a set belonging to a *field of sets.* That is, there is a *null event* represented by the empty set $\emptyset$, and a *universal event* U that contains all other events.[1] Further, new events can be formed from other events in three ways. One way is the *negation* operation, denoted $\sim x$, defined as the set complement. A second way is the *conjunction* operation $x \wedge y$ which is defined by intersection of two sets. A third way is the *disjunction* $x \vee y$ defined as the union of two sets. The events obey the axioms of Boolean algebra, as follows.

B(1) Commutative: $x \vee y = y \vee x$.
B(2) Associative: $x \vee (y \vee z) = (x \vee y) \vee z$.
B(3) Complementation: $x \vee (y \wedge \sim y) = x$.
B(4) Absorption: $x \vee (x \wedge y) = x$.
B(5) Distributive: $x \wedge (y \vee z) = (x \wedge y) \vee (x \wedge z)$.

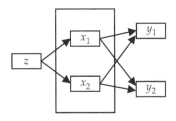

Figure 3.1: The possible measurement outcomes for Condition XY.

---

[1]For simplicity we restrict attention to experiments with only finitely many outcomes; then U can be assumed to be a finite set.

The distributive axiom B(5) is crucial for distinguishing classic probability theory from quantum probability theory.

Classic probabilities. The standard theory of probability, as used throughout the social and behavioral sciences, is based on the Kolmogorov axioms.

K(1) Normalized: $0 \le \Pr(x) \le 1$, $\Pr(\varnothing) = 0$, $\Pr(\mathcal{U}) = 1$.

K(2) Additive: If $x \wedge y = \varnothing$ then $\Pr(x \vee y) = \Pr(x) + \Pr(y)$.

When more than one measurement is involved, the *conditional probability of y given x* is $\Pr(y|x)$, defined by the ratio

$$\Pr(y|x) = \Pr(y \wedge x)/\Pr(x), \tag{3.1}$$

which implies the formula for *joint probabilities*

$$\Pr(y \wedge x) = \Pr(x)\,\Pr(y|x). \tag{3.2}$$

Classic probability distributions

Classically, our simple behavioral experiment is analyzed as follows. Consider first condition XY. We observe $n \times m$ distinct mutually exclusive and exhaustive distinct outcomes, such as $x_i y_j$, which occurs when the pair x_i and y_j are observed. Other events can be formed by union such as the events $x_i = x_i y_1 \vee x_i y_2 \vee \ldots \vee x_i y_m$ and $y_j = x_1 y_j \vee x_2 y_j \vee \ldots \vee x_n y_j$. New sets can also be defined by the intersection operation for sets, such as the event $x_i \wedge y_j = x_i y_j$. These sets obey the axioms of Boolean algebra, and in particular, the distributive axiom B(5) states that

$$y_j = y_j \wedge \mathcal{U} = y_j \wedge (x_1 \vee x_2 \vee \ldots \vee x_n)$$
$$= (y_j \wedge x_1) \vee (y_j \wedge x_2) \vee \ldots \vee (y_j \wedge x_n).$$

Table 3.1 shows all nonzero events for binary-valued measures ($n = m = 2$).

Table 3.1: Events generated by Boolean algebra operators.

Note: $y_1 \wedge (x_1 \vee x_2) = (x_1 \wedge y_1) \vee (x_2 \wedge y_1)$, etc.

Events	y_1	y_2	$y_1 \vee y_2$
x_1	$x_1 \wedge y_1$	$x_1 \wedge y_2$	$x_1 \wedge (y_1 \vee y_2)$
x_2	$x_2 \wedge y_1$	$x_2 \wedge y_2$	$x_2 \wedge (y_1 \vee y_2)$
$x_1 \vee x_2$	$y_1 \wedge (x_1 \vee x_2)$	$y_2 \wedge (x_1 \vee x_2)$	$(x_1 \vee x_2) \wedge (y_1 \vee y_2) = \mathcal{U}$

The Boolean axioms B(1)–B(5) are used in conjunction with the Kolmogorov axioms K(1), K(2) to derive the *law of total probability*:

$$\Pr(y_j) = \Pr(y_j \wedge \mathcal{U}) = \Pr((y_j \wedge (x_1 \vee x_2 \vee \ldots \vee x_n))$$
$$= \Pr((y_j \wedge x_1) \vee (y_j \wedge x_2) \vee \ldots \vee x_n))$$
$$= \sum_i \Pr(x_i \wedge y_j) = \sum_i \Pr(x_i) \Pr(y_j | x_i). \quad (3.3)$$

Thus the marginal probability distribution for Y is determined from the joint probabilities, and this is also true for X. Finally, *Bayes's rule* follows from (3.1), (3.2), and (3.3):

$$\Pr(y_j | x_i) = \frac{\Pr(y_j \wedge x_i)}{\Pr(x_i)} = \frac{\Pr(y_j) \Pr(x_i | y_j)}{\sum_k \Pr(y_k) \Pr(x_i | y_k)}. \quad (3.4)$$

Recall that, in our experiment, under one condition we measure X then Y, but under another condition we measure only variable Y. According to classic probability, there is nothing to prevent us from postulating a joint probability like $\Pr(x_i \wedge y_j)$ for condition Y, which only involves a single measurement. Indeed, the Boolean axioms require the existence of all the events generated by that algebra. Only y_j is observed, but this observed event is assumed to be broken down into counterfactual events,

$$y_j = (y_j \wedge x_1) \vee (y_j \wedge x_2) \vee \ldots \vee (y_j \wedge x_n).$$

In particular, during condition Y, the event $x_i \wedge y_j$ can be considered the counterfactual event that you would have responded at degree of guilt x_i to X if you were asked (but you were not), and responding level of punishment y_j when asked about Y. Thus all of the joint probabilities $\Pr(x_i \wedge y_j | Y)$ are assumed to exist even when we measure only Y. So in the case where only Y is measured, we postulate that the marginal probability distribution, $\Pr(y_j)$, is determined from the joint probabilities such as $\Pr(x_i \wedge y_j)$ according to the law of total probability (3.3). This is actually a big assumption, although it is routinely taken for granted in the social and behavioral sciences.

This critical assumption can be understood more simply using Fig. 3.1. Note that under condition Y, the large box containing X is not observed. However, according to classic probability theory, the probability of starting from z and eventually reaching y_1 is equal to the sum of the probabilities from the two mutually exclusive and exhaustive paths: the joint probability of transiting from z to x_1 and then transiting from x_1 to y_1 plus the joint probability of transiting from z to x_2 and then transiting from x_2 to y_1. How else could one travel from z to y_1 without passing through one of states for x?

If we assume the joint probabilities are the same across conditions, then according to (3.3) we should find $\Pr(y_j|XY) = \Pr(y_j|Y)$. Empirically, however, we often find that $\Pr(y_j|XY) \neq \Pr(y_j|Y)$; the difference is called an *interference effect* (Khrennikov, 2007). Unfortunately, when these effects occur, as they often do in the social and behavioral sciences, classic probability theory does not provide any way to explain them. One is simply forced to postulate a different joint distribution for each experimental condition. This is where quantum probability theory can make a contribution.

Quantum probability theory

Events. Quantum theory assigns probabilities to quantum events (see Hughes, 1989, for an elementary presentation). A *quantum event* (such as L_x representing $X > 4$, or L_y representing $Y < 3$, or the event $z = X + Y = 3$) is defined geometrically as a subspace (*e.g.*, a line or plane or hyperplane, etc.) within a Hilbert space H (*i.e.*, a vector space with *complex numbers*[2] as scalars, and equipped with a *Hermitian inner product* (used to measure length).[3] The *null event* is represented by the zero subspace **0** (containing just the zero vector **0**) of the vector space H, and the *universal event* by H itself. New events can be formed in three ways. One way is the *negation* operation, denoted $L_x^{\perp}$, which is defined as the maximal subspace that is orthogonal to L_x. A second way is the *meet* operation $x \wedge y$ which is defined by intersection of two subspaces: $L_{x \wedge y} = L_x \wedge L_y$. A third way is the *join* operation $x \vee y$ defined as the *span* of two subspaces L_x and L_y. Span is quite different than union, and this is where quantum logic differs from classic logic. It is due to this difference that, although quantum logic obeys axioms B(1)–B(4) of Boolean logic (and therefore all rules of Boolean logic that can be proved from just those axioms), it does not obey the distributive axiom B(5). That is, there are cases in quantum logic where the

---

[2] Complex numbers cannot be avoided in quantum probability; see the section "Why complex numbers?" below, p. 98 *ff.*

[3] For simplicity, we consider only finite dimensional Hilbert spaces. Quantum probability theory includes infinite dimensional spaces, but the basic ideas remain the same for finite and infinite dimensions.

equation $L_z \wedge (L_x \vee L_y) = (L_z \wedge L_x) \vee (L_z \wedge L_y)$ fails to be true (of course it is true in some cases, *e.g.*, if $x = y = z$).

Fig. 3.2 illustrates an example of a violation of the distributive axiom. Suppose H is a 3-dimensional space.[4] This space can be defined in terms of an orthogonal basis formed by the three vectors labeled $|x\rangle$, $|y\rangle$, and $|z\rangle$ corresponding to the three standard coordinate axes (lines) L_x, L_y, L_z.[5] Alternatively, the same space can be defined in terms of an orthogonal basis defined by the three vectors $|u\rangle$, $|v\rangle$, and $|w\rangle$ corresponding to the other three (pairwise perpendicular) lines L_u, L_v, and L_w in Fig. 3.2.[6] Consider the event $(L_u \vee L_w) \wedge (L_x \vee L_y \vee L_z)$. Since x, y, and z are a basis of H, the span of the three lines L_x, L_y, and L_z is all of H: thus the event $L_x \vee L_y \vee L_z$ is equal to the event H. The span of u and w is a plane within H; as an event that plane is $L_u \vee L_w$ (as indicated in Fig. 3.2). Similarly the span of x and y is a plane (the xy-plane), which as an event is

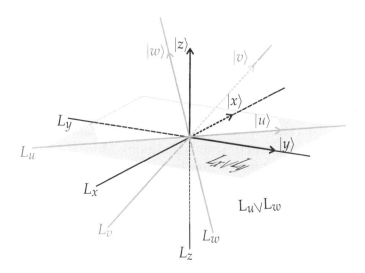

Figure 3.2: Violation of the distributive axiom.

[4]Although Fig. 3.2 is a depiction of ordinary real 3-space $\mathbb{R}^3$ in the usual way, it faithfully reflects the situation for complex 3-space $\mathbb{C}^3$.

[5]Dirac notation is used here. The *ket* $|v\rangle$ corresponds to a column vector, the *bra* $\langle z|$ corresponds to a row vector, the *bra-ket* is an inner product, and $\langle x|P|y\rangle$ is a bra-matrix-ket product.

[6]Precisely, here $|u\rangle = |x\rangle/\sqrt{2} + |y\rangle/\sqrt{2}$, $|v\rangle = |x\rangle/2 + |y\rangle/2 + |z\rangle/\sqrt{2}$, and $|w\rangle = -|x\rangle/2 + |y\rangle/2 + |z\rangle/\sqrt{2}$.

$L_x \vee L_y$. From the definitions, we calculate

$$(L_u \vee L_w) \wedge (L_x \vee L_y \vee L_z) = L_u \vee L_w.$$

If the distributive axiom B(5) were applicable, we would then have

$$(L_u \vee L_w) \wedge (L_x \vee L_y \vee L_z) = (L_u \vee L_w) \wedge ((L_x \vee L_y) \vee L_z)$$
$$= (L_u \vee L_w) \wedge (L_x \vee L_y) \vee (L_u \vee L_w) \wedge L_z. \quad (3.5)$$

Now, $(L_u \vee L_w) \wedge (L_x \vee L_y) = L_u$ because (as shown in Fig. 3.2) the intersection of the two planes is exactly the u-axis, i.e., the event L_u; and $(L_u \vee L_w) \wedge L_z = \mathbf{0}$ because (again, as shown in Fig. 3.2) the intersection of the z-axis and the uw-plane is the single point $\mathbf{0}$. In sum, we find that

$$(L_u \vee L_w) \wedge (L_x \vee L_y \vee L_z) = L_u \vee L_w$$
$$\neq (L_u \vee L_w) \wedge (L_x \vee L_y) \vee (L_u \vee L_w) \wedge L_z = L_u \vee \mathbf{0} = L_u,$$

contradicting (3.5). This example illustrates how quantum logic can violate the distributive axiom B(5) of Boolean logic.

Probabilities. Quantum probabilities are computed using projective rules that involve three steps. First, the probabilities for all events are determined from a *state vector* $|z\rangle \in H$ of unit length (i.e., $\||z\rangle\| = 1$). This state vector depends on the preparation and context (person, stimulus, experimental condition). More is said about this state vector later, but for the time being, assume it is known. Second, to each event L_x there is a corresponding *projection operator* P_x that projects each state vector $|z\rangle \in H$ onto L_x.[7] Finally, probability of an event L_x is equal to the squared length of this projection:

$$\Pr(x) = \|P_x|z\rangle\|^2 = (P_x|z\rangle)^\dagger (P_x|z\rangle)$$
$$= \langle z|P_x^\dagger P_x|z\rangle = \langle z|P_x P_x|z\rangle = \langle z|P_x|z\rangle.$$

Fig. 3.3 illustrates the idea of projective probability. In this figure, the squared length of the projection of $|z\rangle$ onto L_{x_1} is the probability of the event L_{x_1} given the state $|z\rangle$.

---

[7] Projection operators are characterized as being *Hermitian* and *idempotent*. To say P is Hermitian means that $P = P^\dagger$; in matrix terms, for every i and j, the entry $p_{i,j}$ in row i, column j of P and the entry $p_{j,i}$ in row j, column i of P are complex conjugates of each other. To say P is idempotent means that $P^2 = P$.

Quantum probability distributions for a single variable

Consider, for the moment, the measurement of a single variable, say the degree of guilt, X, which can produce one of n distinct outcomes or values, x_i $(i = 1, \ldots, n)$. We will assume that no outcome x_i can be decomposed or refined into other distinguishable parts.

To each distinct outcome x_i we assign a corresponding line or ray L_{x_i} in our Hilbert space H. Corresponding to this subspace is a unit length vector, called a *basis state* and symbolized as $|x_i\rangle$, which generates this ray as the set of its scalar multiples $a|x_i\rangle$.[8] The basis states are assumed to be *orthonormal*: the inner product $\langle x_i|x_j\rangle$ is 0 for all pairs x_i, x_j of states with $i \neq j$, while for each state x_i the length $\|x_i\| = \sqrt{\langle x_i|x_i\rangle}$ equals 1. We can interpret the basis state $|x_i\rangle$ as follows: if the person is put into the initial state $|z\rangle = |x_i\rangle$, then you are certain to observe the outcome x_i.

The projector P_{x_i} projects any point $|z\rangle$ in H into the subspace L_{x_i}. It is constructed from the *outer product* $|x_i\rangle\langle x_i|$; i.e., for all z,

$$P_{x_i}|z\rangle = (|x_i\rangle\langle x_i|)|z\rangle = |x_i\rangle\langle x_i|z\rangle = \langle x_i|z\rangle\,|x_i\rangle,$$

where $\langle x_i|z\rangle$ is the *inner product* (or "bra-ket"; cf. note 5). The inner product $\langle x_i|z\rangle$, in turn, can be interpreted as the probability amplitude[9] of transiting to state $|x_i\rangle$ from state $|z\rangle$. The *probability* of any event L_{x_i} equals the squared projection,

$$\|P_{x_i}|z\rangle\|^2 = \||x_i\rangle\langle x_i|z\rangle\|^2 = \||x_i\rangle\|^2\,|\langle x_i|z\rangle|^2 = 1\cdot|\langle x_i|z\rangle|^2 = |\langle x_i|z\rangle|^2.$$

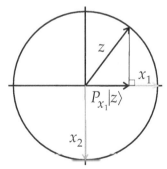

Figure 3.3: Projective probability: $\Pr(x_1) = \|P_{x_1}|z\rangle\|^2$.

---

[8] Actually, both $|x_i\rangle$ and $-|x_i\rangle$ are basis states for x_i; the choice is immaterial.
[9] In general, this can be a complex number.

That is, the probability of transiting to state $|x_i\rangle$ from state $|z\rangle$ equals $|\langle x_i|z\rangle|^2$, the squared magnitude of the probability amplitude.

The probability of the meet $x \wedge y$ of two events x and y is equal to the squared length of the projection of the intersection. For example, if $x = x_i \vee x_j$ and $y = x_i \vee x_k$, then $x \wedge y = x_i$ and

$$\Pr(x \wedge y) = \Pr(x_i) = |\langle x_i|z\rangle|^2.$$

We have $x_i \neq x_j$ for $i \neq j$, so the joint event $x_i \wedge x_j$ is zero, $L_{x_i} \wedge L_{x_j} = \mathbf{0}$, and the projection P_0 onto the zero subspace $\mathbf{0}$ is the zero operator 0; thus the joint probability $\Pr(x_i \wedge x_j)$ is $\|\mathbf{0}\|^2 = 0$.

The join of two events, say $x_i \vee x_j$, is the span $\{|x_i\rangle, |x_j\rangle\}$ of the two basis vectors. The projector for this subspace is

$$P_{x_i \vee x_j} = P_{x_i} + P_{x_j} = |x_i\rangle\langle x_i| + |x_j\rangle\langle x_j|.$$

The probability of the event $x_i \vee x_j$ is thus simply the sum of the separate probabilities,

$$\begin{aligned}
\|P_{x_i \vee x_j}|z\rangle\|^2 &= \|(|x_i\rangle\langle x_i| + |x_j\rangle\langle x_j|)|z\rangle\|^2 \\
&= \||x_i\rangle\langle x_i|z\rangle + |x_j\rangle\langle x_j|z\rangle\|^2 \\
&= |\langle x_i|z\rangle|^2 + |\langle x_i|z\rangle|^2,
\end{aligned}$$

where the final step follows from the orthogonality property.

Finally, for any $|z\rangle$ we have $P_H|z\rangle = |z\rangle$, and so

$$\|P_H|z\rangle\|^2 = \||z\rangle\|^2 = |\langle z|z\rangle|^2 = 1 \text{ and } P_H = \sum_i P_{x_i} = \sum_i |x_i\rangle\langle x_i| = \mathbf{I},$$

where $\mathbf{I}$ is the identity operator $\mathbf{I}|z\rangle = |z\rangle$. From this we see that quantum probabilities obey axioms analogous to the Kolmogorov axioms.

Q(1) $0 \leq \|\Pr(x)|z\rangle\|^2 \leq 1$, $\Pr(\mathbf{0}) = 0$, $\Pr(H) = 1$.

Q(2) If $L_x \wedge L_y = 0$ then $\Pr(L_x \vee L_y) = \Pr(L_x) + \Pr(L_y)$.

The state vector. It is time to return to the problem of defining the state vector $|z\rangle$ prior to the measurement. This vector can be expressed in terms of the basis states as follows:

$$|z\rangle = \mathbf{I}|z\rangle = \left(\sum_i |x_i\rangle\langle x_i|\right)|z\rangle = \sum_i |x_i\rangle\langle x_i|z\rangle = \sum_i \langle x_i|z\rangle |x_i\rangle.$$

Thus the initial state vector is a *superposition* (*i.e.*, linear combination) of the basis states. The inner product $\langle x_i | z \rangle$ is the coefficient (or *component* of the state vector that corresponds to the $|x_i\rangle$ basis state. To be concrete, one can define $|x_i\rangle$ as a column vector with a 0 in every row except for row i, where there is a 1. Then the initial state is a column vector $|z\rangle$ containing coefficient $\langle x_i | z \rangle$ in row i.

The probability of obtaining x_i equals the squared amplitude $|\langle x_i | z \rangle|^2$. Thus we form the initial state by choosing coefficients that have squared amplitudes equal to the probability of the outcome: choose $\langle x_i | z \rangle$ so that $\Pr(x_i) = |\langle x_i | z \rangle|^2$. In short, when only one measurement is made, quantum probability theory is not much different than Kolmogorov probability theory.

Effect of measurement. After one measurement, say X, is taken, and an arbitrary event x is observed, this measurement changes the state from the initial state $|z\rangle$ to a new state $|x\rangle$ which is the normalized projection on the subspace L_x. In fact, one way to prepare an initial state is to take a measurement, after which the person is in a state consistent with the event so obtained. This is called the *state reduction* or *state collapse* assumption of quantum theory. Prior to the measurement, the person was in a superposed state $|z\rangle$, but after measurement the person is in a new state $|x\rangle$. In other words, *measurement changes the person*.

Social and behavioral scientists generally adopt a classical view of measurement that assumes measurement simply records a preexisting reality. In other words, properties exist in the brain at the moment just prior to a measurement, and the measurement simply reveals this preexisting property. Consider condition Y of our experiment, during which only the punishment level is measured. Even though guilt is not measured in this condition, it is still assumed that the criminal case evokes some specific degree of belief in guilt for the person. We just don't bother to measure its specific value. Thus both properties exist even though we measure only one.

The problem with the classical interpretation of measurement can be seen most clearly by reconsidering the example shown in Fig. 3.1 with binary outcomes. If we present a case, then we suppose that it evokes a degree of belief in guilt and a level of punishment. Under condition Y, we measure only the level of punishment. If we measure level y_1, then event $y_1 = y_1 \wedge (x_1 \vee x_2)$ has occurred (here we assume that values x_1, x_2 are mutually exclusive and exhaustive). According to the distributive axiom B(5), this event means that prior to our measurement either the person is in the low guilty state and intends to punish at the low level $x_1 \wedge y_1$ (*i.e.*, the brain experienced the upper path in Fig. 3.1), or the person is in the high guilty state and intends to punish at the low level $x_2 \wedge y_1$ (*i.e.*, the brain experienced the lower path in Fig. 3.1). Condition XY simply resolves the uncertainty about which of these two realities existed at the moment before the measurement.

The classic idea of measurement is rejected in quantum theory (see, *e.g.*, Peres, 1995, p. 14). According to the latter, measurements *create* permanent records that we all can agree upon. To see how this creative process arises in quantum theory, suppose the distributive axiom B(5) fails. Referring again to Fig. 3.1, if we measure punishment state y_1, then event $y_1 \wedge (x_1 \vee x_2)$ has occurred, but from this we cannot infer the existence of any specific degree of belief in guilt: we cannot assume that either $x_1 \wedge y_1$ or $x_2 \wedge y_1$, and not both, existed just prior to measurement (*i.e.*, we cannot assume that either the upper path, or the lower path, is traveled; see Feynman, Leighton, & Sands, 1966, p. 9). On the contrary, if we measure X first in condition XY, then this measurement will *create* a state with a specific belief in guilt before measuring the punishment.

In some ways, quantum systems are more deterministic than classical random error systems. Suppose we measure X twice in succession, and the first measure produces an event x. In a quantum system, the second measure is certain to produce x again, because $\Pr(x|x) = |\langle P_x|x \rangle|^2 = \||x\rangle\|^2 = 1$. Thus the event remains unchanged until a different type of measurement is taken. If a new type of measurement is taken after the first measurement, then the state changes again, and the outcome becomes probabilistic.

In a random error system, the observed values are produced by a true score plus some error perturbation that appears randomly on each trial. In that case, the probability of observing a particular value should change following each and every measurement, regardless of whether or not the same measurement is taken twice in succession.

It is interesting to note that social and behavioral scientists are aware of the quantum principle. When they design experiments to obtain repeated measurements for a particular stimulus, they systematically avoid asking participants to judge the same stimulus back to back. Instead, they insert filler items (other measurements) between presentations (to avoid the deterministic result), and these

filler items disturb the system to generate probabilistic choice behavior for spaced repetitions of the target items.

Quantum probability distributions for several variables

After we have first measured X and observed the event x, the
state changes to $|x\rangle = P_x|z\rangle/\|P_x|z\rangle\|$, where P_x is the projector onto
the subspace L_x. Note that the squared length of the new state
remains equal to one, $|\langle x|x\rangle|^2 = 1$, because of the normalizing factor
in the denominator. This is important to maintain a probability
distribution over outcomes of Y for the next measurement after
measuring X. The probabilities for the next measurement are based
on this new state. If we first measure X and observe the event x,
then the probability of observing y when Y is measured next equals
$\Pr(y|x) = \|P_y|x\rangle\|^2$. This updating process continues for each new
measurement.

When more than one measurement is involved, quantum probability is more general than Kolmogorov probability, and quantum
logic does not have to obey the distributive axiom B(5). In quantum
theory, the analysis of an experimental situation in which more
than one measurement is made, depends on how one represents the
relationship between the measurements. There are two possibilities:
the measures may be *compatible* or *incompatible*.

Compatible measurements. We consider specifically the problem of two measurements, first in the case in which the two measures are compatible. Intuitively, compatibility means that X and
Y can be measured or accessed or experienced simultaneously or
sequentially without interfering with each other. Psychologically
speaking, the two measures can be processed in parallel. If the
measures are compatible, then we form the basis vectors for the
two measurements from all the possible combinations of distinct
outcomes for X and Y of the form $x_i y_j$. The complete Hilbert space
is defined by $n \times m$ orthonormal basis vectors $|x_i y_j\rangle$, $i = 1, \ldots, n$ and
$j = 1, \ldots, m$, spanning a space H of dimension $n \times m$. For example, in
condition XY, the vector $|x_i y_j\rangle$ corresponds to observing x_i from X
and y_j from Y. The orthogonal property implies that $\langle x_i y_j | x_k y_\ell \rangle = 0$,
and the normal property that $\langle x_i y_j | x_i y_j \rangle = 1$. This Hilbert space H is
called the *tensor product space* for the two measures.

Notice that the event x_i is no longer a distinct outcome. Instead,
it is a coarse-grained outcome that can be expressed, using the

tensor decomposition, in terms of more refined parts:

$$L_{x_i} = |x_i y_1\rangle \vee |x_i y_2\rangle \vee \ldots \vee |x_i y_m\rangle.$$

Furthermore, the meet $x_i \wedge y_j$ produces the subspace $L_{x_i} \wedge L_{y_j} = |x_i y_j\rangle$. This implies that, for this tensor decomposition, the distributive axiom B(5) does hold:

$$\begin{aligned} L_{x_i} &= |x_i y_1\rangle \vee |x_i y_2\rangle \vee \ldots \vee |x_i y_m\rangle \\ &= (L_{x_i} \wedge L_{y_1}) \vee (L_{x_i} \wedge L_{y_2}) \vee \ldots \vee (L_{x_i} \wedge L_{y_m}). \end{aligned}$$

Thus Table 3.1 provides an appropriate description of all the relevant events for binary outcomes. In other words, the assumption of compatible measures requires the existence of all joint events, and the individual outcomes can be obtained from the joint events.

The projection operators for the events L_{x_i,y_j}, L_{x_i}, and L_{y_j} are

$$P_{x_i,y_j} = |x_i y_j\rangle\langle x_i y_j|,$$
$$P_{x_i} = \sum_j P_{x_i,y_j} = \sum_j |x_i y_j\rangle\langle x_i y_j|,$$
$$P_{y_j} = \sum_i P_{x_i,y_j} = \sum_i |x_i y_j\rangle\langle x_i y_j|.$$

The orthogonality properties then imply

$$|x_i y_j\rangle\langle x_i y_j| = \left(\sum_j |x_i y_j\rangle\langle x_i y_j|\right)\left(\sum_i |x_i y_j\rangle\langle x_i y_j|\right) \tag{3.6}$$
$$= P_{x_i} P_{y_j} \tag{3.7}$$
$$= P_{y_j} P_{x_i}. \tag{3.8}$$

(3.6) implies that the projection for the joint event L_{x_i,y_j} can be viewed as a series of two successive measurements, and vice versa. (3.7) and (3.8) show that the projectors for X *commute with* the projectors for Y: that is, the order of projection does not matter— both orders project onto the same final subspace. In general, given operators (in particular, projectors) A and B, their combination $AB - BA$ is called the *commutator* of A and B. We have shown that the commutator of compatible measures is always zero.

Now let us consider a series of two measurements. Using the reduction principle, if X is measured first and x_i is observed, then the new state after measurement is $|x_i\rangle = P_{x_i}|z\rangle/\|P_{x_i}|z\rangle\|$; similarly if Y is measured first and we observe y_j, then the new state after measurement is $|y_j\rangle = P_{y_j}|z\rangle/\|P_{y_j}|z\rangle\|$. Consider again the probability of the event L_{x_i,y_j}, viewed as a series of projections.

$$P_{x_i,y_j}|z\rangle = P_{x_i}(P_{y_j}|z\rangle) = P_{x_i}|y_j\rangle \|P_{y_j}|z\rangle\|$$

$$P_{y_j,x_i}|z\rangle = P_{y_j}(P_{x_i}|z\rangle) = P_{y_j}|x_i\rangle \|P_{x_i}|z\rangle\|$$

$$Pr(x_i \wedge y_j) = \|P_{x_i,y_j}|z\rangle\|^2$$

$$= \|P_{x_i}|y_j\rangle\|^2 \|P_{y_j}|z\rangle\|^2 = Pr(x_i|y_j)Pr(y_j) \qquad (3.9)$$

$$Pr(y_j \wedge x_i) = \|P_{y_j,x_i}|z\rangle\|^2$$

$$= \|P_{y_j}|x_i\rangle\|^2 \|P_{x_i}|z\rangle\|^2 = Pr(y_j|x_i)Pr(x_i). \qquad (3.10)$$

From (3.9) and (3.10) we obtain the conditional probability axioms for quantum probabilities:

$$Pr(x_i|y_j) = \|P_{x_i}|y_j\rangle\|^2 = \|P_{x_i,y_j}|z\rangle\|^2/\|P_{y_j}|z\rangle\|^2, \qquad (3.11)$$

$$Pr(y_j|x_i) = \|P_{y_j}|x_i\rangle\|^2 = \|P_{x_i,y_j}|z\rangle\|^2/\|P_{x_i}|z\rangle\|^2. \qquad (3.12)$$

In general $\|P_{x_i}|z\rangle\|^2 \neq \|P_{y_j}|z\rangle\|^2$ and so also $Pr(y_j|x_i) \neq Pr(x_i|y_j)$.

The projection onto L_{x_i} is $P_{x_i}|z\rangle = \sum_j |x_iy_j\rangle\langle x_iy_j|z\rangle$ and the probability of this event equals

$$Pr(x_i) = \sum_j |\langle x_iy_j|z\rangle|^2 = \sum_j \|P_{x_i}|y_j\rangle\|^2 \|P_{y_j}|z\rangle\|^2. \qquad (3.13)$$

(3.13) is the law of total probability for quantum probabilities. From (3.12) and (3.13), we can derive a quantum analogue of Bayes's rule (3.4):

$$Pr(y_j|x_i) = \|P_{y_j}|x_i\rangle\|^2 = \frac{\|P_{x_i}|y_j\rangle\|^2 \|P_{y_j}|z\rangle\|^2}{\sum_k \|P_{x_i}|y_k\rangle\|^2 \|P_{y_k}|z\rangle\|^2}.$$

Let us re-examine the initial state vector $|z\rangle$ for the case of two compatible measurements. As before, this state vector can be described in terms of the basis vectors:

$$|z\rangle = \mathbf{I}|z\rangle = \left(\sum_i \sum_j |x_iy_j\rangle\langle x_iy_j|\right)|z\rangle = \sum_i \sum_j \langle x_iy_j|z\rangle |x_iy_j\rangle.$$

Once again, we see that the initial state is a superposition of the basis states. The inner product $\langle x_iy_j|z\rangle$ is the coefficient of the state vector corresponding to the $|x_iy_j\rangle$ basis state. The probability of obtaining the joint event x_iy_j equals the squared amplitude of the corresponding coefficient, $|\langle x_i y_j | z \rangle|^2$. Thus we form the initial state by choosing coefficients that have squared amplitudes equal to the probability of the joint outcome: choose $\langle x_i y_j | z \rangle$ so that $\Pr(x_i y_j) = \Pr(x_i \wedge y_j) = |\langle x_i y_j | z \rangle|^2$.

In sum, all of these results exactly correspond to the classic probability axioms. In short, quantum probability theory reduces to classic probability theory for compatible measures. If all measures were compatible, then quantum probability would produce exactly the same results as classical probability.[10]

Incompatible measurements. Incompatibility means that X and Y cannot be measured or accessed or experienced simultaneously. Psychologically speaking, the two measures must be processed se-rially, and measurement of one variable interferes with the other. This implies that X produces n distinct outcomes x_i ($i = 1, \ldots, n$) that cannot be decomposed into more refined parts, because we can't simultaneously measure Y. Similarly, Y produces n distinct outcomes y_i ($i = 1, \ldots, n$) that cannot be decomposed into more refined parts, because we can't simultaneously measure X. In this case, we assume that the outcomes from the measure X produce one orthonormal set of basis states, $|x_i\rangle$ ($i = 1, \ldots, n$), and that the outcomes of Y produce another orthonormal set of basis states $|y_j\rangle$ ($j = 1, \ldots, n$). To account for the fact that one measure influences the other, it is assumed that one set of basis states is a (non-identity) linear transformation of the other. Thus we now have two *different* bases for the *same* n-dimensional Hilbert space. This idea is illus-trated in Fig. 3.4. In this figure, we assume that the outcomes are binary. The outcomes of the first measure (regarding the guilt) are represented by the basis vectors $|x_1\rangle$ and $|x_2\rangle$, and the outcomes of the second measure (regarding the punishment) by the basis vectors $|y_1\rangle$ and $|y_2\rangle$. Note that the basis vectors for the Y measure are a linear transformation—specifically, an orthogonal rotation—of the basis vectors for the X measure (and vice versa). One can use either the $|x_1\rangle$, $|x_2\rangle$ basis or the $|y_1\rangle$, $|y_2\rangle$ basis to describe the state vector $|z\rangle$, but one cannot use both bases at the same time.

One cannot experience or measure both variables X and Y si-multaneously. If one measures X first, then one needs to project the state $|z\rangle$ onto the X basis, not the Y basis, and if one the value x_1 is

---

[10]This is not quite true. We are only focusing on change caused by measurement, and disregarding change caused by dynamic laws.

observed then the outcome for the next measurement of Y must be uncertain: $\Pr(y_j) = |\langle y_j|x_1\rangle|^2$ ($j = 1, 2$). Similarly, if one measures Y first, then the Y basis must be used, and if the value y_1 is observed, then the outcome for the next measurement on X must be uncertain: $\Pr(x_i) = |\langle x_i|y_1\rangle|^2$ ($i = 1, 2$). It is impossible to be certain about both values simultaneously! Therefore, it is impossible to completely and correctly measure all the values of the system. This is essentially the idea behind the famous Heisenberg uncertainty principle (Peres, 1995, Chap. 2).

The distributive axiom B(5) of Boolean logic is violated by incompatible measures. For example, considering Fig. 3.4, note that $|x_i\rangle \wedge |y_j\rangle = 0$ for all i and j, and therefore we have

$$L_{y_1} = L_{y_1} \wedge (L_{x_1} \vee L_{x_2}) \neq (L_{y_1} \wedge L_{x_1}) \vee (L_{y_1} \wedge L_{x_2}) = 0 \vee 0 = 0,$$

a violation of distributivity. In this example, because of incompatibility the event each of the events $L_{y_1} \wedge L_{x_1}$ and $L_{y_2} \wedge L_{x_2}$ is impossible, yet clearly the event L_{y_1} is possible. This is where quantum probability deviates from classic probability. Table 3.2 shows the events for incompatible measures, including clear violations of the distributive axiom B(5).

To get a deeper understanding of the violation of the distributive axiom B(5), let us return to Fig. 3.1 again. Suppose only Y is measured, and we observe y_1. How does the person go from the initial state $|z\rangle$ to this observed state $|y_1\rangle$? We *cannot* say 'The person traveled one of two paths—either $|z\rangle \to |x_1\rangle \to |y_1\rangle$ or $|z\rangle \to |x_2\rangle \to |y_1\rangle$— but we are uncertain about which path was taken.' In other words, if the person intends to punish at the low level, then we cannot say he or she reached that decision having first concluded that the

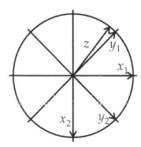

Figure 3.4: Rotated basis vectors for incompatible measurements.

person was guilty at a low degree or that the person was guilty at a high degree. Because we do not measure guilt, we cannot assume that the person is definitely in one of these two guilt states; on the contrary, in the quantum probability model, the person is indefinite (or superposed) between these two states. When we do not observe what happens, quantum theory allows for a type of uncertainty regarding state changes that is more general than classical probability theory.

The fact that there are two different bases for the same Hilbert space implies that the same state vector has two different descriptions in terms of those bases:

$$|z\rangle = \mathbf{I}\,|z\rangle = \left(\sum_i |x_i\rangle\langle x_i|\right)|z\rangle = \sum_i |x_i\rangle\langle x_i|z\rangle,$$
$$|z\rangle = \mathbf{I}\,|z\rangle = \left(\sum_i |y_i\rangle\langle y_i|\right)|z\rangle = \sum_i |y_i\rangle\langle y_i|z\rangle.$$

If the X basis is used to describe the state vector $|z\rangle$, then the inner products $\langle x_i|z\rangle$ form the coordinates for $|z\rangle$. In this basis, we can represent the initial state vector by a column vector Ψ with $\langle x_i|z\rangle$ in row i, and the marginal probability distribution for X is $\Pr(x_i) = \|\Psi_i\|^2 = |\langle x_i|z\rangle|^2$. But if the Y basis is used to describe the state vector $|z\rangle$, then the inner products $\langle y_i|z\rangle$ form the coordinates for $|z\rangle$; in this basis, we can represent the initial state vector by a column vector Φ with $\langle y_i|z\rangle$ in row i, and the marginal probability distribution for Y is $\Pr(y_j) = \|\Phi_i\|^2 = |\langle y_j|z\rangle|^2$. No joint distribution exists, but both marginal distributions are derived from a common state vector $|z\rangle$. The equality of the two representations implies

$$\sum_i |x_i\rangle\langle x_i|z\rangle = \sum_i |y_i\rangle\langle y_i|z\rangle,$$
$$\implies \langle x_j|\sum_i |x_i\rangle\langle x_i|z\rangle = \langle x_j|\sum_i |y_i\rangle\langle y_i|z\rangle,$$
$$\implies \sum_i \langle x_j|x_i\rangle\langle x_i|z\rangle = \sum_i \langle x_j|y_i\rangle\langle y_i|z\rangle,$$
$$\implies \langle x_j|z\rangle = \sum_i \langle x_j|y_i\rangle\langle y_i|z\rangle,$$

Table 3.2: Events generated by incompatible measures.

Note: $L_{y_1} = L_{y_1} \wedge (L_{x_1} \vee L_{x_2}) \neq (L_{y_1} \wedge L_{x_1}) \vee (L_{y_1} \wedge L_{x_2}) = \mathbf{0}$, etc.

Events	y_1	y_2	$y_1 \vee y_2$
x_1	0	0	$x_1 \wedge (y_1 \vee y_2)$
x_2	0	0	$x_2 \wedge (y_1 \vee y_2)$
$x_1 \vee x_2$	$y_1 \wedge (x_1 \vee x_2)$	$y_2 \wedge (x_1 \vee x_2)$	$(x_1 \vee x_2) \wedge (y_1 \vee y_2)$

which is the linear transformation that maps coefficients of the state described by the Y basis into coefficients of the state described by the X basis. The inner product $\langle x_j|y_i \rangle = \langle y_i|x_j \rangle^*$ is the probability amplitude of transiting to the $|x_j\rangle$ state from the $|y_i\rangle$ state;[11] its square $|\langle x_j|y_i \rangle|^2$ equals the probability of observing x_j on the next measurement of X given that y_i was obtained from a previous measurement of Y. A similar argument produces

$$\sum_i \langle y_j|x_i \rangle\langle x_i|z \rangle = \langle y_j|z \rangle,$$

which is the linear transformation that maps coefficients of the state described by the X basis into coefficients of the state described by the Y basis. The inner product $\langle y_j|x_i \rangle$ is the probability amplitude of transiting to the $|y_j\rangle$ state from the $|x_i\rangle$ state; its square $|\langle y_j|x_i \rangle|^2$ equals the probability of observing y_j on the next measurement of Y given that x_i was obtained from a previous measurement of X.

In sum, one constructs (a) the first marginal distribution from the inner products like $\langle x_i|z \rangle$ that relate the initial state to the states for the first basis, and (b) the second marginal distribution from the inner products like $\langle y_j|x_i \rangle$ that relate the states from the first basis to the states of the second basis.

The inner products relating one basis to another must satisfy several important constraints.

First, the fact that $|\langle x_j|y_i \rangle|^2 = |\langle y_i|x_j \rangle|^2$ implies that (even) incompatible measurements must satisfy

$$Pr(x_j|y_i) = Pr(y_i|x_j), \tag{3.14}$$

the so-called *law of reciprocity* (Peres, 1995, p. 34). Of course, classic probability is not subject to this constraint. It is important to note that (3.14) need hold only for transitions between basis states, not for more general (coarse-grained) events.

Second, consider the matrix U of coefficients with $\langle y_i|x_j \rangle$ (representing the transition to state $|y_j\rangle$ from state $|x_i\rangle$) in row i and column j. Then the column vector Ψ (which describes the initial state in terms of the X basis) is related to the column vector Φ

---

[11]The notation ζ^* stands for the complex conjugate of the complex number ζ (recall that, in general, probability amplitudes like $\langle y_i|x_j \rangle$ can be complex numbers). That $\langle y_i|x_j \rangle^*$ always equals $\langle x_j|y_i \rangle$ is ensured by the assumption (p. 80) that the bra-ket inner product on the Hilbert space H is Hermitian.

(which describes the initial state in terms of the Y) by the linear transformation $\Phi = U\Psi$; and similarly, Φ is related to Ψ by the linear transformation $\Psi = U^\dagger \Phi$. Notice that therefore

$$\Phi = UU^\dagger \Phi, \quad \Psi = U^\dagger U \Psi$$

and this is true (with the same U) no matter what the initial state (represented in terms of the two bases by Φ and Ψ) happens to be. It follows that

$$UU^\dagger = \mathbf{I} = U^\dagger U;$$

in other words, U is a *unitary* matrix. Unitarity of U guarantees that U preserves the lengths of the vectors before and after transformation, and implies that the *transition matrix T*, which has $|\langle y_i|x_j\rangle|^2$ in row i and column j, must be *doubly stochastic*: each row and each column of T must sum to unity. This is called the *doubly stochastic law* (Peres, 1995, p. 33). In classic probability theory, the transition matrix must be stochastic (each column sums to unity) but need not be doubly stochastic.

Thus, for incompatible measures, quantum probabilities must obey two laws that are not required by classic probability: the law of reciprocity (3.14) and the doubly stochastic law. On the other hand, classic probability must obey the law of total probability (3.3), which is not required by quantum probability for incompatible measures. These three properties can be used to distinguish quantum models from classical models empirically.

The projector for the event L_{x_i} is $P_{x_i} = |x_i\rangle\langle x_i|$, and that for the event L_{y_j} is $P_{y_j} = |y_j\rangle\langle y_j|$. It is interesting to compare the composition of projections produced by measuring Y first, then X,

$$P_{x_i} P_{y_j} = |x_i\rangle\langle x_i| \, |y_j\rangle\langle y_j| = \langle x_i|y_j\rangle |x_i\rangle\langle y_j|,$$

with that produced by measuring X first, then Y,

$$P_{y_j} P_{x_i} = |y_j\rangle\langle y_j| \, |x_i\rangle\langle x_i| = \langle y_j|x_i\rangle |y_j\rangle\langle x_i|.$$

In contrast to the case of compatible measures, where the commutator is always zero (*cf.* p. 88), here the assumption of incompatibility ensures that the commutator

$$P_{x_i} P_{y_j} - P_{y_j} P_{x_i} = \langle x_i|y_j\rangle |x_i\rangle\langle y_j| - \langle y_j|x_i\rangle |y_j\rangle\langle x_i|$$

is nonzero for some i and j. This implies that different orders of measurement can produce different final projections and thus different probabilities. In other words, quantum probability provides a theory for explaining *order effects* on measurements, a pervasive phenomenon throughout the social and behavioral sciences.

Let us now examine the event probabilities in the case of incompatible measures. Here we have to give separate careful analyses of the different possible experimental conditions.

First consider condition XY. In this case we have

$$\Pr(y_j \wedge x_i | XY) = \| P_{y_j} P_{x_i} | z \rangle \|^2 = |\langle y_j | x_i \rangle|^2 |\langle x_i | z \rangle|^2$$
$$= \Pr(y_j | x_i, XY) \Pr(x_i | XY),$$

so that

$$\Pr(y_j | XY) = \sum_i |\langle x_i | z \rangle|^2 |\langle y_j | x_i \rangle|^2, \tag{3.15}$$

similar to the situations for compatible measurements in both classic and quantum probability.

To get a more intuitive idea, refer again to Fig. 3.1. The probability $\Pr(x_1 | XY)$ of responding x_1 to question X on the first measure is equal to $|\langle x_1 | z \rangle|^2$, the squared probability amplitude of transiting from the initial state $|z\rangle$ to the basis vector $|x_1\rangle$. Given that the first measurement produces x_1, and the state now equals $|x_1\rangle$, the probability $\Pr(y_1 | x_1, XY)$ of responding $Y = y_1$ to the second question is equal to $|\langle y_1 | x_1 \rangle|^2$, the squared probability amplitude of transiting from $|x_1\rangle$ to $|y_1\rangle$. The probability of observing $X = x_1$ on the first test followed by $Y = y_1$ on the second test equals

$$\Pr(x_1 | XY) \Pr(y_1 | x_1, XY) = |\langle x_1 | z \rangle|^2 |\langle y_1 | x_1 \rangle|^2.$$

A similar analysis produces

$$\Pr(x_2 | XY) \Pr(y_1 | x_2, XY) = |\langle x_2 | z \rangle|^2 |\langle y_1 | x_2 \rangle|^2$$

for the probability of observing $X = x_2$ on the first test followed by $Y = y_1$ on the second test. Thus the probability of observing $Y = y_1$ on the second test, given the XY condition, equals

$$\Pr(y_1 | XY) = |\langle x_1 | z \rangle|^2 |\langle y_1 | x_1 \rangle|^2 + |\langle x_2 | z \rangle|^2 |\langle y_1 | x_2 \rangle|^2.$$

Next consider the probability of responding to question Y alone, not preceded by question X. The projection of the initial state onto the L_{y_j} event is $P_{y_j}|z\rangle = |y_j\rangle\langle y_j||z\rangle = |y_j\rangle\langle y_j|z\rangle$, and so $\Pr(y_j|Y) = \langle z|y_j\rangle\langle y_j|y_j\rangle\langle y_j|z\rangle = |\langle y_j|z\rangle|^2$. More intuitively, this is obtained from the squared amplitude of transiting from the initial state $|z\rangle$ to the basis vector $|y_j\rangle$ without measuring or knowing anything about the first question. Expansion of the identity operator produces the following interesting result:

$$\Pr(y_j|Y) = |\langle y_j|z\rangle|^2 = |\langle y_j|\mathbf{I}|z\rangle|^2 =$$
$$\left|\sum_i \langle y_j|(|x_i\rangle\langle x_i|)|z\rangle\right|^2 = \left|\sum_i \langle y_j|x_i\rangle\langle x_i|z\rangle\right|^2. \quad (3.16)$$

Comparing (3.16) with (3.15), we see that $\Pr(y_j|Y)$ and $\Pr(y_j|XY)$ need not be equal. This difference can explain interference effects. Let us analyze the interference effect in more detail for the special case shown in Fig. 3.1, with only two outcomes for each measure.

$$|\langle y_1|z\rangle|^2 = (\langle y_1|x_1\rangle\langle x_1|z\rangle + \langle y_1|x_2\rangle\langle x_2|z\rangle)$$
$$(\langle y_1|x_1\rangle\langle x_1|z\rangle + \langle y_1|x_2\rangle\langle x_2|z\rangle)^*$$
$$= |\langle y_1|x_1\rangle\langle x_1|z\rangle|^2 + |\langle y_1|x_2\rangle\langle x_2|z\rangle|^2 +$$
$$\langle y_1|x_1\rangle\langle x_1|z\rangle\langle y_1|x_2\rangle^*\langle x_2|z\rangle^* +$$
$$\langle y_1|x_2\rangle\langle x_2|z\rangle\langle y_1|x_1\rangle^*\langle x_1|z\rangle^*$$
$$= |\langle y_1|x_1\rangle\langle x_1|z\rangle|^2 + |\langle y_1|x_2\rangle\langle x_2|z\rangle|^2 +$$
$$|\langle y_1|x_1\rangle||\langle x_1|z\rangle||\langle y_1|x_2\rangle||\langle x_2|z\rangle|$$
$$(e^{i(\langle y_1|x_1\rangle\langle x_1|z\rangle\langle y_1|x_2\rangle\langle x_2|z\rangle)} + e^{-i(\langle y_1|x_1\rangle\langle x_1|z\rangle\langle y_1|x_2\rangle\langle x_2|z\rangle)})$$
$$= |\langle y_1|x_1\rangle\langle x_1|z\rangle|^2 + |\langle y_1|x_2\rangle\langle x_2|z\rangle|^2 +$$
$$|\langle y_1|x_1\rangle||\langle x_1|z\rangle||\langle y_1|x_2\rangle||\langle x_2|z\rangle|$$
$$(\cos(\theta) + i\sin(\theta) + \cos(\theta) - i\sin(\theta))$$
$$= |\langle y_1|x_1\rangle\langle x_1|z\rangle|^2 + |\langle y_1|x_2\rangle\langle x_2|z\rangle|^2 +$$
$$2|\langle y_1|x_1\rangle||\langle x_1|z\rangle||\langle y_1|x_2\rangle||\langle x_2|z\rangle|\cos(\theta),$$

where θ is the angle in the complex plane of the complex number $\langle y_1|x_1\rangle\langle x_1|z\rangle\langle y_1|x_2\rangle^*\langle x_2|z\rangle^*$ (see Fig. 3.5). If we restrict the probability amplitudes to real numbers, then we are restricted to the horizontal line in Fig. 3.5, so that $\theta = 0$ or $\theta = \pi$ and $\cos(\theta) = \pm 1$.

Note that the first two terms in the expression just derived for $\Pr(y_1|Y)$ exactly match those found when computing $\Pr(y_1|XY)$. If the cosine in the third term is zero, then $\Pr(y_1|Y) - \Pr(y_1|XY) = 0$ and there would be no interference. Thus the difference $\Pr(y_1|Y) -$

$\Pr(y_1|XY)$ is contributed solely by the cosine term, which is called the *interference term*. Here we see the uniquely quantum prediction of interference effects for incompatible measures.

Quantum probability provides a more coherent and elegant explanation of interference effects than classic probability theory. The former uses a single interference coefficient θ to relate the two marginal distributions, $\Pr(y_1|Y)$ and $\Pr(y_1|XY)$, whereas the latter postulates two separate joint probability distributions and then derives the marginals for each condition from these separate joint distributions.

It is also instructive to compare the probabilities of the binary-valued responses for condition XY with those for YX:

$$\Pr(x_1 \wedge y_1|XY) = |P_{y_1}P_{x_1}|z\rangle|^2 = |\langle x_1|z\rangle|^2\,|\langle y_1|x_1\rangle|^2,$$
$$\Pr(y_1 \wedge x_1|YX) = |P_{x_1}P_{y_1}|z\rangle|^2 = |\langle y_1|z\rangle|^2\,|\langle x_1|y_1\rangle|^2.$$

Note that $|\langle y_1|x_1\rangle|^2 = |\langle x_1|y_1\rangle|^2$ and so

$$\Pr(x_1 \wedge y_1|XY) - \Pr(y_1 \wedge x_1|YX) = |\langle y_1|z\rangle|^2),$$

which differs from zero as long as $|\langle x_1|z\rangle|^2 \neq |\langle y_1|z\rangle|^2$. An illustration of these two different projections appears in Fig. 3.6. Once again, quantum theory provides a direct explanation for the relation between the distributions produced by the two conditions, whereas classic probability theory needs to assume an entirely new probability distribution for each condition.

Finally it is interesting to re-examine the conditional probabilities for incompatible measures.

$$\Pr(y_1|x_1, YX) = |\langle y_1|x_1\rangle|^2 = |\langle x_1|y_1\rangle|^2 = \Pr(x_1|y_1, YX).$$

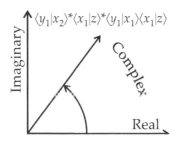

Figure 3.5: The angle between probability amplitudes.

This law of reciprocity places a very strong constraint on quantum probability theory. This relation only holds, however, for complete measures that involve transitions from one basis state to another. It is no longer true for coarse measurements that are disjunctions of several basis vectors.

Why complex numbers?

Consider again Fig. 3.1 which involves binary outcomes for each measure. If we are restricted to real valued probability amplitudes, then we obtain the following simplification of our basic theoretical result for incompatible measures:

$$\Pr(y_1|Y) = |\langle y_1|x_1\rangle|^2|\langle x_1|z\rangle|^2 + |\langle y_1|x_2\rangle|^2|\langle x_2|z\rangle|^2$$
$$\pm 2|\langle y_1|x_1\rangle||\langle x_1|z\rangle||\langle y_1|x_2\rangle||\langle x_2|z\rangle|.$$

The interference term is now simply determined by the sign and magnitude of $|\langle u|x\rangle||\langle x|z\rangle||\langle u|y\rangle||\langle y|z\rangle|$. Complex probability amplitudes can be shown to be needed under the following conditions and results. Suppose we can perform variations on our basic experiment by changing some experimental factor F, and that we find that changing the experimental factor from level F_1 to level F_2 produces an interference effect of the same sign (+ or −), but increases the magnitude of the interference:

$$|\Pr(y_1|Y, F_2) - \Pr(y_1|XY, F_2)| > |\Pr(y_1|Y, F_1) - \Pr(y_1|XY, F_1)|.$$

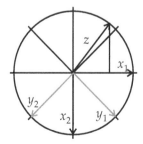 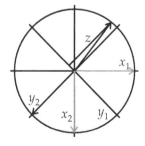

Figure 3.6: Projections of the initial state on basis vectors from two different orthonormal bases.

Suppose also that this same manipulation does not change the joint probabilities, so that

$$Pr(x_1 \wedge y_1 | XY, F_1) = |\langle y_1|x_1\rangle|^2 |\langle x_1|z\rangle|^2$$
$$= Pr(x_1 \wedge y_1 | XY, F_2), \qquad (3.17)$$
$$Pr(x_2 \wedge y_1 | XY, F_1) = |\langle y_1|x_2\rangle|^2 |\langle x_2|z\rangle|^2$$
$$= Pr(x_2 \wedge y_1 | XY, F_2). \qquad (3.18)$$

Together, (3.17) and (3.18) imply that changes in this factor F leave $|\langle y_1|x_1\rangle||\langle x_1|z\rangle||\langle y_2|x_2\rangle||\langle x_2|z\rangle|$ constant and vary $\cos(\theta)$ instead.

Consider the example from physics called the paradox of recombined beams (French & Taylor, 1978, pp. 295–296; cf. also Fig. 3.1). In this experiment, a plane polarized photon z is shot through a quarter wave plate to produce a circularly polarized photon. There are two possible channel outputs for the quarter wave plate, a left-circular or right-circular rotation (to which we assign, respectively, the labels x_1 and x_2 in Fig. 3.1). A final detector determines whether the output from the quarter wave plate can be detected (symbolized y_1 in Fig. 3.1) by a linear polarized detector rotated at angle φ with respect to the original state of the photon. In this situation, the critical factor F that is manipulated is the angle φ between the initial and final linear polarization.

The two channel outputs from the quarter wave plate form an orthonormal basis of two vectors, $|x_1\rangle$ and $|x_2\rangle$, in terms of which the state is represented. The probability amplitude of transiting from the initial state to the final state equals

$$\langle y_1|z\rangle = \langle y_1|\mathbf{I}|z\rangle = \langle y_1| \left(|x_1\rangle\langle x_1| + |x_2\rangle\langle x_2| \right) |z\rangle$$
$$= \langle y_1|x_1\rangle\langle x_1|z\rangle + \langle y_1|x_2\rangle\langle x_2|z\rangle.$$

When the right channel is closed, the then probability of passing through the left channel is $|\langle x_1|z\rangle|^2 = 1/2$, and the probability of detection is also $|\langle y_1|x_1\rangle|^2 = 1/2$. The same is true when the left channel is closed: then the probability of passing through right channel is $|\langle x_2|z\rangle|^2 = 1/2$ and the probability of detection is $|\langle y_1|x_2\rangle|^2 = 1/2$. Further, when both channels are open, the probability of detection is $\cos(\varphi)$. Therefore, we have five equations in the four unknowns $\langle x_1|z\rangle$, $\langle y_1|x_1\rangle$, $\langle x_2|z\rangle$, $\langle y_1|x_2\rangle$:

$$|\langle x_1|z\rangle| = |\langle y_1|x_1\rangle| = |\langle x_2|z\rangle| = |\langle y_1|x_2\rangle| = 1/\sqrt{2},$$
$$\langle y_1|x_1\rangle\langle x_1|z\rangle + \langle y_1|x_2\rangle\langle x_2|z\rangle = \cos(\varphi).$$

The first four equations do not depend on φ, but the last one does. This forces us to find a solution using complex numbers. In this case, the solutions are

$$\langle x_1|z \rangle = 1/\sqrt{2} = \langle x_2|z \rangle, \langle y_1|x_1 \rangle = e^{-i\varphi}/\sqrt{2}, \langle y_2|x_2 \rangle = e^{i\varphi}/\sqrt{2}.$$

What is the difference between superposition and mixture?

A superposition state is a linear combination of the basis states for a measurement. The initial state $|z\rangle$ is not restricted to just one of the basis states. According to quantum logic, if L_{x_1} is an event corresponding to the observation of x_1 and L_{x_2} is another event corresponding to the observation x_2, then we can form a new disjunction event $L_{x_1} \vee L_{x_2}$ which is the set of all linear combinations $|z\rangle = a|x_1\rangle + b|x_2\rangle$, where the coefficients a and b are complex numbers with $|a|^2 + |b|^2 = 1$. In the above case, the initial state, $|z\rangle$, would be in a superposition state with respect to the basis states for measure A. In this case we observe the value x_1 with probability $|\langle x_1|z\rangle|^2 = |a|^2$, and the value x_2 with probability $|\langle x_2|z\rangle|^2 = |b|^2$.

It is difficult to interpret the superposition state. There is no well agreed upon psychological interpretation of superposition—indeed, the interpretation of this concept has produced great controversy (Schrödinger's cat problem). Intuitively, a superposition seems to be something like a *fuzzy* and *uncertain* representation of a state. It is tempting, but invalid, to interpret superposition is meaning that immediately before measurement, you are either in state $|x_1\rangle$ with probability $|a|^2$ or in state $|x_2\rangle$ with probability $|b|^2$. In fact, that is a description of a *mixed state* (either classical or quantum), not a *superposition state* (quantum only). Superposition quantum states and mixed quantum states are distinguishable by their probability predictions, as the following example shows.

Again, we start from Fig. 3.1 with binary outcomes, this time letting the relationship between the two bases be given by

$$|y_1\rangle = (|x_1\rangle + |x_2\rangle)/\sqrt{2}, |y_2\rangle = (|x_1\rangle - |x_2\rangle)/\sqrt{2},$$
$$|x_1\rangle = (|y_1\rangle + |y_2\rangle)/\sqrt{2}, |x_2\rangle = (|y_1\rangle - |y_2\rangle)/\sqrt{2}$$

(as in Fig. 3.6). After a measurement of $Y = y_1$, we are in the super-position state $|y_1\rangle = (|x_1\rangle + |x_2\rangle)/\sqrt{2}$, and we have the probabilities

$$Pr(x_1) = Pr(x_2) = 1/2, Pr(y_1) = 1, Pr(y_2) = 0. \qquad (3.19)$$

On the other hand, consider the mixed state in which the basis states $|x_1\rangle$ and $|x_2\rangle$ each have probability 1/2. Whichever of the two basis states you are in, y_1 and y_2 are equally likely measurements for Y, so the mixed state produces the probabilities

$$Pr(x_1) = Pr(x_2) = Pr(y_1) = Pr(y_2) = 1/2,$$

differing dramatically from (3.19). In sum, an equal mixture of $|x_1\rangle$ and $|x_2\rangle$ produces different results from an equally weighted superposition of $|x_1\rangle$ and $|x_2\rangle$: but this difference is only revealed by obtaining probabilities from both X and an incompatible measure Y.

Concluding comments

Quantum probability was discovered by physicists in the early 20th century solely for applications to physics. But von Neumann axiomatized the theory and discovered that it implied a new logic, quantum logic, and a new probability, quantum probability. Just as the mathematics of differential equations spread from purely physical applications in Newtonian mechanics to applications throughout the social and behavioral sciences, it is very likely that the mathematics of quantum probability will also see new applications in the social and behavioral sciences.

Quantum probability reduces to classical probability when all the measures are compatible. But quantum probability departs dramatically from classical probability when the measures are incompatible. In particular, quantum probabilities do not have to obey the law of total probability as required by classical probabilities. Thus one can view quantum probability as a generalization of classical probability with the inclusion of incompatible measures. However, there are several important restrictions on quantum probabilities for incompatible measures. In this case the quantum probabilities must obey the law of reciprocity and the doubly stochastic law, which classical probabilities do not have to obey.

There are several advantages for using a quantum probability approach over a classical probability approach. First, the quantum approach does not always require or need to assume a joint probability space to derive and relate marginal probabilities from different measures. Marginal probabilities from different measures can all be derived from a common state vector without postulating a common joint distribution. Second, quantum probability theory provides an explanation for order effects on measurements, which is a pervasive problem in the social and behavioral sciences. Third, quantum probability provides an explanation for the interference effect that one measure has on another measure, which is another pervasive problem of measurements in the social and behavioral sciences. Finally, quantum probabilities allow for deterministic as well as probabilistic behavior, which matches human behavior better than random error theories.

Quantum probability theory is a new and exciting field of mathematics with many interesting and potentially useful applications to the social and behavioral sciences. The intention of this chapter was to show the simplicity, coherence, and generality of quantum probability theory.

References

Bruza, P. D., Busemeyer, J. R., & Gabora, L. (2009). Introduction to the special issue on quantum cognition. *Journal of Mathematical Psychology, 53*(5), 303–305. (Special Issue: Quantum Cognition)

Bruza, P. D., Lawless, W., van Rijsbergen, C. J., & Sofge, D. (Eds.). (2007). *Proceedings of the AAAI Spring Symposium on Quantum Interaction, March 27–29, Stanford University.* AAAI Press.

Bruza, P. D., Lawless, W., van Rijsbergen, C. J., & Sofge, D. (Eds.). (2008). *Proceedings of the Second Conference on Quantum Interactions, March 26–28, Oxford University.* AAAI Press.

Feynman, R. P., Leighton, R. B., & Sands, M. (1966). *The Feynman Lectures on Physics: Vol. III.* Reading, MA: Addison Wesley.

French, A. P., & Taylor, E. F. (1978). *An Introduction to Quantum Physics.* New York, NY: W. W. Norton.

Hughes, R. I. G. (1989). *The Structure and Interpretation of Quantum Mechanics.* Cambridge, MA: Harvard University Press.

Khrennikov, A. Yu. (2007). Can quantum information be processed by macroscopic systems? *Quantum Information Theory, 6*(6), 401–429.

Peres, A. (1995). *Quantum Theory: Concepts and Methods.* Dordrecht, NL: Kluwer Academic.

4

The Sorites Paradox: A Behavioral Approach

Ehtibar N. Dzhafarov and Damir D. Dzhafarov

The issues discussed in this chapter can be traced back to the Greek philosopher Eubulides of Miletus. He lived in the 4th century B.C.E., a contemporary of Aristotle whom he, according to Diogenes Laërtius (Yonge, 1901, pp. 77–78), "was constantly attacking". Eubulides belonged to what is known as the Megarian school of philosophy, founded by a pupil of Socrates named Euclid(es). Besides his quarrels with Aristotle we know from Diogenes Laërtius that Eubulides was the target of an epigram referring to his "false arrogant speeches", and that he "handed down a great many arguments in dialectics", mostly trivial sophisms of the kind ridiculed by Socrates in Plato's Dialogues. For example, the Horned Man argument asks you to agree that 'whatever you haven't lost you have', and points out that then you must have horns since you have not lost them.

Two of the "arguments in dialectics" ascribed to Eubulides, however, are among the most perplexing and solution-resistant puzzles in history. The first is the Liar paradox, which demonstrates the impossibility of assigning a truth value to the statement 'This statement is false'. Eubulides's second 'serious' paradox, the Heap, is the subject of this chapter. It can be stated as follows. (1) A single grain of sand does not form a heap, but many grains (say 1,000,000) do. (2) If one has a heap of sand, then it will remain a heap if one removes a single grain from it. (3) But, by removing from a

Keywords: closeness, comparison, connectedness, matching, regular mediality, regular minimality, sorites, supervenience on stimuli, tolerant responses, vagueness.

heap of sand one grain at a time sufficiently many times, one can eventually be left with too few grains to form a heap. This argument is traditionally referred to by the name 'sorites' (from the Greek σορός [*soros*] meaning 'heap'), with the adjective 'soritical' used to indicate anything 'sorites-related'. Thus, the Bald Man paradox which Diogenes Laërtius lists as yet another argument of Eubulides in dialectics is a 'soritical argument', because it follows the logic of the sorites but applies it to the example of the number of hairs forming or not forming a full head of hair.

Two varieties of sorites

This chapter is based on Dzhafarov & Dzhafarov(2010a,b), in which we proposed to treat sorites as a behavioral issue, with 'behavior' broadly understood as the relationship between stimuli acting upon a system (the 'system' being a human observer, a digital scale, a set of rules, or anything whatever). We present here a sketch of this treatment, omitting some of the more delicate philosophical points. Examples of behavioral questions pertaining to sorites include: Can a person consistently respond by different characterizations, such as 'is 2 meters long' and 'is not 2 meters long', to visually presented line segments a and b which only differ by one billionth of one percent? Is the person bound to say that these segments, a and b, look 'the same' when they are presented as a pair? But soritical questions can also be directed at non-sentient systems: Can a crude two-pan balance at equilibrium be upset by adding to one of the pans a single atom? Can the probability that this balance will remain at equilibrium change as a result of adding to one of the pans a single atom?

Sorites, viewed behaviorally, entails two different varieties of problems. The first, *classificatory sorites*, is about the identity or nonidentity of responses, or some properties thereof, to stimuli that are 'almost identical', 'differ only microscopically'. The second, *comparative sorites*, concerns 'match/not match'-type responses to *pairs of stimuli*, or more generally, response properties interpretable as indicating whether the two stimuli in a pair 'match' or 'do not match'. A prototypical example would be visually presented pairs of line segments with 'matching' understood as 'appearing the same in length'. Perhaps surprisingly, the two varieties of sorites turn out to be very different. The classificatory sorites is a logical impossibility, and our contribution to its analysis consists in demonstrating this impossibility by explicating its underlying assumptions on arguably the highest possible level of generality (using the mathematical language of Maurice Fréchet's proto-topological V-spaces). The comparative sorites (also called 'observational' in the philosophical literature) is, by contrast, perfectly possible: one can construct abstract and even physically realizable systems which exhibit soritical behavior of the comparative variety. This, however, by no means *has to be* the case for any system with 'match/not match'-type response properties. Most notably, we will argue that contrary to the widespread view this is not the case for the human comparative judgments, where comparative sorites contradicts a certain regularity principle supported by all available empirical evidence, as well as by the practice and language of the empirical research dealing with perceptual matching.

The compelling nature of the view that human comparative judgments are essentially soritical is apparent in the following quotation from R. Duncan Luce (who used this view to motivate the introduction of the important algebraic notion of a semiorder).

> It is certainly well known from psychophysics that if "preference" is taken to mean which of two weights a person believes to be heavier after hefting them, and if "adjacent" weights are properly chosen, say a gram difference in a total weight of many grams, then a subject will be indifferent between any two "adjacent" weights. If indifference were transitive, then he would be unable to detect any weight differences, however great, which is patently false. (Luce, 1956, p. 179)

One way of conceptualizing this quotation so that it appears to describe a 'paradox' is this. (1) If the two weights being hefted and compared are the same, x and x, they 'obviously' match perceptually. (2) If one adds to one of the two weights a 'microscopic' amount ε (say, the weight of a single atom), the human's response to x and $x + \varepsilon$ cannot be different from that to x and x, whence x and $x + \varepsilon$ must still match perceptually. (3) But by adding ε to one of the weights many times one can certainly obtain a pair of weights x and $x + n\varepsilon$ that are clearly different perceptually.

This reasoning follows the logic of the *classificatory* sorites, and is thus logically invalid. This reasoning, however, seems intuitively compelling, which explains both why the philosophers call the comparative ('observational') sorites a sorites, and why psychophysicists need to deal with the classificatory sorites even if they are interested primarily in human comparative judgments.

Luce definitely did not imply a connection to the classificatory sorites of Eubulides. Rather, he simply presented as "well known" the impossibility of telling apart very close stimuli. This being a logically tenable position, we will see that the "well known" fact in question is in reality a theoretical belief not founded in empirical evidence. Almost everything in it contradicts or oversimplifies what we know from modern psychophysics. Judgments like 'x weighs the same as y' or 'x is heavier than y' given by human observers in response to stimulus pairs cannot generally be considered predicates on the set of stimulus pairs, as these responses are not uniquely determined by these stimulus pairs: an indirect approach is needed to define a matching relation based on these inconsistent judgments. When properly defined, the view represented by the quotation from Luce (1956) loses its appearance of self-evidence.

Classificatory sorites

The classificatory sorites is conceptually simpler than the comparative one and admits a less technical formal analysis. Within the framework of the behavioral approach we view the elements that the argument is concerned with (such as collections of grains of sand) as stimuli 'acting' upon a system and 'evoking' its responses. Thus, the stimuli may be electric currents passing through a digital ammeter which responds by displaying a number on its indicator; or the stimuli may be schematic drawings of faces visually presented to a human observer who responds by saying that the face is 'nice' or 'not nice'; or the stimuli may be appropriately measured weather conditions in May to which a flock of birds reacts by either migrating north or not. This is a very general framework which many examples can be molded to fit. In Eubulides's original argument, the stimuli are collections of sand grains, presented visually or described verbally, and the system responding by either 'form(s) a heap' or 'do(es) not form a heap' may be a human observer, if one is interested in factual classificatory behavior, or a system of linguistic rules, if one is interested in the normative use of language.

Supervenience, tolerance, and connectedness

To get our analysis off the ground, we would like to identify some properties which characterize stimulus-effect systems amenable to (classificatory) soritical arguments. Consider the following.

Supervenience assumption (Sup). There is a certain property of the system's responses to stimuli that—all else being equal—cannot have different values for different instances (replications) of one and the same stimulus. That is, there is a function π such that a certain property of the response of the system to stimulus x is $\pi(x)$. We call π the *stimulus-effect function*, and its values *stimulus effects*.[1]

Tolerance assumption (Tol). The stimulus-effect function $\pi(x)$ is 'tolerant to microscopic changes' in stimuli: if $x' \neq x$ is chosen sufficiently close to x, then $\pi(x') = \pi(x)$.

Connectedness assumption (Con). The stimulus set S contains at least one pair of stimuli a, b with $\pi(a) \neq \pi(b)$ such that one can find a finite chain of stimuli $a = x_1, \ldots, x_i, x_{i+1}, \ldots, x_n = b$ leading from a to b 'by microscopic steps': x_{i+1} is arbitrarily or maximally close to but different from x_i for $i = 1, \ldots, n - 1$.

We have, of course, yet to define what precisely we mean by 'closeness' and 'connectedness by microscopic steps' here, but deferring that for the moment, it is not difficult to see that the conjunction Sup $\wedge$ Tol $\wedge$ Con is sufficient and necessary for formulating the classificatory sorites 'paradox'.

Classificatory Sorites. There exists a stimulus-effect system satisfying Sup, Tol, and Con.

It is clear that this statement is false: the three assumptions in question are mutually inconsistent. Indeed, by Sup and Con we can fix

---

[1]The response itself, *e.g.*, 'heap' or 'not heap', may be viewed as a response property (namely, its identity or content),] and this property may or may not be a stimulus effect. Other candidates for being stimulus effects can be such response properties as response time, response probability, the probability with which response time falls within a certain interval, etc.

a pair of stimuli a, b with $\pi(a) \neq \pi(b)$, connectable by a *classificatory soritical sequence* $x_1, \ldots, x_n$ with $a = x_1$, $b = x_n$, and x_{i+1} only 'microscopically' different from x_i for each i. By Tol, $\pi(x_i) = \pi(x_{i+1})$ for $i = 1, \ldots, n-1$, whence $\pi(a) = \pi(b)$, a contradiction. Therein lies the classificatory sorites.

Notice that if Sup is not satisfied, then Tol and Con simply cannot be formulated as above, as these formulations make use of a stimulus-effect function $\pi(x)$. Once Sup is accepted, simple mathematical examples can be constructed to witness the independence of Tol and Con. It is natural to ask whether Tol and Con can be formulated without an explicit reference to Sup, but this can readily be seen not to be an option, at least not without making Tol 'automatically' false. Indeed, if a property π_x of a response to stimulus x is not a function of x, then π_x will generally be different from π_y even if y is a replication of x, let alone close to but different from x.

The fact that Sup is indispensable for the formulability of the classificatory sorites leads one to reject the philosophical tradition of relating soritical issues to 'vague predicates'. A vague predicate is defined as one whose truth value is not determinable at least for some objects to which the predicate applies, whether the structure of its truth values is dichotomous (true, false), trichotomous (true, false, not known), or the entire interval between 0 (false) and 1 (true). Here we see a major advantage of the behavioral approach to sorites. One may very well argue about the truth value structure of the predicate 'form(s) a heap', and one may suggest that for certain values of x this predicate's truth value is indeterminate when applied to a collection of x grains of sand. However, a statement like "in this trial, this observer responded to x grains of sand by saying 'they form a heap'" or "in this trial, this observer did not produce a response to x grains of sand when asked to choose between 'they form a heap' and 'they do not form a heap'" is true or false in the simplest sense, with no controversy involved. The 'observer' in these examples can very well be replaced with a set of linguistic rules or the group of expert language users if one is interested in the normative rather than factual use of language.[2]

From the behavioral point of view (and in agreement with the

---

[2]In the philosophical literature, Varzi (2003) comes close to the behavioral approach by arguing that soritical issues are essentially non-semantic and are not confined to linguistic phenomena.

position stated with admirable clarity in the dictionary article by Peirce, 1901/1960), a 'vague predicate' is merely a special case of an inconsistent response. If we call the predicate 'form(s) a heap' vague, this is because the choice of an allowable response associated with this predicate (such as 'yes, it is true', 'possibly', or 'I don't know') is not uniquely determined by the number x of the sand grains to which it applies. Thus, a human observer is likely to classify one and the same collection of x grains sometimes as forming and sometimes as not forming a heap (and sometimes neither if this is an option); and in a group of competent speakers of the language some will choose one response, and some another. This implies a violation of Sup and the impossibility of formulating the soritical argument with this predicate. The behavioral scientist in a situation like this would likely redefine the stimulus-effect function $\pi = \pi(x)$ as the probability distribution on the set of all allowable responses, the hypothesis being that this probability distribution is now uniquely determined by stimuli. Thus, if the allowable responses are 'form(s) a heap' and 'do(es) not form a heap', then the hypothesis is that for some probability function $p(x)$, called a *psychometric function* in psychophysics,

$$\pi(x) = p(x) = \Pr['a \text{ collection of } x \text{ grains of sand is a heap'}].$$

Of course, to say that a probability p of a response to x is an effect of the stimulus x amounts to treating probabilities as occurring at individual instances of x 'within' the system responding to x, rather than characterizing patterns of the system's behaviors over a potential infinity of instances of x. While this view may encounter philosophical misgivings, it is routine in the established conceptual schemes of probability theory, physics, and behavioral sciences. Our analysis is not critically based on accepting this 'probabilistic realism', but the class of physically realizable response properties uniquely determined by stimuli may get precariously small if one rejects it.[3] Without allowing probability distributions over responses to function as legitimate stimulus effects one would often have to declare sorites altogether unformulable and hence automatically dissolved, or would have to seek additional factors to include in the description of stimuli.

---

[3]In the context of comparative sorites (p. 117 *ff.*), Hardin (1988) argued for the necessity of taking into account the probabilistic nature of responses to stimuli.

There are a number of avenues for redefining stimuli in order to achieve the compliance of some response property (deterministic or probabilistic) with Sup. Thus, one might think it important to take into account sequential effects, that is, to make the response property in question dependent on a sequence of previously presented stimuli, or even on both the previous stimuli and the responses given to each of them. In either case we deal with some form of compound stimuli, the space of which we can endow (not necessarily in a unique way) with a closeness structure based on that of the original space of stimuli. For instance, if in the sequence $x_0, x_1, x_2, \ldots$ of stimuli each x_{i+1} differs from x_i 'microscopically', then the same can be said of x^*_{i+1} and x^*_i in the sequence $x^*_0, x^*_1, x^*_2, \ldots$ of compound stimuli $x^*_i = \{x_j\}_{j \le i}$. If now a stimulus-effect function π, such as the probability of saying 'form(s) a heap', is uniquely determined by such finite sequences of successive stimuli, the characterizations Sup, Tol, and Con will be formulable for this function on the redefined stimulus set, and our analysis applies with no modifications.

To see the generality of our approach, one may even consider a radical redefinition of stimuli (by no means feasible for scientific purposes) which consists in taking stimulus instances as part of stimulus identities, so that each stimulus is formally characterized by a pair (x, t) where x is the stimulus's physical value, and t designates an 'instance' at which it occurs, say, a trial number. With this redefinition no stimulus (x, t) is replicable, so every response to (x, t) can be viewed as a function of (x, t), a stimulus effect. Given an 'initial' closeness measure between stimulus values x and y, and the conventional distance $|t - t'|$ between time moments t and t', it is easy to see that Tol in this situation means that (x, t) and (y, t') evoke identical responses if x and y are sufficiently close and $|t - t'|$ sufficiently small; similarly, Con means that for some (a, t) and (b, t') which evoke different responses one can find a sequence $(a, t) = (x_1, t_1), \ldots, (x_n, t_n) = (b, t')$ whose successive elements fall within the sphere of Tol. As Sup here is satisfied 'automatically', our analysis again applies with no modifications.

Closeness and connectedness

It is apparent from our formulation of Tol and Con that we need conceptual means of saying that two distinct stimuli x and y can be chosen 'as close as one wishes' or 'as close as possible'. The meaning is clear in the case the stimulus set is endowed with a metric. Thus, if stimulus values (say, time intervals) are represented by real numbers, y can be chosen arbitrarily close to x by making the difference $|x - y|$ arbitrarily small. If, as is the case with grains of sand, stimulus values are represented by integers, y is as close as possible to x (without being equal to it) if $|x - y| = 1$. The requirement of a full-fledged metric, however, is too stringent in general, and it is moreover unnecessary. The analysis of the classificatory sorites can be carried out at a much higher (arguably, the highest possible) level of generality using the following concept due to Fréchet (1918).

Definition 1. A *V-space* on a nonempty set S is a pair $\{S, \{V_x\}_{x \in S}\}$ where V_x, for each $x \in S$, is a collection of subsets of S satisfying (1) $V_x \neq \emptyset$, (2) if $V \in V_x$ then $x \in V$. For each $x \in S$, any element V of V_x is called a *vicinity* of x. Any set of vicinities obtained by choosing one element of V_x for every $x \in S$ is called a *V-cover* of S.

For each $x \in S$, each $V \in V_x$ represents the stimuli which are close to x *in some sense*, namely, in the sense of belonging to V. In particular, since x belongs to each of its vicinities, x is 'close' to itself in all possible senses. (Fig. 4.1 illustrates 'closeness' in a V-space.)

The notion of V-space obviously generalizes that of a topological space, and in particular, a V-space on any metric space (S, d) can be obtained by letting V_x consist of all open balls

$$B_x(\varepsilon) = \{u \in S : d(x, u) < \varepsilon\}$$

where $\varepsilon > 0$. In general, however, V-spaces provide for a notion of closeness that does not have to be numerical.

Next we need to use this notion of closeness to define connectedness. Eubulides's Heap riddle would not be perplexing were it not for the fact that by adding one grain of sand at a time one can obtain from a very small collection of grains of sand a collection large enough to form a heap. The key to defining connectedness in the general language of V-spaces is in replacing the notion of being connectable 'by microscopic steps' (which requires a quantitative measure of closeness) with the notion of being connectable, from each choice of vicinities covering the space, by a chain of sequentially intersecting vicinities.

Definition 2. A point $a \in S$ is *V-connected* to a point $b \in S$ in a V-space $\{S, \{V_x\}_{x \in S}\}$ if for any V-cover $\{V_x\}_{x \in S}$ of S one can find a finite chain of points $x_1, x_2, \ldots x_{n-1}, x_n \in S$ such that (1) $a = x_1$, (2) $b = x_n$, (3) $V_{x_i} \cap V_{x_{i+1}} \neq \emptyset$ for $i = 1, \ldots n - 1$. A V-space $\{S, \{V_x\}_{x \in S}\}$ is V-*connected* if any two points in S are V-connected in $\{S, \{V_x\}_{x \in S}\}$.

Consider the following example. Let $S = \mathbb{N}$ be the set of all natural numbers $0, 1, 2, \ldots$, and give it a V-space structure by defining, for each $n, k \in \mathbb{N}$, $V_{n,k} = \{n, n+1, \ldots, n+k\}$ and $\mathcal{V}_n = \{V_{n,k} : k > 0\}$. Then any two numbers $a < b$ in this space are V-connected because for any V-cover, the chain $x_1, x_2, \ldots, x_{b-a+1}$ with $x_i = a + i - 1$ satisfies properties (1)–(3) of Definition 2. If, however, we take instead $\mathcal{V}_n = \{V_{n,k} : k \geq 0\}$, then no two elements can be V-connected, as witnessed by the V-cover $\{V_{n,0} : n \in \mathbb{N}\}$.

It is easy to see that the relation 'is connected to' is an equivalence relation, whence we immediately have the following lemma.

Lemma 1. *For any V-space* $\{S, \{V_x\}_{x \in S}\}$*, the set S is a union* $\bigcup S_\gamma$ *of pairwise disjoint nonempty subsets (the V-components of S) such that*

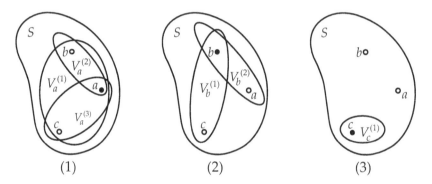

(1)   (2)   (3)

Figure 4.1: An example of a V-space $\{S, \{V_x\}_{x \in S}\}$. The set S consists of all points within the large outlined area, including the points a, b, and c shown. The vicinities $V_a^{(1)}$, $V_a^{(2)}$, $V_a^{(3)}$ of point a that together comprise $\mathcal{V}_a$ are indicated by the ovals in (1), and similarly for $\mathcal{V}_b = \{V_b^{(1)}, V_b^{(2)}\}$ in (2) and $\mathcal{V}_c = \{V_c^{(1)}\}$ in (3). These three sets of vicinities determine the closeness relations among a, b, and c. Thus, c is close to a in the sense of belonging to $V_a^{(1)}$ and $V_a^{(3)}$; c is close to b in the sense of belonging to $V_b^{(1)}$; and a and b are not close to c in any sense (*i.e.*, they are 'not close at all' to c) as they do not belong to any vicinity of c. Note that, at this level of generality, closeness is not a symmetric relation. Any V-cover of S will contain $V_c^{(1)}$, one and only one of $V_b^{(1)}$, $V_b^{(2)}$, and one and only one of $V_a^{(1)}$, $V_a^{(2)}$, $V_a^{(3)}$.

any two points in every V-component are V-connected and no two points belonging to different V-components are V-connected.

We also need the following definition to formalize Sup and Tol.

Definition 3. Given a V-space $\{S, \{V_x\}_{x \in S}\}$ and an arbitrary set R, any function $\pi: S \to R$ is a *stimulus-effect function* (and R a set of *stimulus effects*). A stimulus effect function π is called *tolerant at $x \in S$* in $\{S, \{V_x\}_{x \in S}\}$ if there is a vicinity $V_x \in \mathcal{V}_x$ on which π is constant; π is *tolerant* if it is tolerant at every point.

Thus R can be the two-element set {'form(s) a heap', 'do(es) not form a heap'}; or the set $[0, 1]$ representing the probabilities of choosing the response 'form(s) a heap' over 'do(es) not form a heap'; or the set of all probability distributions on $[0, 1]$, with $x \in [0, 1]$ representing the degree of confidence with which a stimulus is judged to form a heap.

Dissolving the classificatory sorites 'paradox'

The reward for formulating the soritical concepts on such a high level of generality is that the classificatory sorites 'paradox' can now be dissolved by means of a simple mathematical theorem.

Theorem 1 (No-tolerance theorem). *Let $\{S, \{V_x\}_{x \in S}\}$ be a V-space and $\pi: S \to R$ a stimulus-effect function, such that S contains two V-connected elements a, b with $\pi(a) \neq \pi(b)$. Then π is not tolerant: for at least one $x \in S$, π is non-constant on every vicinity of x ('however small').*[4]

N.B. The words 'however small' are added for emphasis only and do not imply that the vicinities have numerical sizes.

Proof. Assume π is tolerant: every x has a vicinity V_x^* such that π is constant on V_x^*. The set $\{V_x^*\}_{x \in S}$ is a V-cover of S, and a, b being V-connected, one can form a sequence $V_{x_1}^*, \ldots, V_{x_n}^*$ satisfying (1)–(3) of Definition 2. Then, denoting by y_i an arbitrary element of $V_{x_i}^* \cap V_{x_{i+1}}^*$ for $i = 1, \ldots n - 1$, we would have $\pi(x_i) = \pi(y_i) = \pi(y_{i+1}) = $

---

[4]One may recognize in this formulation a generalized version of what is known as the epistemic dissolution of the classificatory sorites, proposed by Sorensen (1988a, 1988b) and Williamson (1994, 2000), except that they apply this dissolution to vague predicates. This would only be legitimate if vague predicates were assigned to stimuli consistently, but then they would not be considered vague.

$\pi(x_{i+1})$ (since $y_i, y_{i+1} \in V^A_{x_{i+1}}$), whence $\pi(a) = \pi(x_1) = \pi(x_n) = \pi(b)$, contradicting the premise $\pi(a) \neq \pi(b)$.   □

Corollary 1. *No non-constant function on a V-connected V-space is tolerant.*

One can easily recognize in Theorem 1 the spelled-out version of the classical (classificatory) sorites, taken as a *reductio ad absurdum* proof of the incompatibility of Sup, Tol, and Con. More precisely, this incompatibility is formulated as Sup ∧ Con ⟹ ~Tol. In fact, when discussing classificatory sorites, Sup and Con are almost always assumed implicitly, although Sup is sometimes mentioned as an innocuous premise. The 'paradox' (in these cases) is thus dissolved by pointing out the inescapable truth of the assertion our intuition often finds hard to accept:

Non-tolerance principle (~Tol). If Sup and Con hold, then there is at least one point $x_0 \in S$ in every vicinity of which ('however small'), the stimulus-effect function $\pi(x)$ is non-constant.

Note that it may very well be that every single point in S complies with ~Tol, and this is probably the case in a host of situations with continuously varying stimulus effect (as a mathematical example, consider the identity function, mapping each stimulus to itself).

To prevent misunderstanding, ~Tol does not imply that the responding system can be used to measure some of the stimulus values with absolute precision. The situation here is very much like the one with a stopped clock: it shows the correct time twice a day, but one cannot determine when. In order to distinguish a stimulus x from its arbitrarily close neighbors x' by means of a stimulus-effect function $\pi(x)$ a human researcher must know that x being presented on two different instances is indeed one and the same x, and that x' presented on another instance is not the same as x. This amounts to having a system identifying stimuli $\iota_S(x)$ being presented and a system identifying stimulus effects $\iota_R(\pi(x))$ being recorded, hence to having a stimulus-effect function besides $\pi(x)$ whose values react to arbitrarily small differences from precisely the same stimulus x (something not impossible but definitely not deliberately construable unless the stimuli have been identified by some $\iota'_S(x)$, which assumption leads to an infinite regress).

For completeness, we should mention that there is another way of approaching the inconsistency of Sup ∧ Tol ∧ Con, formulable as

Sup ∧ Tol ⟹ ~Con. We can define vicinities as constant-response areas of the stimulus set S; we call them 'pi-vicinities' since we denote the stimulus-effect function by π. This makes the stimulus-effect function 'automatically' tolerant, and one sees subsequently that no two points in S are V-connected unless they map into one and the same stimulus effect.

Definition 4. Given a nonempty set S, an arbitrary set R, and a stimulus-effect function $\pi: S \to R$, the *pi-vicinity* of $x \in S$ is the set P_x of all $x' \in S$ such that $\pi(x') = \pi(x)$. The pair $\{S, \{P_x\}_{x \in S}\}$ is called the *pi-space associated to* π.

Lemma 2. *Any pi-space $\{S, \{P_x\}_{x \in S}\}$ uniquely corresponds to the V-space on S in which the only vicinity of $x \in S$ is P_x. The collection of the sets $\{P_x\}_{x \in S}$ is the only V-cover of S in this V-space.*

Proof. Clearly $\{S, \{V_x\}_{x \in S}\}$ with $V_x = \{P_x\}$ satisfies Definition 1.   □

Lemma 3. (1) *The pi-space $\{S, \{P_x\}_{x \in S}\}$ associated to $\pi: S \to R$ is uniquely determined by π.* (2) *π is tolerant in the corresponding V-space $\{S, \{V_x = \{P_x\}\}_{x \in S}\}$.*

Proof. Immediate consequences of Definition 4, Lemma 2, and Definition 3.   □

This yields the following alternative dissolution of the classificatory sorites.

Theorem 2 (No-connectedness theorem). *Given a stimulus-effect function $\pi: S \to R$ and its associated pi-space $\{S, \{P_x\}_{x \in S}\}$, elements $a, b \in S$ are V-connected in the corresponding V-space $\{S, \{V_x = \{P_x\}\}_{x \in S}\}$ if and only if $\pi(a) = \pi(b)$.*

Proof. An immediate consequence of Definition 2 and the fact that either $P_x = P_y$ or $P_x \cap P_y = \varnothing$, for any $x, y \in S$.   □

Comparative sorites

The *comparative* sorites pertains to situations in which a system responds to *pairs* of stimuli (x, y), and there is some binary response whose values are interpretable as two complementary relations, 'x is matched by y' and 'x is not matched by y'. A prototypical example would be an experiment in each trial of which a human observer is shown a pair of line segments and asked whether they are or are not of the same length, or two circles of light and asked which of them is brighter. The matching relation in both cases is computed from probabilities of the observer's answers, as discussed later. Thus, we regard the comparative sorites as pertaining to what in psychophysics would be called *discrimination* or *pairwise comparison* tasks.

Comparing stimuli

The comparative sorites 'paradox' can be intuitively stated.

Comparative Sorites. A set of stimuli S acting upon a system and presentable in pairs may contain a finite sequence of stimuli $x_1, \ldots, x_n$ such that 'from the system's point of view' x_i is matched by x_{i+1} for $i = 1, \ldots, n-1$, but x_1 is not matched by x_n.

We call $x_1, \ldots, x_n$ as above a *comparative soritical sequence*.
    At the outset, the comparative sorites may seem very similar to the classificatory sorites, and it is natural to ask whether the former may be only a special case of the latter. This turns out not to be the case. It is true that nothing prevents one from redefining a pair of stimuli (x_i, x_j) into a single 'bipartite' stimulus x_{ij}, and treating 'match' and 'not match' as classificatory responses to x_{ij}. However, given a sequence $x_{12}, x_{23}, \ldots, x_{n-1,n}, x_{1n}$, we can only apply to it the rationale of the classificatory sorites provided each term in this sequence is only 'microscopically' different from its successor. While this may be the case for $x_{i,i+1}$ and $x_{i+1,i+2}$ for each $i = 1, \ldots, n-2$ (perhaps because x_i is very close to x_{i+1}, which in turn is very close to x_{i+2}), there is in general no reason at all to think that $x_{n-1,n}$ is only 'microscopically' different from x_{1n}, assuming a reasonable notion of closeness between the two can be formulated at all.[5] Thus,

---

[5]Nor can one consider the classificatory sorites a special case of the comparative one. Given a purported classificatory soritical sequence $x_1, \ldots, x_n$ under some stimulus-effect function π, we can indeed recast $\pi(x_i)$ into a function $f(x_i, x_i + 1)$ loosely interpretable as a 'comparison' of x_i with x_{i+1} for each $i = 1, \ldots, n-1$. But then the logic of the classificatory sorites would lead to the 'comparison' of x_{n-1} with x_n being different from that of x_1 with x_2 (while in the comparative sorites the two pairs produce the same effect, 'match'), with nothing in this logic necessitating

the existence of comparative soritical sequences is not automatically precluded by the analysis of the classificatory sorites given in the previous section. Indeed, it is easy to construct simple mathematical examples of such sequences,

Example 1. Fix $\varepsilon > 0$. Let the stimulus set S be the set $\mathbb{R}$ of reals, and define

$$'y \text{ matches } x' \longleftrightarrow |y - x| \le \varepsilon.$$

Then any sequence $0, \delta, 2\delta$ with $\varepsilon/2 < \delta \le \varepsilon$ is a comparative soritical sequence.

This example also precludes the possibility of reducing the comparative sorites to the classificatory one by postulating the existence of a stimulus-effect function $\pi(x)$ such that

$$'x \text{ is matched by } y' \Longleftrightarrow \pi(x) = \pi(y).$$

If such a function could always be found, the comparative sorites would indeed be obtained as a 'logical consequence' of the classificatory one and would then be ruled out together with the latter. In the preceding example, however, it is readily seen that given any non-constant stimulus-effect function π, there exists an x with $\pi(x) \ne \pi(y)$ for some y with $x < y \le x + \varepsilon$, even though $|y - x| \le \varepsilon$ and so 'y matches x'.

We see that, on the one hand, the comparative sorites cannot be ruled out as a logical inconsistency, as it was in the case of the classificatory one. On the other hand, the comparative sorites 'paradox' is rarely if ever presented as an exercise in constructing abstract mathematical examples like the one above, and is generally assumed to apply to systems which, in their responses, resemble human comparative judgments. That human comparative judgments are soritical is often considered self-evident and well-known. As we shall see in the next subsection, this assumption contradicts a certain psychophysical principle (Regular Mediality/Minimality) proposed for comparative judgments. In systems similar to human comparative judgments, for which it is plausible that this principle holds, the 'matching' relation must satisfy (a certain form of) transitivity, thus compelling the system to behave more in accordance with the following example than with Example 1.

---

x_1 to be 'compared' with x_n (while in the comparative sorites this comparison is critical).

Example 2. Fix $\lambda > 0$. Let the stimulus set be $S = \mathbb{R}_{\geq 0}$ (non-negative reals), and define

$$\text{'}y \text{ matches } x\text{'} \iff \lfloor \lambda x \rfloor = \lfloor \lambda y \rfloor,$$

where $\lfloor a \rfloor$ denotes the floor of a, *i.e.*, the greatest integer $\leq a$. Then the relation 'matches' is reflexive, symmetric, and transitive, whence no comparative soritical sequence involving this relation is possible.[6]

Stimulus areas

It is clear from the comparison of the two examples above that transitivity or lack thereof is at the heart of dealing with comparative soritical sequences. We will see, however, that with the recognition of the fact that two stimuli being compared belong to distinct 'stimulus areas' (the notion we explain next), the notion of transitivity should be approached with some caution, and the transitivity sometimes has to be formulated differently from the familiar, 'triadic' way (x is matched by y and y is matched by z, hence x is matched by z). For the same reason ('stimulus areas') the same degree of caution should be exercised in approaching the properties of reflexivity and symmetry, which most writers seem to take for granted.[7]

The notion of distinct 'stimulus areas' is both simple and fundamental. To say meaningfully that two physically identical stimuli, x and x, are judged as being the same or different, the two xs have to designate identical values of two otherwise different stimuli. If not for this fact we would have a single stimulus rather than two stimuli with identical values, and we would not be able to speak of pairwise comparisons. Thus, one of the two stimuli can be presented on the left and another on the right from a certain point, or one presented chronologically first and the other second. Stimuli, therefore, should be referred to by both their *values* (for example, length) and their

---

[6]This agrees with the obvious fact that $\lfloor \lambda x \rfloor$ can be viewed as a stimulus-effect function defined on individual number-stimuli, so comparative soritical sequences here are ruled out by the nonexistence of classificatory ones.

[7]Thus, Goodman (1951/1997), Armstrong (1968), Dummett (1975), and Wright (1975) view it as obvious that perceptual matching is intransitive, Jackson and Pinkerton (1973) and Graff (2001) argue for its transitivity (in the conventional, 'triadic' sense), and all of them consider it self-evident that perceptual matching is reflexive and symmetric.

stimulus areas (for example, left and right).[8] The complete reference here is then (x, left) and (y, right), or more briefly, $x^{(l)}$ and $y^{(r)}$.

Stimulus areas need not be defined only by spatiotemporal positions of stimuli. Thus, two line segments compared in their length may be of two different fixed orientations, and two patches of light compared in their brightness may be of two different fixed colors (in addition to occupying different positions in space or time). Nor should there be only two distinct observation areas: pairs of light patches compared in brightness can appear in multiple pairs of distinct spatial positions, and can be of various colors.

The sets to which we shall address our formal analysis will consequently be of the form $S \times \Omega$, where S is a set of stimulus *values* and Ω a set of stimulus *areas*, both containing at least two elements. (We will continue to use the more convenient notation $x^{(\omega)}$ for the stimulus $(x, \omega) \in S \times \Omega$.) The most basic property of the matching relation M is then

$$x_1^{(\omega)} M x_2^{(\omega')} \implies \omega \neq \omega',$$

that is, that we do not compare stimulus values from the same stimulus area. This implies, in particular, that M is antireflexive:

$$\sim x^{(\omega)} \, M \, x^{(\omega)}$$

holds for all $x^{(\omega)}$.[9]

For $\omega, \omega' \in \Omega$, the sets $S \times \{\omega\}$ and $S \times \{\omega'\}$ may simply be viewed as sets with different, further unanalyzable elements. We could, in fact, replace $S \times \Omega$ with $\{S_\omega\}_{\omega \in \Omega}$ treating thereby Ω as an indexing set for a collection of sets otherwise unrelated to each other. This is an important point in view of situations where one would want to speak of matching between entities of different natures, *e.g.*, abilities of examinees and difficulties of the tests offered to them (as is routinely done in psychometric models). We prefer, however, an intermediate notational approach: we write $S \times \Omega$ as a reminder

---

[8]The term used in Dzhafarov (2002), where the concept was introduced in a systematic fashion, was 'observation area', but 'stimulus area' seems preferable if the present analysis is also to apply to non-perceptual responses.

[9]This should not be confused with the statement that $\sim x^{(\omega)} \, M \, x^{(\omega')}$ holds for all $x^{(\omega)}$ and $x^{(\omega')}$. We are perfectly free to compare two stimuli with the same value, so long as they belong to different stimulus areas, and these stimuli may match or not match.

that stimuli in different stimulus areas may have identical values, but we treat $x^{(\omega)}$ 'holistically', saying, *e.g.*,

for any $a^{(\omega)}, b^{(\omega')} \in S \times \Omega$, the stimuli $a^{(\omega)}, b^{(\omega')}$ are...

instead of

for any $\omega, \omega' \in \Omega$ and $a, b \in S$, the stimuli $a^{(\omega)}, b^{(\omega')}$ are... .

Psychophysics of matching

Let us begin with a situation involving only two stimulus areas, say, $\Omega = \{l, r\}$, standing for 'left' and 'right'. In modern psychophysics (*cf.* Dzhafarov, 2002, 2003), matching relations on a given system $S \times \{l, r\}$ are defined from discrimination probability functions, as follows. For an observer asked to say whether two stimuli are the same or different, either with respect to a specified subjective property or overall, but ignoring the conspicuous difference in the stimulus areas, we can form a 'probability of being judged to be different' function,

$$\psi(x^{(l)}, y^{(r)}) = \Pr[x^{(l)} \text{ and } y^{(r)} \text{ are judged to be different}].$$

If the stimuli are compared with respect to a specified property, and if this property is linearly ordered (as in the cases of length, brightness, attractiveness, etc.), then the question can also be formulated in terms of which of the two stimuli has a greater amount of this property, and we can form a 'probability of being judged to be greater' function,

$$\gamma(x^{(l)}, y^{(r)}) = \Pr[y^{(r)} \text{ is judged to be greater than } x^{(l)}].$$

We can then define the 'matching' relation, henceforth denoted by M, either by

$$x^{(l)} \text{ M } y^{(r)} \text{ iff } \psi(x^{(l)}, y^{(r)}) = \min_z \psi(x^{(l)}, z^{(r)})$$
$$y^{(r)} \text{ M } x^{(l)} \text{ iff } \psi(x^{(l)}, y^{(r)}) = \min_z \psi(z^{(l)}, y^{(r)})$$

if dealing with ψ, or by

$$x^{(l)} \text{ M } y^{(r)} \text{ iff } \gamma(x^{(l)}, y^{(r)}) = 1/2$$
$$y^{(r)} \text{ M } x^{(l)} \text{ iff } \gamma(x^{(l)}, y^{(r)}) = 1/2$$

if dealing with γ. (No claim is being made that we get the same notion of matching in both cases.)

We now present two properties of the matching relation M that are critical for our treatment of comparative sorites. These properties, formulated and developed in Dzhafarov (2002, 2003) and Dzhafarov and Colonius (2006), constitute a principle which is called *Regular Mediality* or *Regular Minimality*, according as it is applied to (a function like) γ or to ψ. These properties are formulated here in a form better suited to the present context of studying the comparative sorites, but their formulations in psychophysics are entirely unrelated to and unmotivated by soritical issues.

Regular Mediality/Minimality, Part 1 (RM1). For every stimulus in either of the two stimulus areas, one can find a stimulus in the other stimulus area such that if $x^{(l)}$ and $y^{(r)}$ are the stimuli in question then $x^{(l)} \, M \, y^{(r)}$ and $y^{(r)} \, M \, x^{(l)}$.

To formulate the second property, we need the following notion. We call two stimuli in a given stimulus area *equivalent* if they match exactly the same stimuli in the other stimulus area. So, $x_1^{(l)}$ and $x_2^{(l)}$ are equivalent, in symbols $x_1^{(l)} \, E \, x_2^{(l)}$, if

$$y^{(r)} \, M \, x_1^{(l)} \Longleftrightarrow y^{(r)} \, M \, x_2^{(l)}$$

for every $y^{(r)}$; and $y_1^{(r)} \, E \, y_2^{(r)}$ if

$$x^{(l)} \, M \, y_1^{(r)} \Longleftrightarrow x^{(l)} \, M \, y_2^{(r)}$$

for every $x^{(l)}$.[10]

Regular Mediality/Minimality, Part 2 (RM2). Two stimuli in one stimulus area are equivalent if there is a stimulus in the other area which matches both of them, *i.e.,*

$$\text{if } x_1^{(l)} \, M \, y^{(r)} \text{ and } x_2^{(l)} \, M \, y^{(r)} \text{ then } x_1^{(l)} \, E \, x_2^{(l)},$$

$$\text{if } y_1^{(r)} \, M \, x^{(l)} \text{ and } y_2^{(r)} \, M \, x^{(l)} \text{ then } y_1^{(r)} \, E \, y_2^{(r)}.$$

---

[10]In Dzhafarov and Colonius (2006) the equivalence is defined in a stronger way: $x_1^{(l)} \, E \, x_2^{(l)}$ if $\psi(x_1^{(l)}, y^{(r)}) = \psi(x_2^{(l)}, y^{(r)})$ for all $y^{(r)}$, and $y_1^{(r)} \, E \, y_2^{(r)}$ if $\psi(x^{(l)}, y_1^{(r)}) = \psi(x^{(l)}, y_2^{(r)})$ for all $x^{(l)}$.

In regard to γ, the relations $x^{(l)} \, M \, y^{(r)}$ and $y^{(r)} \, M \, x^{(l)}$ mean one and the same thing, $\gamma(x^{(l)}, y^{(r)}) = 1/2$. For γ, then, RM1 requires only that, for every $x_0^{(l)}$, the function $y \mapsto \gamma(x_0^{(l)}, y^{(r)})$ reaches the median level $1/2$ at some point $y_0^{(r)}$, and then it follows that the function $x \mapsto \gamma(x^{(l)}, y_0^{(r)})$ reaches the median level at $x_0^{(l)}$. In regard to ψ, however, RM1 is more restrictive, requiring not only that the functions $y \mapsto \psi(x_0^{(l)}, y^{(r)})$ and $x \mapsto \psi(x^{(l)}, y_0^{(r)})$ reach their minima at some points, but also that, for every x_0, if y_0 minimizes the function $y \mapsto \psi(x_0^{(l)}, y^{(r)})$ then x_0 minimizes the function $x \mapsto \psi(x^{(l)}, y_0^{(r)})$. Unlike with γ, RM1 imposes nontrivial restrictions on the properties of the function ψ (Kujala & Dzhafarov, 2008).

To understand RM2, note that it is satisfied trivially if every match is determined uniquely, i.e., if for each $y^{(r)}$ there is only one $x^{(l)}$ satisfying $x^{(l)} \, M \, y^{(r)}$, and for each $x^{(l)}$ there is only one $y^{(r)}$ satisfying $y^{(r)} \, M \, x^{(l)}$. However, this is not generally the case, and whether it is the case in specific cases depends on one's choice of the physical description of stimuli. The most familiar example is that of matching isoluminant colors. A given color on the right, $y^{(r)}$, usually matches a single color on the left, $x^{(l)}$, provided that $x^{(l)}$ is identified, say, by its CIE coordinates. But if $x^{(l)}$ is identified by its radiometric spectrum, then $y^{(r)}$ matches an infinite multitude of $x^{(l)}$, all of them mapped into a single CIE point. All these different versions of $x^{(l)}$, however, are equivalent: they all match precisely the same colors on the right. Another example: a stimulus of luminance level L_1 and size s_1 can have the same (subjective) brightness as a stimulus of some other luminance L_2 and size s_2, regardless of whether the two stimuli belong to the 'left' or 'right' stimulus area. One would expect then that $(L_2, s_2)^{(l)} \, M \, (L, s)^{(r)}$ if and only if $(L_1, s_1)^{(l)} \, M \, (L, s)^{(r)}$. That is, a given right-hand stimulus would match more than one left-hand stimulus. It would be reasonable to expect, however, that then $(L_1, s_1)^{(l)}$ and $(L_2, s_2)^{(l)}$ are equivalent, i.e., that it is impossible for one of them to match and the other to not match one and the same stimulus on the right. This is precisely what RM2 posits: the uniqueness of matches *up to equivalence*.

The following proposition contains two most important consequences of RM1-RM2.

Proposition 1. *Assuming* M *satisfies RM1 and RM2, we have*

(1) $a^{(l)} \, M \, b^{(r)} \iff b^{(r)} \, M \, a^{(l)}$,

(2a) $a^{(l)} \, \mathrm{M} \, b^{(r)} \wedge b^{(r)} \, \mathrm{M} \, c^{(l)} \wedge c^{(l)} \, \mathrm{M} \, d^{(r)} \implies a^{(l)} \, \mathrm{M} \, d^{(r)}$,

(2b) $a^{(r)} \, \mathrm{M} \, b^{(l)} \wedge b^{(l)} \, \mathrm{M} \, c^{(r)} \wedge c^{(r)} \, \mathrm{M} \, d^{(l)} \implies a^{(r)} \, \mathrm{M} \, d^{(l)}$

for all $a^{(l)}, b^{(r)}, c^{(r)}, d^{(l)}$.

Proof. To prove (1), suppose $a^{(l)} \, \mathrm{M} \, b^{(r)}$. By RM1, there exists some $e^{(l)}$ with $e^{(l)} \, \mathrm{M} \, b^{(r)}$ and $b^{(r)} \, \mathrm{M} \, e^{(l)}$. By RM2, the first of these relations implies that $e^{(l)} \, \mathrm{E} \, a^{(l)}$, and by definition of E) the second implies that $b^{(r)} \, \mathrm{M} \, a^{(l)}$. Thus $a^{(l)} \, \mathrm{M} \, b^{(r)} \implies b^{(r)} \, \mathrm{M} \, a^{(l)}$; the reverse implication is proved symmetrically.

We now prove (2a), the proof of (2b) being symmetric. Assume that $a^{(l)} \, \mathrm{M} \, b^{(r)}$, $b^{(r)} \, \mathrm{M} \, c^{(l)}$, and $c^{(l)} \, \mathrm{M} \, d^{(r)}$. By symmetry of M we have $d^{(r)} \, \mathrm{M} \, c^{(l)}$, and since $b^{(r)} \, \mathrm{M} \, c^{(l)}$ RM2 implies $b^{(r)} \, \mathrm{E} \, d^{(r)}$. Since $a^{(l)} \, \mathrm{M} \, b^{(r)}$, $a^{(l)} \, \mathrm{M} \, d^{(r)}$ by definition of E.   □

The proposition says that if one accepts RM1 and RM2 (which are in agreement with, or at least do not contradict, what we know about human comparative judgments),

then the (idealized) 'matching' relation designed to generalize psychophysical matching ought to be *symmetric* in the sense of (1) and satisfy the notion of *tetradic transitivity*, (2). Let us briefly comment on each of these.

Regarding symmetry, it is very important to note that the values x and y in the expression $x^{(l)} \, \mathrm{M} \, y^{(r)} \iff y^{(r)} \, \mathrm{M} \, x^{(l)}$ remain in their respective stimulus areas on both sides. The symmetry condition does not necessarily allow for the exchange of values between the two stimulus areas,

$$x^{(l)} \, \mathrm{M} \, y^{(r)} \iff y^{(l)} \, \mathrm{M} \, x^{(r)}.$$

The naïve notion of symmetry represented by this statement is definitely not a general rule.[11] The symmetry in the sense of (1), on the other hand, is supported by all available empirical evidence (Dzhafarov, 2002; Dzhafarov & Colonius, 2006) and underlies the

---

[11]Thus, in one of the same-different discrimination experiments described in Dzhafarov and Colonius (2005), a right hand segment of length r happens to match a left-hand segment of length $x - 2$, if measured in minutes of arc. So, a $17^{(r)}$ min arc segment and a $15^{(l)}$ min arc one match each other. But, clearly, a $15^{(r)}$ min arc and a $17^{(l)}$ min arc do not match: rather the former of the two is matched by a $13^{(l)}$ min arc. One can list a host of such illustrations involving what in psychophysics is called *constant error*.

very language of psychophysical research dealing with matching-type relations. A psychophysicist is likely to consider the description

> the observer adjusted the right-hand stimulus until its appearance matched that of the fixed stimulus on the left

as saying precisely the same as

> the observer adjusted the right-hand stimulus until the fixed stimulus on the left matched its appearance

and precisely the same as

> the observer adjusted the right-hand stimulus until its appearance and that of the fixed stimulus on the left matched each other.

Regarding now the tetradic form of transitivity in Proposition 1, it is easily seen that the 'ordinary', triadic transitivity is simply false (or unformulable) if one deals with two stimulus areas: if $a^{(l)} \, \mathrm{M} \, b^{(r)}$ and $b^{(r)} \, \mathrm{M} \, c^{(l)}$, it is never true that $a^{(l)} \, \mathrm{M} \, c^{(l)}$ since only stimuli from different stimulus areas can be compared. Thus M can be said to be transitive in the tetradic sense but *antitransitive* in the triadic sense. It is easy to see, and will be rigorously demonstrated in the next subsection, that the tetradic transitivity is all one needs to rule out the existence of comparative soritical sequences in the case involving just two distinct stimulus areas (not necessarily the 'left' and 'right' used here for concreteness only).

If the number of stimulus areas is greater than two, the analysis above does apply, of course, to any two of them. In addition, however, for any three distinct stimulus areas (let us denote them 1, 2, and 3) one can formulate the familiar triadic transitivity property,

$$a^{(1)} \, \mathrm{M} \, b^{(2)} \wedge b^{(2)} \, \mathrm{M} \, c^{(3)} \implies a^{(1)} \, \mathrm{M} \, dc^{(3r)}.$$

This property can be derived from appropriate reformulations of RM1 and RM2 for three distinct stimulus areas. We forgo this task, however, as it is subsumed by the formal treatment presented next, which applies to an arbitrary set of stimulus areas.

Formal theory of regular well-matched stimulus spaces

We work throughout with a set $S \times \Omega$ in which at least two stimulus values are paired with at least two stimulus areas. We endow $S \times \Omega$ with binary relations M and E such that

(1) $\sim a^{(\omega)}$ M $b^{(\omega)}$ for all $a^{(\omega)}, b^{(\omega)} \in S \times \Omega$, and

(2) $a^{(\omega)}$ E $b^{(\omega')}$ if and only if

$$c^{(\iota)} \text{ M } a^{(\omega)} \iff c^{(\iota)} \text{ M } b^{(\omega')}$$

for all $a^{(\omega)}, b^{(\omega')}, c^{(\iota)} \in S \times \Omega$.

Since E is uniquely defined in terms of M, we refer to the space $(S \times \Omega, M, E)$ by the more economical $(S \times \Omega, M)$.[12] We omit the simple proof that E is an equivalence relation on $S \times \Omega$.

Definition 5. A sequence $x_1^{(\omega_1)}, \ldots, x_n^{(\omega_n)}$ in a space $(S \times \Omega, M)$ is called

(1) *chain-matched* if $x_i^{(\omega_i)}$ M $x_{i+1}^{(\omega_{i+1})}$ for $i = 1, \ldots, n-1$;

(2) *well-matched* if $\omega_i \neq \omega_j \implies x_i^{(\omega_i)}$ M $x_j^{(\omega_j)}$ for all $i, j \in \{1, \ldots n\}$;

(3) *soritical* if it is chain-matched, $\omega_1 \neq \omega_n$ and $\sim x_1^{(\omega_1)}$ M $x_n^{(\omega_n)}$.

In (3) we easily recognize a formal version of what we earlier defined as comparative soritical sequences. Soritical sequences are clearly always chain-matched and never well-matched.[13] It is also clear that there are no soritical sequences with just two elements, and that all soritical sequences consisting of three elements are of the form $a^{(\alpha)}, b^{(\beta)}, c^{(\gamma)}$ with $\{\alpha, \beta, \gamma\}$ pairwise distinct. Longer soritical sequences, as it turns out, can always be reduced to one of two types (illustrated in Fig. 4.2): three-element sequences like the one just mentioned, and four-element sequences with two alternating stimulus areas, $a^{(\alpha)}, b^{(\beta)}, c^{(\alpha)}, d^{(\beta)}$.

Lemma 4. *If $x_1^{(\omega_1)}, \ldots, x_n^{(\omega_n)}$ in a space $(S \times \Omega, M)$ is a soritical sequence, then it contains either a triadic soritical subsequence $a^{(\alpha)}, b^{(\beta)}, c^{(\gamma)}$ or a tetradic soritical subsequence $a^{(\alpha)}, b^{(\beta)}, c^{(\alpha)}, d^{(\beta)}$.*

---

[12]In dealing with both E and M, and treating them as two interrelated but different relations, we follow Goodman (1951/1997).

[13]This shows that chain-matchedness does not imply well-matchedness. That the reverse implication also does not hold can be seen by considering any sequence of the form $x_1^{(\omega)}, \ldots, x_n^{(\omega)}$. Thus, chain-matchedness and well-matchedness are logically independent conditions.

Proof. Let $x_{i_1}^{(\omega_{i_1})}, \ldots, x_{i_m}^{(\omega_{i_m})}$ be a soritical subsequence of our sequence having the shortest possible length. If there exists an ℓ such that $1 < \ell < m$ and $\omega_{i_1} \neq \omega_{i_\ell} \neq \omega_{i_m}$ then it must be that $x_{i_1}^{(\omega_{i_1})} \mathrm{M} x_{i_\ell}^{(\omega_{i_\ell})}$: otherwise $x_{i_1}^{(\omega_{i_1})}, \ldots, x_{i_\ell}^{(\omega_{i_\ell})}$ would be yet a shorter soritical subsequence of the original sequence. Similarly, it must be that $x_{i_\ell}^{(\omega_{i_\ell})} \mathrm{M} x_{i_m}^{(\omega_{i_m})}$. Hence,

$$(a^{(\alpha)}, b^{(\beta)}, c^{(\gamma)}) = (x_{i_1}^{(\omega_{i_1})}, x_{i_\ell}^{(\omega_{i_\ell})}, x_{i_m}^{(\omega_{i_m})})$$

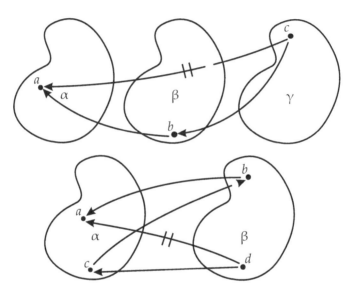

Figure 4.2: Examples of a triadic soritical sequence (top) and a tetradic soritical sequence (bottom). According to Lemma 4, at least one of these can be found as a subsequence in any soritical sequence. The outlined areas represent stimulus areas: α, β, γ in the top illustration and α, β in the bottom one. An arrow (resp., interrupted arrow) drawn from one point to another indicates that the latter point is matched (resp., not matched) by the former. Thus, in the top illustration, $b^{(\beta)}$ matches $a^{(\alpha)}$, $c^{(\gamma)}$ matches $b^{(\beta)}$, but $c^{(\gamma)}$ does not match $a^{(\alpha)}$. In the bottom illustration, $b^{(\beta)}$ matches $a^{(\alpha)}$, $c^{(\alpha)}$ matches $b^{(\beta)}$, $d^{(\beta)}$ matches $c^{(\alpha)}$, but $d^{(\beta)}$ does not match $a^{(\alpha)}$. According to Theorem 3 (p. 130), neither of these scenarios (and hence no soritical sequence) is possible in a regular well-matched space (Definition 6).

is a triadic subsequence of the kind desired. If no such ℓ exists, then it must be that $m \geq 4$ and that $\omega_{i_1} = \omega_{i_3} \neq \omega_{i_2} = \omega_{i_m}$. Again, the choice of $x_{i_1}^{(\omega_{i_1})}, \ldots, x_{i_m}^{(\omega_{i_m})}$ as the shortest soritical subsequence ensures that $x_{i_3}^{(\omega_{i_3})} \, M \, x_{i_m}^{(\omega_{i_m})}$, so in this case

$$(a^{(\alpha)}, b^{(\beta)}, c^{(\alpha)}, d^{(\beta)}) = (x_{i_1}^{(\omega_{i_1})}, x_{i_\ell}^{(\omega_{i_2})}, x_{i_j}^{(\omega_{i_3})}, x_{i_m}^{(\omega_{i_m})})$$

is a tetradic soritical subsequence of our sequence.   □

Our goal being to dissolve the comparative sorites 'paradox' in systems resembling human comparative judgments, we would like to restrict our interest from arbitrary spaces to those in which we can formalize (suitably generalized versions of) the conditions RM1 and RM2 discussed in the previous subsection. To this end, we introduce the following two types of spaces.

Definition 6. We call $(S \times \Omega, M)$ a
(1) *well-matched space* if, for any sequence $\alpha, \beta, \gamma \in \Omega$ and any $a^{(\alpha)} \in S \times \Omega$, there is a well-matched sequence $a^{(\alpha)}, b^{(\beta)}, c^{(\gamma)}$;
(2) *regular space* if, for any $a^{(\alpha)}, b^{(\alpha)}, c^{(\beta)} \in S \times \Omega$ with $\alpha \neq \beta$,

$$a^{(\alpha)} \, M \, c^{(\beta)} \wedge b^{(\alpha)} \, M \, c^{(\beta)} \implies a^{(\alpha)} \, E \, b^{(\alpha)}.$$

In the concept of regularity, we clearly see the formalization of RM2. The following lemma shows that, similarly, the notion of well-matchedness suffices for the formalization of RM1. It also shows that in such spaces the relation E behaves as we expect it to.

Lemma 5. Let $(S \times \Omega, M)$ be a well-matched space. (1) For any $\alpha, \beta \in \Omega$ and any $a^{(\alpha)} \in S \times \Omega$, there exists $b^{(\beta)} \in S \times \Omega$ such that $a^{(\alpha)} \, M \, b^{(\beta)}$ and $b^{(\beta)} \, M \, a^{(\alpha)}$. (2) For any $a^{(\alpha)}, b^{(\beta)} \in S \times \Omega$, if $a^{(\alpha)} \, E \, b^{(\beta)}$ holds then $\alpha = \beta$.

Proof. For (1), consider a sequence α, β, β and apply Definition 6(1) (notice that no assumption is made in that definition that α, β, γ need to be pairwise distinct). To prove (2), suppose for a contradiction that $a^{(\alpha)} \, E \, b^{(\beta)}$ and $\alpha \neq \beta$. Then by part (1), we can find some $c^{(\alpha)}$ with $c^{(\alpha)} \, M \, b^{(\beta)}$ and $b^{(\beta)} \, M \, c^{(\alpha)}$. By definition of E, this implies that $c^{(\alpha)} \, M \, a^{(\alpha)}$, which contradicts the definition of M.   □

It follows that if comparative judgments satisfying RM1 and RM2 are to serve as the model for our formal analysis, then we should apply our analysis to spaces which are both well-matched and regular.[14] The following lemma provides us with additional information about the structure of such spaces. Part (1) is the analogue of Proposition 1(1); parts (2) and (3) are obvious and are stated mostly for convenience of reference in the next subsection.

Lemma 6. *Let* $(S \times \Omega, M)$ *be a regular well-matched space, and let* $a^{(\alpha)}$, $b^{(\beta)}, c^{(\alpha)}, d^{(\beta)} \in S \times \Omega$. *Then*

(1) $a^{(\alpha)} \, M \, b^{(\beta)} \iff b^{(\beta)} \, M \, a^{(\alpha)}$;

(2) *if either* $a^{(\alpha)} \, M \, b^{(\beta)} \wedge c^{(\alpha)} \, M \, b^{(\beta)}$ *or* $b^{(\beta)} \, M \, a^{(\alpha)} \wedge b^{(\beta)} \, M \, c^{(\alpha)}$, *then* $a^{(\alpha)} \, E \, c^{(\alpha)}$;

(3) *if* $a^{(\alpha)} \, E \, c^{(\alpha)} \wedge b^{(\beta)} \, E \, d^{(\beta)}$ *then* $a^{(\alpha)} \, M \, b^{(\beta)} \iff c^{(\alpha)} \, M \, d^{(\beta)}$.

Proof. Part (1) is proved exactly as Proposition 1(1), only replacing references to RM1 by those to Lemma 5(1) and replacing references to RM2 by those to the regularity of the space. Parts (2) and (3) are straightforward consequences of Definition 2 and part (1).   □

Dissolving the comparative sorites 'paradox'

The main result of the section is in the theorem presented next. As mentioned above, soritical sequences are always chain-matched and never well-matched, so the theorem in particular implies that soritical sequences do not exist in regular well-matched spaces.

Theorem 3. *Let* $(S \times \Omega, M)$ *be a regular well-matched space. Then any chain-matched sequence in this space is well-matched.*

Proof. In view of Lemma 4, it suffices to prove the result for all soritical sequences of the form $a^{(\alpha)}, b^{(\beta)}, c^{(\gamma)}$ and $a^{(\alpha)}, b^{(\beta)}, c^{(\alpha)}, d^{(\beta)}$. We begin with the former.

Suppose $a^{(\alpha)}, b^{(\beta)}, c^{(\gamma)}$ is a chain-matched sequence. This means $a^{(\alpha)} \, M \, b^{(\beta)} \wedge b^{(\beta)} \, M \, c^{(\gamma)}$. All we have to prove is that then $a^{(\alpha)} \, M \, c^{(\gamma)}$, as the rest of the matches in $a^{(\alpha)}, b^{(\beta)}, c^{(\gamma)}$ then obtain by symmetry of M. By Definition 6, there exists a well-matched sequence $x^{(\alpha)}, b^{(\beta)}, y^{(\gamma)}$. Since $a^{(\alpha)} \, M \, b^{(\beta)} \wedge x^{(\alpha)} \, M \, b^{(\beta)}$, it follows by Lemma 6(2) that $a^{(\alpha)} \, E \, x^{(\alpha)}$. Since $b^{(\beta)} \, M \, c^{(\gamma)} \wedge b^{(\beta)} M y^{(\gamma)}$, it follows similarly that $c^{(\gamma)} \, E \, y^{(\gamma)}$. Since $x^{(\alpha)}, b^{(\beta)}, y^{(\gamma)}$ is well-matched, we have $x^{(\alpha)} \, M \, y^{(\gamma)}$, and so, by Lemma 6(3), $a^{(\alpha)} \, M \, c^{(\gamma)}$.

---

[14]These are independent assumptions. It is easy to construct toy examples demonstrating that a well-matched space need not be regular, and vice versa.

Now suppose $a^{(\alpha)}, b^{(\beta)}, c^{(\alpha)}, d^{(\beta)}$ is a chain-matched sequence, so $a^{(\alpha)} \text{ M } b^{(\beta)} \wedge b^{(\beta)} \text{ M } c^{(\alpha)} \wedge c^{(\alpha)} \text{ M } d^{(\beta)}$. We have to show that $a^{(\alpha)} \text{ M } d^{(\beta)}$. By Definition 6, there exists a well-matched sequence $x^{(\alpha)}, b^{(\beta)}, y^{(\alpha)}, z^{(\beta)}$. Since $b^{(\beta)} \text{ M } c^{(\alpha)} \wedge b^{(\beta)} \text{ M } y^{(\alpha)}$, it follows by Lemma 6(2) that $c^{(\alpha)} \text{ E } y^{(\alpha)}$. Then we should have $y^{(\alpha)} \text{ M } d^{(\beta)}$ (by Lemma 6(3), because $c^{(\alpha)} \text{ M } d^{(\beta)}$). Since $y^{(\alpha)} \text{ M } d^{(\beta)} \wedge y^{(\alpha)} \text{ M } z^{(\beta)}$, we have by Lemma 6(2) that $d^{(\beta)} \text{ E } z^{(\beta)}$. By the same lemma, we also have $a^{(\alpha)} \text{ E } x^{(\alpha)}$, because $a^{(\alpha)} \text{ M } b^{(\beta)} \wedge x^{(\alpha)} \text{ M } b^{(\beta)}$. But now $a^{(\alpha)} \text{ E } x^{(\alpha)}$ and $d^{(\beta)} \text{ E } z^{(\beta)}$, so from the fact that $x^{(\alpha)} \text{ M } z^{(\beta)}$ it follows that $a^{(\alpha)} \text{ M } d^{(\beta)}$, by Lemma 6(3). □

Corollary 2. *Any chain-matched sequence in a regular well-matched space is well-matched: one cannot form a soritical sequence in such a space.*

We conclude with a method of merging the antireflexive relation M and the equivalence relation E into a single identity relation EM. To this end, we define the notion of 'canonical labeling'. The idea is very simple: given a regular well-matched space $S \times \Omega$, any two equivalent stimuli $a^{(\omega)}$ and $b^{(\omega)}$ in any stimulus area ω can be assigned one and the same *label* (say, x). Then every new label in any one stimulus area will match and be matched by one and only one label in any other stimulus area—and then it is possible to assign the same label x to all stimuli in all stimulus areas which match (and are matched by) $x^{(\omega)}$. The resulting simplicity is the reward: for any two stimulus areas ω and ω', any 'relabeled stimulus' $x^{(\omega)}$ matches the 'relabeled stimulus' $x^{(\omega')}$ and none other; and in any given stimulus area any $x^{(\omega)}$ is only equivalent to itself. If one now 'merges' the relations E and M into a single relation EM, the latter is simply the indicator of the equality of labels and is therefore reflexive, symmetric, and transitive:

$$\mathbf{a}^{(\alpha)} \text{ EM } \mathbf{b}^{(\beta)} \Longleftrightarrow \mathbf{a} = \mathbf{b},$$

where α and β need not be distinct. The formal procedure described below effects the canonical (re)labeling by means of a single function cal (from 'canonical labeling') applied to all stimuli in the space.

Canonical representation is not a return to the naive idea that every stimulus matches 'itself'. One cannot dispense with the notion of a stimulus area: for distinct ω and ω' and one and the same label x, the original identities (*e.g.*, conventional physical descriptions) of $x^{(\omega)}$ and $x^{(\omega')}$ are generally different. In fact, $x^{(\omega)}$ and $x^{(\omega')}$

designate two equivalence classes of stimuli, whose values may be non-overlapping (partially or completely).

Definition 7. A surjective function

$$\mathrm{cal} : S \times \Omega \to \mathbf{S}$$

is called a *canonical labeling* of a regular well-matched space $(S \times \Omega, M)$ (and $\mathbf{S}$ is called a *set of canonical labels*) if, for any $a^{(\alpha)}, b^{(\beta)} \in S \times \Omega$,

$$\mathrm{cal}(a^{(\alpha)}) = \mathrm{cal}(b^{(\beta)}) \Longleftrightarrow \begin{cases} a^{(\alpha)} \, \mathrm{E} \, b^{(\beta)} & \text{if } \alpha = \beta \\ a^{(\alpha)} \, \mathrm{M} \, b^{(\beta)} & \text{if } \alpha \neq \beta \end{cases}.$$

Theorem 4. *A canonical labeling function* cal *exists for any regular well-matched space* $(S \times \Omega, M)$. *If* cal: $S \times \Omega \to \mathbf{S}$ *and* cal*: $S \times \Omega \to \mathbf{S}^*$ *are such functions, then* cal$^* \equiv h \circ$ cal, *where* h *is a bijection* $\mathbf{S} \to \mathbf{S}^*$.

Proof. For each $x^{(\omega)} \in S \times \Omega$, let

$$N_{x^{(\omega)}} = \{ y^{(\iota)} \in S \times \Omega : x^{(\omega)} \, \mathrm{M} \, y^{(\iota)} \vee x^{(\omega)} \, \mathrm{E} \, y^{(\iota)} \}.$$

Since $(S \times \Omega, M)$ is regular and well-matched, it is easy to see that if $N_{x^{(\omega)}} \cap N_{y^{(\iota)}} \neq \varnothing$ for some $x^{(\omega)}, y^{(\iota)} \in S \times \Omega$, then $N_{x^{(\omega)}} = N_{y^{(\iota)}}$. Setting $\mathbf{S} = \{ N_{x^{(\omega)}} : x^{(\omega)} \in S \times \Omega \}$, we define cal: $S \times \Omega \to \mathbf{S}$ by cal$(x^{(\omega)}) = N_{x^{(\omega)}}$, and this function clearly satisfies Definition 7. If now cal*: $S \times \Omega \to \mathbf{S}^*$ is another canonical labeling function for $(S \times \Omega, M)$, define $h: \mathbf{S} \to \mathbf{S}^*$ by $h(N_{x^{(\omega)}}) = $ cal$^*(x^{(\omega)})$. Since cal$^*(x^{(\omega)}) = $ cal$^*(y^{(\iota)})$ if and only if $x^{(\omega)} \mathrm{M} y^{(\iota)}$ or $x^{(\omega)} \, \mathrm{E} \, y^{(\iota)}$, it follows that h is well-defined and injective, while its surjectivity follows immediately from that of cal*. Clearly, cal$^* - h \circ$ cal, whence the proof is complete. $\qquad \square$

Consider now the set $\mathbf{S} \times \Omega$. As before, let us use the notation $\mathbf{x}^{(\omega)}$ for $(\mathbf{x}, \omega)$.

Definition 8. Given a regular well-matched space $(S \times \Omega, M)$ and a canonical labeling function cal: $S \times \Omega \to \mathbf{S}$, for any $\mathbf{a}, \mathbf{b} \in \mathbf{S}$ and $\alpha, \beta \in \Omega$, we say that $\mathbf{b}^{(\beta)}$ *matches* $\mathbf{a}^{(\alpha)}$, and write $\mathbf{a}^{(\alpha)} \, \mathrm{EM} \, \mathbf{b}^{(\beta)}$, if

$$a^{(\alpha)} \, \mathrm{M} \, b^{(\beta)} \vee a^{(\alpha)} \, \mathrm{E} \, b^{(\beta)}$$

for some $a^{(\alpha)}$ and $b^{(\beta)}$ in $S \times \Omega$ such that cal$(a^{(\alpha)}) = \mathbf{a}$ and cal$(b^{(\beta)}) = \mathbf{b}$. We refer to $(\mathbf{S} \times \Omega, \mathrm{EM})$ as a *canonical comparison space*.

Theorem 5. *Given any regular well-matched space* $(S \times \Omega, M)$ *and any canonical labeling function* $\mathrm{cal} \colon S \times \Omega \to \mathbf{S}$, *for any* $\mathbf{a}, \mathbf{b} \in \mathbf{S}$ *and* $\alpha, \beta \in \Omega$ *we have*

$$\mathbf{a}^{(\alpha)} \, \mathrm{EM} \, \mathbf{b}^{(\beta)} \iff \mathbf{a} = \mathbf{b}.$$

Hence EM *is an equivalence relation on* $\mathbf{S} \times \Omega$.

Proof. By Definition 8, $\mathbf{a}^{(\alpha)} \, \mathrm{EM} \, \mathbf{b}^{(\beta)}$ means that for some $a^{(\alpha)}$ with $\mathrm{cal}(a^{(\alpha)}) = \mathbf{a}$ and $b^{(\beta)}$ with $\mathrm{cal}(b^{(\beta)}) = \mathbf{b}$, either $a^{(\alpha)} \, \mathrm{M} \, b^{(\beta)}$ (which implies $\alpha \neq \beta$) or $a^{(\alpha)} \, \mathrm{E} \, b^{(\beta)}$ (implying $\alpha = \beta$). But by Definition 7,

$$\left. \begin{array}{ll} a^{(\alpha)} \, \mathrm{E} \, b^{(\beta)} & (\alpha = \beta) \\ a^{(\alpha)} \, \mathrm{M} \, b^{(\beta)} & (\alpha \neq \beta) \end{array} \right\} \iff \mathrm{cal}(a^{(\alpha)}) = \mathrm{cal}(b^{(\beta)}).$$

This proves the theorem. □

Corollary 3. *With* EM *in place of* M *and* $\mathbf{S}$ *in place of* S, *any canonical comparison space* $(\mathbf{S} \times \Omega, \mathrm{EM})$ *is a regular well-matched space.*

Conclusion

We have approached soritical arguments and (allegedly) soritical phenomena within a broadly understood behavioral framework, in terms of stimuli acting upon a system (such as a biological organism, a group of people, or a set of normative linguistic rules) and the responses they evoke.

The classificatory sorites, dating back to Eubulides of the Megarian school, is about the identity of or difference between the effects of stimuli which differ 'only microscopically'. We have formulated the notions and assumptions underlying this variety of sorites in a highly general mathematical language, and we have shown that the 'paradox' is dissolved on grounds unrelated to vague predicates or other linguistic issues traditionally associated with it. If stimulus effects are properly defined (*i.e.*, if they are uniquely determined by stimuli), and if the space of the stimuli is endowed with appropriate closeness and connectedness properties, then this space must contain points in every vicinity of which, 'however small', the stimulus effect is not constant. This conclusion clashes with the common but nonetheless false intuition that a 'macroscopic' system cannot 'react differently' to two 'microscopically different' stimuli. In fact, a non-constant stimulus effect upon a system can only be insensitive to small differences in stimuli if the closeness structure which is used to define very close stimuli does not render the space of stimuli appropriately connected, and in this case we have no 'paradox'. The 'paradox' cannot even be formulated using response properties that are not true stimulus effects, *i.e.*, are not uniquely determined by stimuli. This is the reason the classificatory sorites is not related to the issue of vagueness in human responses to stimuli: 'vague predicates' are always assigned inconsistently, whatever other properties they may be thought to have.

The comparative sorites (also known in the literature as 'observational') is very different from the classificatory one. Here, it has been discussed in terms of a system mapping pairs of stimuli into a binary response characteristic whose values are uniquely determined by stimulus pairs and are interpretable as the complementary relations 'match' and 'do not match' (overall or in some respect). The comparative sorites is about hypothetical sequences of stimuli in which every two successive elements are mapped into the relation 'match', while the pair comprising the first and the last elements of the sequence is mapped into 'do not match'. Although soritical sequences of this kind are logically possible, we have argued that insofar as human comparative judgments are concerned, their existence is far from being a well-known, let alone obvious, empirical fact. Rather it is a naïve theoretical idea that overlooks the fundamental notion of stimulus areas and the necessity of defining the matching relation so that it is uniquely determined by stimulus pairs (which is critical in view of the probabilistic nature of comparative judgments in humans). Moreover, the comparative soritical sequences are excluded by the principle of Regular Mediality/Minimality proposed for human comparative judgments in a context unrelated to soriticial issues. In this chapter we have generalized this principle into the mathematical notion of regular well-matched spaces of stimulus values paired with stimulus areas. The matching relation in such spaces is irreflexive, symmetric, and transitive in the tetradic or triadic sense, according as we deal with two or more than two stimulus areas, respectively.

Acknowledgments

The first author's work was supported by NSF grant SES 0620446 and AFOSR grants FA9550-06-1-0288 and FA9550-09-1-0252. The second author's work was partially supported by an NSF Graduate Research Fellowship.

References

Armstrong, D. M. (1968). *A Materialist Theory of the Mind*. London, GB: Routledge and Kegan Paul.

Dummett, M. (1975). Wang's paradox. *Synthese, 30*, 301–324.

Dzhafarov, E. N. (2002). Multidimensional Fechnerian scaling: Pairwise comparisons, regular minimality, and nonconstant self-similarity. *Journal of Mathematical Psychology, 46*, 583–608.

Dzhafarov, E. N. (2003). Thurstonian-type representations for 'same-different' discriminations: Deterministic decisions and independent images. *Journal of Mathematical Psychology, 47*, 208–228.

Dzhafarov, E. N., & Colonius, H. (2005). Psychophysics without physics: A purely psychological theory of Fechnerian scaling in continuous stimulus spaces. *Journal of Mathematical Psychology, 49*(1), 1–50.

Dzhafarov, E. N., & Colonius, H. (2006). Regular minimality: A fundamental law of discrimination. In H. Colonius & E. N. Dzhafarov (Eds.), *Measurement and Representation of Sensations* (pp. 1–46). Mahwah, NJ: Lawrence Erlbaum Associates, Inc.

Dzhafarov, E. N., & Dzhafarov, D. D. (2010a). Sorites without vagueness I: Classificatory sorites. *Theoria, 76*(1), 4–24.

Dzhafarov, E. N., & Dzhafarov, D. D. (2010b). Sorites without vagueness II: Comparative sorites. *Theoria, 76*(1), 25–53.

Fréchet, M. (1918). Sur la notion de voisinage dans les ensembles abstrait [On the notion of neighborhood in abstract sets]. *Bulletin des Sciences Mathématiques, 42*, 138–156.

Goodman, N. (1997). *The Structure of Appearance*. Dordrecht, NL: Reidel. (Original work published 1951)

Graff, D. (2001). Phenomenal continua and the sorites. *Mind, 110*, 905–935.

Hardin, C. L. (1988). Phenomenal colors and sorites. *Noûs, 22*, 213–234.

Jackson, F., & Pinkerton, R. J. (1973). On an argument against sensory items. *Mind, 82*, 269–272.

Kujala, J. V., & Dzhafarov, E. N. (2008). On minima of discrimination functions. *Journal of Mathematical Psychology*, *52*, 116–127.

Luce, R. D. (1956). Semiorders and a theory of utility discrimination. *Econometrica*, *24*, 178–191.

Peirce, C. S. (1960). Vague. In J. M. Baldwin (Ed.), *Dictionary of Philosophy and Psychology* (p. 748). New York, NY: Macmillan. (Original work published 1901)

Sorensen, R. A. (1988a). *Blindspots*. Oxford, GB: Clarendon Press.

Sorensen, R. A. (1988b). Vagueness, blurriness, and measurement. *Synthese*, *75*, 45–82.

Varzi, A. C. (2003). Cut-offs and their neighbors. In J. C. Beall (Ed.), *Liars and Heaps* (pp. 24–38). Oxford, GB: Oxford University Press.

Williamson, T. (1994). *Vagueness*. London, GB: Routledge.

Williamson, T. (2000). *Knowledge and its Limits*. Oxford, GB: Oxford University Press.

Wright, C. (1975). On the coherence of vague predicates. *Synthese*, *30*, 325–365.

Yonge, C. D. (1901). *The Lives and Opinions of Eminent Philosophers, by Diogenes Laërtius*. London, GB: G. Bell & Sons.

5

On an Intensity Attribute—Loudness

Che Tat Ng

This chapter is devoted to discussion and investigation of an open problem in the psychophysical theory of *intensity perception* as formulated by R. D. Luce in a series of papers (Luce, 2002, 2004, 2005; see also Steingrimsson & Luce, 2005a, 2005b, 2006, 2007). Luce's theory follows the classical model for measurements of physical attributes like mass, with a behavioral adaptation to psychophysical intensity attributes like loudness. The material presented here includes a compact outline of part of this theory, derived from lectures by Professor Luce in 2005 (at the joint Summer Meeting of the Canadian Mathematical Society and the Canadian Society for History and Philosophy of Mathematics, June 4–6) and 2007 (at the University of Waterloo, June 14), in which he derived some representation theorems on loudness, and stated the problem to be considered here—it is to characterize the precise mathematical form that a certain psychophysical function takes.

The first two sections of this chapter, closely based on my notes of Professor Luce's lectures, end with a statement of the open problem. They are followed by discussion, including statements and proofs of several theorems. As to the open problem, I conclude that by correctly (re)interpreting the role of *bisymmetry* in the model, the theory elaborated by Luce (2002, 2004, 2005) and Steingrimsson and Luce (2005a, 2005b) can be put in much better agreement with the experimental data of Steingrimsson and Luce (2006, 2007).

Keywords: functional equations, joint presentation of tones, loudness, magnitude production, measurements, ordering structure, perceived intensity, proportion representation, psychophysical functions, tone intensity.

Introduction—a physical model

In his 2005 and 2007 lectures, Luce used a natural science model for the physical measures of mass as a lead-in to introduce his subject. That model contains a few primitives and axioms. We list some.

- A *mass, m,* of a homogeneous substance is an ordered pair (s, v) where s labels the substance and v labels the volume container.
- There are two manipulations of mass: first, a *concatenation,* $v \oplus v'$, of two volumes v, v' of a homogeneous substance s (instantiated by, say, placing them together on the same pan of a pair of scales):

$$(s, v) \oplus (s, v') := (s, v \oplus v');$$

  and second, a *trade-off* of volumes and substances that induces on a set of masses a weak ordering $\gtrsim$ and an equivalence relation $\approx := \gtrsim \cap \lesssim$.

- Some extensive axioms of Hölder yield a numerical representation Ψ that preserves order

$$m \gtrsim m' \Leftrightarrow \Psi(m) \geq \Psi(m')$$

  and also preserves concatenation

$$\Psi(m \oplus m') = \Psi(m) + \Psi(m').$$

- A second measure of mass based on multiplicative conjoint axioms yields a representation Φ that is order-preserving and multiplicative:

$$\Phi(s, v) = \phi(s)\psi(v).$$

- Luce pointed out that having two distinct measures of mass (Ψ and Φ) of unknown relation (other than just being monotonically related) is scientifically unacceptable. He proposed a distribution law to link them up:

$$(r, v) \approx (r', v'), (r, u) \approx (r', u') \Rightarrow (r, v \oplus u) \approx (r', v' \oplus u').$$

  This distribution law, together with certain structural axioms, yields the relation

$$\Psi = \gamma \Phi^\beta.$$

Luce's psychophysical theories follow this framework.

A psychophysical model—sound intensity and loudness

This section lays out the general model pioneered by R. D. Luce. It follows my notes of his lectures, with minor notational adjustments (to improve consistency with notation used in his publications) and some small additions. For the general theory of measurements, the classic reference is Krantz, Luce, Suppes, and Tversky (1971).

Monaural and binaural presentation

Pure tones, of the same frequency and in phase, are presented to the two ears of a listener. The sounds used are bounded from below by thresholds and from above by the IRB restrictions on research with human subjects (currently the maximum allowable sound pressure level is 85 dB). Let (x,u) denote physical intensity x presented to the left ear and physical intensity u to the right ear. Such pairs (x,u), with intensities $x > 0$ and $u > 0$ respectively to the left and right ears, is called a *binaural presentation*.

A *monaural presentation* is a binaural presentation $(x,0)$ or $(0,u)$, where 0 denotes the respective threshold.

Some primitives

The model under study is stated in terms of two binary relations, $\succsim$ and $\approx$, and two binary operations, $\oplus_s$ and $\circ_p$ (the latter depending on a parameter p) together with some variants. First we describe the primitives, then some axioms that will be imposed on them.

- *Loudness* is an attribute in the perception of pure tones received. It is expressed as a weak order relation

$$(x,u) \succsim (y,v) \tag{5.1}$$

over the range of all binaural presentations (x,u), idealized to be $[0,\infty[\times[0,\infty[$ (*i.e.*, not bounded from above as in reality). The meaning of (5.1) is that the (binaural) presentation (x,u) is perceived to be *louder* than or equally loud as (y,v). As in Luce's physical example above, and writing $\precsim$ for the *inverse relation* of $\succsim$ (*i.e.*, $(y,v) \precsim (x,u)$ if and only if $(x,u) \succsim (y,v)$), the assumption that $\precsim$ is a weak order implies that the relation $\approx := \succsim \cap \precsim$ is an equivalence relation: that is, $(x,u) \approx (y,v)$ means that the two presentations are indifferent in perceived loudness, so '$\approx$' can be read 'sounds equally loud as'. It is assumed that monaural loudness orderings agree with that of physical intensity. So the loudness ordering is strictly monotonic in the variables x and u separately, i.e., if (say) u, the physical intensity of sound presented to the right ear, is held constant and sounds of two physical intensities x_1, x_2, with $x_1 > x_2$, are presented to the left ear, then

$$(x_1, u) \gtrsim (x_2, u) \text{ and not } (x_1, u) \approx (x_2, u)$$

and similarly if x is held constant.

- *Symmetric matching* across two ears is represented using a binary operation $\oplus_s$ on physical intensities. By definition

$$x \oplus_s u = z \text{ if and only if } (x, u) \approx (z, z) \qquad (5.2)$$

  and it is assumed that for each given (x, u), some such z exists; its uniqueness then follows from the assumption of strict monotonicity on the loudness order. It also follows, from the assumed strict monotonicity in x and u separately, that

$$\min(x, u) \leq x \oplus_s u \leq \max(x, u),$$

  i.e., that $\oplus_s$ is a *strict mean*. In particular, $x \oplus_s x = x$ for all x.
- *Ratio production* is a generalization of what Stevens (1973) called *magnitude production*. For any given positive real p, and physical intensities $x > y$,

$$x \circ_p y \qquad (5.3)$$

  is by definition the (unique) physical intensity z that the respondent produces so as to make the respondent's perception of the subjective 'interval' from (y, y) to (z, z) have the ratio $p : 1$ to the subjective interval from (y, y) to (x, x). (Stevens's magnitude production is the special case $y = 0$.)

Note that clarity may be gained, at some cost in conciseness, by rewriting the primitive $x \circ_p y$ as $(x, x) \circ_p (y, y)$. Once we go beyond symmetric matching and start talking about ratio production using (only) the left or right ear, this longer form might seem actually indispensable. However, as Steingrimsson and Luce (2005a, p. 292; see also 2005b, section 1.1) point out, under "several background assumptions" previously stated by Luce (2002, 2004), "there is no loss of generality in studying $\circ_p$ in the symmetric case" only. The "background assumptions" come down simply to saying that $\circ_p$ is a binary operation on the indifference classes of intensities.

The evolution of $\circ_p$ from a binary operation between two intensities x and y applied to one ear, as introduced in Luce (2002), first to (5.3) with intensities applied symmetrically to both ears, and subsequently to a binary operation between two joint presentations and their indifference classes, is developed in Luce (2004). Moving ahead, at times I retain the notations $\circ_{\ell,p}$, $\circ_{r,p}$, and $\circ_{s,p}$ to distinguish the descriptive interpretations for left, right, and symmetric presentations, respectively. In this notation, (5.3) becomes $x \circ_{s,p} y$.

Some behavioral axioms

As already stated, the psychophysical theories set forth by R. D. Luce follow the classical physical framework. A key emphasis is the axiomatization and the derivations of the various associated numerical representations. Some of the axioms are behavioral, and some are mathematically technical for the derivations. Here we list some of the behavioral properties employed in the discussions. The list is by no means complete, but it contains the major axioms.

- *Thomsen Condition:*[1]

$$(x, w) \approx (z, v) \text{ and } (z, u) \approx (y, w) \Rightarrow (x, u) \approx (y, v). \qquad (5.4)$$

- *i-Bisymmetry* $(i \in \{\ell, r, s\})$:

$$(x \oplus_i y) \oplus_i (u \oplus_i v) \approx (x \oplus_i u) \oplus_i (y \oplus_i v). \qquad (5.5)$$

- *Proportion Commutativity*:

$$((x, x) \circ_p (y, y)) \circ_q (y, y) \approx ((x, x) \circ_q (y, y)) \circ_p (y, y). \qquad (5.6)$$

---

[1][The reader who finds this condition mysterious may find it helpful to pretend temporarily—in the face of all psychophysical evidence—that the perceived loudness of a binaural presentation (s, t) is identical to the (ordinary, arithmetical) sum $s + t$ of the two physical intensities it comprises (measured in, say, dB). Then $(q, r) \approx (s, t)$ is true if and only if $q + r = s + t$, and (5.4) becomes

$$x + w = z + v \text{ and } z + u = y + w \Rightarrow x + u = y + v,$$

which it is easy algebra to verify. Thus (5.4) may be viewed as an attempt to salvage a minimal amount of algebraic simplicity even after casting overboard the simplistic identification of loudness with physical intensity. Similar remarks apply to (5.5) and (5.6). (Ed.)]

Representations associated with the axioms

In his lectures, Luce stated several results. Details of their derivations are in his cited papers and will not be elaborated here.

- *Additive and p-additive representations.* The Thomsen Condition together with an Archimedean axiom and some structural axioms imply the p-*additive* representation:

$$\Psi(x,u) = \Psi(x,0) + \Psi(0,u) + \delta\Psi(x,0)\Psi(0,u) \qquad (5.7)$$

for some order-preserving function Ψ and some constant $\delta \geq 0$. Here Ψ is *singular at* 0, i.e., $\Psi(0,0) = 0$.

It was also stated that, assuming (5.7), s-bisymmetry is equivalent to $\delta = 0$ in (5.7), leading to the purely *additive* representation

$$\Psi(x,u) = \Psi(x,0) + \Psi(0,u). \qquad (5.8)$$

However, this last statement is incorrect. A slightly different statement appears in Steingrimsson and Luce (2005b, p. 310): "Luce [...] showed that, under the assumptions of the theory, $\gamma \neq 1$ and $\delta = 0$ are equivalent to" i-bisymmetry (5.5) for $i = \ell$, r, and s. (Here γ is as defined below in (5.17).) This statement is also incorrect, as shown below, p. 145 *ff.*

- *Representation Φ with a distortion function W.* The property known as *proportion commutativity,* studied by Narens (1996), combined with some structural assumptions, implies the existence of an order-preserving function Φ and an increasing numerical *distortion function W* mapping $[0, \infty[$ onto $[0, \infty[$ such that

$$W(p) = \frac{\Phi\big((x,x) \circ_p (y,y)\big) - \Phi(y,y)}{\Phi(x,x) - \Phi(y,y)} \qquad (x > y \geq 0). \qquad (5.9)$$

Luce (2004, p. 449) explains that, because of (5.2), (5.9) is equivalent to

$$W(p) = \frac{\Phi\big((x,u) \oplus_s (y,v)\big) - \Phi(y,v)}{\Phi(x,u) - \Phi(y,v)}, \qquad (x,u) > (y,v). \qquad (5.10)$$

Linking Ψ and Φ

Luce links Ψ and Φ by using three main axioms.

- *Left segregation*:

$$u \oplus_s (x \circ_{s,p} y) = (u \oplus_s x) \circ_{s,p} (u \oplus_s y). \tag{5.11}$$

- *Right segregation*:

$$(x \circ_{s,p} y) \oplus_s u = (x \oplus_s u) \circ_{s,p} (y \oplus_s u). \tag{5.12}$$

- *Simple joint presentation decomposition*:

$$(x \oplus_s y) \circ_{s,p} 0 = (x \circ_p 0) \oplus_s (y \circ_p 0). \tag{5.13}$$

Luce proves that (5.11), (5.12), and (5.13) imply

$$\Phi = k\Psi \tag{5.14}$$

for some constant $k > 0$. Thereafter, without loss of generality, it is assumed that the p-additive representation Φ and the order-preserving function Ψ are identical.

Representation with the link $\Phi = \Psi$

Assuming the linking axioms which give rise to a pyschophysical function Ψ representing both structures via

$$\Psi(x, u) = \Psi(x, 0) + \Psi(0, u) + \delta\Psi(x, 0)\Psi(0, u) \quad (\delta \geq 0), \tag{5.15}$$

$$W(p) = \frac{\Psi\big((x, x) \circ_p (y, y)\big) - \Psi(y, y)}{\Psi(x, x) - \Psi(y, y)} \quad (x > y \geq 0), \tag{5.16}$$

various extra assumptions are made to further determine the functions $\Psi(x, 0)$, $\Psi(0, u)$, and the constant δ. Some, but not all, lead to a linear relationship

$$\Psi(x, 0) = \gamma\Psi(0, x) \quad (\gamma > 0) \tag{5.17}$$

which we call *constant bias*.[2] Adding more assumptions leads to the specific form (obviously satisfying (5.17))[3]

$$\Psi(x, u) = x^\beta + \gamma u^\beta. \tag{5.18}$$

---

[2][Algebraic manipulation shows that a function $\Psi(x, u)$ of two variables is a simultaneous solution of the two equations (5.15) and (5.17) if, and only if, $\Psi(x, u) = \gamma\psi(x) + \psi(u) + \gamma\delta\psi(x)\psi(u)$ for some function $\psi(x)$. (Ed.)]

[3][This is more algebraic manipulation. Assume (5.18); then (5.15) becomes

$$x^\beta + \gamma u^\beta = (x^\beta + 0) + (0 + \gamma u^\beta) + \delta(x^\beta + 0)(0 + \gamma u^\beta) = x^\beta + \gamma u^\beta + \delta\gamma x^\beta u^\beta,$$

whence $\delta\gamma x^\beta u^\beta = 0$ for all x and u. By assumption $\gamma > 0$, so $\delta = 0$. (Ed.)]

Experimental evaluations have been carried out by Steingrimsson and Luce (2005a, 2005b, 2006, 2007).

An open question raised by Luce

We can now state Luce's open question: it is to *characterize the form*

$$\Psi(x,u) = \gamma_\ell x^{\beta_\ell} + \gamma_r u^{\beta_r}. \tag{5.19}$$

This form, a strict generalization of (5.18) (for $\beta_\ell \neq \beta_r$), still satisfies (5.15). The motivation for generalizing (5.18), by allowing
the powers β_ℓ and β_r to differ, is that constant bias (5.17)—which
(as just noted) is one of the assumptions that leads to (5.18)—
has not been sustained at all in the laboratory experiments that
have been run so far.

Discussion of the open question

To comprehend the open problem better, I re-read my lecture notes
(as summarized above) and compared them with the details found
in the various cited papers. Some points need clarification.

The binary primitives $\oplus_s$, $\oplus_\ell$, and $\oplus_r$

My interpretation is that taking the symmetric matching $\oplus_s$ to be a
primitive indicates that it is (i) intrinsic to a particular respondent,
and (ii) independent of any numerical representation.

The meaning of (i) is that some complex process executed in the
particular respondent's nervous system yields $\oplus_s$; the process is not
revealed to the investigator, only the resulting judgments of relative
loudness.

The meaning of (ii) is that, if Ψ is any numerical representation
(order-preserving and singular at 0), then by virtue of the definition
of $\oplus_s$, we have

$$\Psi(x,u) = \Psi(x \oplus_s u, x \oplus_s u), \tag{5.20}$$

and conversely: any order-preserving Ψ determines $\oplus_s$, by the identity (5.20). More precisely, if h is a homeomorphism of the interval

$[0, \infty[,^4$ then $\Psi' := h(\Psi)$ is also an order-preserving representation; applying h on both sides of (5.20), we have

$$\Psi'(x, u) = \Psi'(x \oplus_s u, x \oplus_s u).$$

In this sense we see that Ψ and Ψ' represent, or give rise to, the same primitive structure $\oplus_s$. A property of $\oplus_s$ that can be inferred from a given representation Ψ should also be susceptible to being inferred from the transformed representation $\Psi' = h(\Psi)$, because the two functions define the same $\oplus_s$.

We clarify this with a general example. Suppose that a representation Ψ has the form

$$\Psi(x, u) = f(x) + f(u) \tag{5.21}$$

where f is a homeomorphism of $[0, \infty[$. One can infer the symmetry of $\oplus_s$, that $x \oplus_s u = u \oplus_s x$, from (5.21). Then $\Psi' = h(\Psi)$ also yields the symmetry of $\oplus_s$, although we don't anticipate that the form (5.21) remains intact, that is, it is not to be expected that $\Psi'(x, u) = f'(x) + f'(u)$ for any f'. Simply put, if $\oplus_s$ is seen to satisfy an identity, say $x \oplus_s u = u \oplus_s x$, then that fact stands its ground regardless of whether Ψ or Ψ' is used to represent the order structure.

Asymmetric matches by the left ear and the right ear (Steingrimsson & Luce, 2005a) are defined by

$$(x, u) \approx (x \oplus_\ell u, 0) \approx (0, x \oplus_r u). \tag{5.22}$$

Both operations, $\oplus_\ell$ and $\oplus_r$, are also primitives, and similar observations can be made.

An issue: p-additivity *vs.* additivity and bisymmetry

The existence of a p-additive representation is equivalent to the existence of an additive representation. To see this, suppose that we have a p-additive representation Ψ with $\delta > 0$. Then the function $\widetilde{\Psi}$ defined by

$$\widetilde{\Psi} := \log(1 + \delta\Psi) \tag{5.23}$$

---

[4] [This hypothesis on h means that (a) $h(0) = 0$, (b) if $0 \le s < t$ then $0 \le h(s) < h(t)$, and (c) $\lim_{x \to \infty} h(x) = \infty$. In the language of structures and their automorphisms (see Chaps. 9 and 13), h is an automorphism of the ordered set $[0, \infty[$. (Ed.)]

is again order-preserving and singular at 0, and (5.7) can be rewritten as a (purely) additive representation (5.8):

$$\widetilde{\Psi}(x,u) = \widetilde{\Psi}(x,0) + \widetilde{\Psi}(0,u). \tag{5.24}$$

The equivalence between the existence of a p-additive representation with $\delta > 0$ and the existence of a purely additive representation (*i.e.*, with $\delta = 0$) has been noted in the cited papers.

According to the discussion of representation independence, point (ii) on p. 144, whether the operation $\oplus_s$ satisfies or fails to satisfy bisymmetry can not depend on whether inference is made through $\widetilde{\Psi}$ or through Ψ. Thus the assertion (see Steingrimsson & Luce, 2005b, p. 310) about (5.7), that bisymmetry separates the case of $\delta > 0$ from that of $\delta = 0$, is apparently an error. I find it necessary to understand the role of bisymmetry before moving ahead; my findings are given below, on pp. 147–153.

Discussion of the linkage between Ψ and Φ

The linkage (5.14) is also in contradiction with point (ii), since Ψ and $\widetilde{\Psi}$ that differ by more than just a scalar factor can still give equivalent representations for $\oplus_s$. Apparently the relation (5.14) is confined to the case of an additive Ψ.

A sharper statement is found in Luce (2004, pp. 451–452). The result stated there is that a certain set of axioms and conditions, centering around those described on p. 142, gives rise to the following representation with a single psychophysical function $\Psi = \Phi$ that represents both structures $\oplus_s$ and $\circ_p$:

$$\Psi = \Phi,$$
$$\Psi(x,u) = \Psi(x,0) + \Psi(0,u), \tag{5.25}$$
$$W(p) = \frac{\Phi\big((x,u) \circ_p (y,v)\big) - \Phi(y,v)}{\Phi(x,u) - \Phi(y,v)} \quad \text{for } (x,u) > (y,v). \tag{5.26}$$

Equations (5.26) and (5.10) are identical.

I explore the linkage a bit further, using simple joint presentation decompositions, on pp. 153–159.

Monaural ratio productions

The special case of $(x,u) \circ_p (y,v)$ with $(y,v) = (0,0)$, called *ratio production*, was studied by Stevens (1973). In this case, (5.26) implies

$$\Phi\big((x,u)\circ_p(0,0)\big) = W(p)\Phi(x,u). \tag{5.27}$$

There are two monaural ratio productions, $x\circ_{\ell,p}0$ by the left ear and $u\circ_{r,p}0$ by the right ear. The descriptive definition of $x\circ_{\ell,p}0$ is that it is an intensity which, if received by the left ear, will be judged by the respondent as being "p times as loud" as the input x, and similarly for $u\circ_{\ell,p}0$. These monaural ratio productions are tied to the more general $\circ_p$ by

$$(x\circ_{\ell,p}0,0) \approx (x,0)\circ_p(0,0), \quad (0,u\circ_{r,p}0) \approx (0,u)\circ_p(0,0). \tag{5.28}$$

It is supposed that each production has a *separable* representation:

$$\phi_\ell(x\circ_{\ell,p}0) = W_\ell(p)\phi_\ell(x), \tag{5.29}$$

$$\phi_r(u\circ_{r,p}0) = W_r(p)\phi_r(u), \tag{5.30}$$

where ϕ_ℓ, ϕ_r, W_ℓ and W_r are strictly monotonic functions mapping $[0,\infty[$ onto itself (see Luce, 2002, pp. 521 and 523). The existence of some separable representation is implied by (5.27), using $\phi_\ell(x) := \Phi(x,0)$, $\phi_r(u) := \Phi(0,u)$, $W_\ell = W_r := W$.

The notion "p times as loud" is subject to interpretation. It has a linguistic aspect. One natural question to test that interpretation of the respondent could be: if (for you, the respondent) y is p times as loud as x, and z is q times as loud as y, would z be pq times as loud (for you) as x? For a respondent who conforms to that notion, it is reasonable to postulate that W satisfies $W(pq) = W(p)W(q)$, and is therefore a *power function*. This aspect has been discussed quite fully in the works of Luce and of Narens (*op. cit.*).

Based on the studies of Stevens, the *power law* appears to be approximately sustained in the labs:

$$\Phi(x,0) = \gamma_\ell x^{\beta_\ell}, \quad \Phi(0,u) = \gamma_r u^{\beta_r}. \tag{5.31}$$

However, studies of individuals suggest that this is indeed only *approximately* true. The proposed form (5.19) evidently is tied to (5.31).

Bisymmetry in a p-additive representation

In this section, we state and prove our main theorem, giving necessary and sufficient conditions for bisymmetry in p-additive representations. The proof makes use of previously published theorems of the author, stated below but not proved here.

Preliminary theorems on functional equations

Let X and Y be real intervals and $T\colon X \times Y \to \mathbb{R}$ a given continuous function. Consider the functional equation

$$f(x) + g(y) = h(T(x,y)), \quad x \in X, y \in Y \tag{5.32}$$

in unknown functions $f\colon X \to \mathbb{R}$, $g\colon Y \to \mathbb{R}$, $h\colon T(X \times Y) \to \mathbb{R}$.

Theorem 1. Regularity (Ng, 1973a). *If (f,g,h) is any solution of (5.32) with f and g locally bounded and non-constant, then f and g are necessarily continuous.* □

Theorem 2. Uniqueness (Ng, 1973b). *If (f_0, g_0, h_0) is any particular solution of (5.32) with f_0 and g_0 continuous and non-constant, then the general solution (f, g, h) with f and g continuous is given by*

$$f = \gamma f_0 + \beta_1, \ g = \gamma g_0 + \beta_2, \ h = \gamma h_0 + \beta_1 + \beta_2,$$

where γ, β_1, and β_2 are real constants. □

There is also a multiplicative version of Theorem 2. Let X and Y be real intervals and $T\colon X \times Y \to \mathbb{R}$ a given continuous function. Consider the functional equation

$$f(x)g(y) = h(T(x,y)), \quad x \in X, y \in Y \tag{5.33}$$

in unknown functions $f\colon X \to\ ]0,\infty[$, $g\colon Y \to\ ]0,\infty[$, $h\colon T(X \times Y) \to\ ]0,\infty[$.

Theorem 3. *If (f_0, g_0, h_0) is any particular solution of (5.33) with f_0 and g_0 continuous and non-constant, then the general solution (f, g, h) with f and g continuous is given by*

$$f = \beta_1 f_0^{\gamma}, \ g = \beta_2 g_0^{\gamma}, \ h = \beta_1 \beta_2 h_0^{\gamma},$$

where $\beta_1 > 0$, $\beta_2 > 0$, and γ are arbitrary constants. □

Statement of the main theorem

Given a p-additive representation Ψ, we distinguish the cases

$$\Psi(x,u) = \Psi(x,0) + \Psi(0,u), \qquad\qquad (5.34a)$$
$$\Psi(x,u) = \Psi(x,0) + \Psi(0,u) + \delta\Psi(x,0)\Psi(0,u) \text{ with } \delta > 0 \quad (5.34b)$$

as indicated.

Theorem 4. *The following statements are equivalent.*
(i) $\oplus_\ell$ *is bisymmetric.*
(ii) $\oplus_r$ *is bisymmetric.*
(iii) $\oplus_s$ *is bisymmetric.*
(iv)   *In case* (5.34a) *there exists some constant* $\gamma > 0$ *such that*

$$\Psi(x,0) = \gamma\Psi(0,x) \qquad\qquad (5.35a)$$

and in case (5.34b) *there exists some constant* $\gamma > 0$ *such that*

$$1 + \delta\Psi(x,0) = (1 + \delta\Psi(0,x))^\gamma. \qquad\qquad (5.35b)$$

We defer the proof of Theorem 4 briefly, in order to place it more firmly in the context of the open problem.

Relationship of the main theorem to the open problem

With regard to the status of the open problem, I find the following statements in papers by Luce and by Steingrimsson and Luce.

> Steingrimsson and Luce (2004) give an argument based on [equation (5.16)] that leads to the representation [given in equation (5.19)]. The constant-ratio condition excludes this except for $\beta_\ell = \beta_r$ [...]. It would be desirable to verify empirically the constant-ratio condition, but I do not really know how to do so very effectively. No purely behavioral condition equivalent to it has yet been found. (Luce, 2004, p. 449; our equation numbering)

> For symmetric matches, [equation (5.25)] with $\delta = 0$ and [equation (5.16)] both hold, but the theory for symmetric matches does not predict the constant bias of [equation (5.17)]. (Steingrimsson & Luce, 2005a, p. 304; our equation numbering)

Steingrimsson and Luce (2005b) give tests for bisymmetry.

> These results provide good initial support for the bisym-
> metric property. Within the context of the theory, this
> means that either $\gamma = 1$ or, when $\gamma \neq 1$, $\delta = 0$ [...] we are
> justified in assuming that $\delta = 0$ [...]. (Steingrimsson &
> Luce, 2005b, p. 313)

In short, the data collected by Steingrimsson and Luce provide good
support for the bisymmetry property, strongly reject property (5.17),
and to a fairly good approximation seem to accord with

$$\Psi(x,0) = \gamma_\ell x^{\beta_\ell} \text{ and } \Psi(0,x) = \gamma_r x^{\beta_r} \tag{5.36}$$

with $\beta_\ell \neq \beta_r$. The form $\Psi(x,u) = \gamma_\ell x^{\beta_\ell} + \gamma_r u^{\beta_r}$ proposed in (5.19),
with $\beta_\ell \neq \beta_r$, is apparently driven by the desire to stay away from
(5.17) and at the same time stay in accord with (5.36). The question
becomes, is (5.19) (with $\beta_\ell \neq \beta_r$) consistent with bisymmetry? Unfortunately, Theorem 4 shows that it is not. In that sense, it might
be said that interest in the open problem has faded.

Proof of the main theorem

Proof. To clarify the exposition of this and later proofs, we adopt
the abbreviated notations

$$\psi_s(x) := \Psi(x,x), \quad \psi_\ell(x) := \Psi(x,0), \quad \psi_r(x) := \Psi(0,x) \tag{5.37}$$

Note that, because Ψ is order-preserving, so are ψ_s, ψ_ℓ, and ψ_r.
    We first show that in the additive case (5.34a), each of (i), (ii),
and (iii) implies and is implied by (5.35a).
    (i) $\Rightarrow$ (5.35a): Assume (i), *i.e.*,

$$(x \oplus_\ell y) \oplus_\ell (u \oplus_\ell v) \approx (x \oplus_\ell u) \oplus_\ell (y \oplus_\ell v). \tag{5.38}$$

Using (5.37), we rewrite (5.34a) as

$$\psi_\ell(x \oplus_\ell u) = \psi_\ell(x) + \psi_r(u). \tag{5.39}$$

Applying ψ_ℓ to both sides of (5.38), then using (5.39) once, we get

$$\psi_\ell(x \oplus_\ell y) + \psi_r(u \oplus_\ell v) = \psi_\ell(x \oplus_\ell u) + \psi_r(y \oplus_\ell v).$$

Using (5.39) again for $\psi_\ell(x \oplus_\ell y)$ and $\psi_\ell(x \oplus_\ell u)$, and then canceling the resulting common term $\psi_\ell(x)$, we get

$$\psi_r(y) + \psi_r(u \oplus_\ell v) = \psi_r(u) + \psi_r(y \oplus_\ell v). \tag{5.40}$$

Setting the variable y to 0 in (5.40), writing $f(v) := \psi_r(0 \oplus_\ell v)$, and using $\psi_r(0) = 0$, we get

$$\psi_r(u \oplus_\ell v) = \psi_r(u) + f(v)$$

which, upon renaming the variables, becomes

$$\psi_r(x \oplus_\ell u) = \psi_r(x) + f(u). \tag{5.41}$$

We now have two representations for $\oplus_\ell$, (5.39) and (5.41). According to Theorem 2 (uniqueness), there exist constants γ, b_1, b_2 relating these representations as follows:

$$\psi_\ell(x) = \gamma\psi_r(x) + b_1, \tag{5.42}$$
$$\psi_r(u) = \gamma f(u) + b_2, \tag{5.43}$$
$$\psi_\ell(t) = \gamma\psi_r(t) + b_1 + b_2. \tag{5.44}$$

Comparing (5.42) and (5.44), we get $b_2 = 0$. Because $\psi_\ell(0) = \psi_r(0) = 0$, (5.42) now gives us $b_1 = 0$, and (5.42) becomes $\psi_\ell(x) = \gamma\psi_r(x)$. Since both ψ_ℓ and ψ_r are homeomorphisms on $[0, \infty[$, $\gamma > 0$. By (5.37),

$$\Psi(x, 0) = \psi_\ell(x) = \gamma\psi_r(x) = \gamma\Psi(0, x) \tag{5.45}$$

which is a mere rewriting of (5.35a).

(i) $\Leftarrow$ (5.35a): It is straightforward to check, using the fact that the functions defined in (5.37) are order-preserving, that all the transformations applied to get from (5.38) to (5.45) are invertible; thus (5.35a), rewritten as (5.45), implies (i).

(ii) $\Longleftrightarrow$ (5.35a): The argument is completely parallel to those for $\oplus_\ell$; we simply transpose the variables in Ψ.

(iii) $\Rightarrow$ (5.35a): Assume (iii), *i.e.*,

$$(x \oplus_s y) \oplus_s (u \oplus_s v) \approx (x \oplus_s u) \oplus_s (y \oplus_s v). \tag{5.46}$$

As before, we use (5.37) to rewrite (5.34a), getting

$$\psi_s(x \oplus_s u) = \psi_\ell(x) + \psi_r(u). \tag{5.47}$$

Applying ψ_s to (5.46) and using (5.47) once, we get

$$\psi_\ell(x \oplus_s y) + \psi_r(u \oplus_s v) = \psi_\ell(x \oplus_s u) + \psi_r(y \oplus_s v).$$

Fixing $u = v = 0$ and noting $0 \oplus_s 0 = 0$, $\psi_r(0) = 0$, we get

$$\psi_\ell(x \oplus_s y) = \psi_\ell(x \oplus_s 0) + \psi_r(y \oplus_s 0)$$

which becomes

$$\psi_\ell(x \oplus_s u) = \psi_\ell(x \oplus_s 0) + \psi_r(u \oplus_s 0)$$

upon renaming y as u. As before, we now have two representations for $\oplus_s$, to which we can apply Theorem 2, concluding that there exist constants c, b_1, and b_2 relating the representations as follows:

$$\psi_\ell(x) = c\psi_\ell(x \oplus_s 0) + b_1, \tag{5.48}$$
$$\psi_r(u) = c\psi_r(u \oplus_s 0) + b_2, \tag{5.49}$$
$$\psi_s(t) = c\psi_\ell(t) + b_1 + b_2. \tag{5.50}$$

Because $\psi_\ell(0) = \psi_s(0) = 0$, (5.50) is reduced to

$$\psi_s = c\psi_\ell. \tag{5.51}$$

Because $x \oplus_s x = x$, setting $x = u$ in (5.47) gives

$$\psi_s = \psi_\ell + \psi_r.$$

Feeding this back into (5.51) gives us $c\psi_\ell = \psi_\ell + \psi_r$, so $\psi_r = (c-1)\psi_\ell$. Since the functions ψ_ℓ and ψ_r are homeomorphisms, $c - 1 > 0$. Let $\gamma := 1/(c-1)$; then $\gamma > 0$ and we get

$$\Psi(x,0) = \psi_\ell(x) = \gamma\psi_r(x) = \gamma\Psi(0,x)$$

and thus (5.35a).

(iii) $\Leftarrow$ (5.35a): In this case, we cannot simply reverse the steps of the converse implication. Instead, we note that (5.35a) can be rewritten as

$$\psi_s = (\gamma + 1)\psi_r$$

so that (5.47) becomes

$$(\gamma + 1)\psi_r(x \oplus_s u) = \gamma\psi_r(x) + \psi_r(u). \tag{5.52}$$

(5.52) states that $\oplus_s$ is a *weighted quasi-arithmetic mean* generated by ψ_r, i.e.,

$$x \oplus_s u = \psi_r^{-1}[\lambda \psi_r(x) + (1-\lambda)\psi_r(u)]$$

where $\lambda = \gamma/(\gamma+1)$ is a constant in the open interval $]0,1[$. It is known that such a mean satisfies the bisymmetry equation (iii) (a recent reference is Daróczy, 2009).

This completes the proof of the main theorem in the additive case (5.34a). The non-additive, p-additive case (5.34b), is reduced to the additive case as follows: if Ψ is a p-additive representation satisfying

$$\Psi(x,u) = \Psi(x,0) + \Psi(0,u) + \delta\Psi(x,0)\Psi(0,u)$$

for some $\delta > 0$, then $\widetilde{\Psi} := \log(1 + \delta\Psi)$ is an additive representation; and Ψ satisfies (5.35b) if and only if $\widetilde{\Psi}$ satisfies (5.35a).   □

Further results on the linkage between Ψ and Φ

In this section we state and prove several results relating magnitude production, p-additive representation, and simple joint presentation decomposition. They are somewhat more technical than the results in earlier sections, and the proofs leave more to the reader.

A theorem on symmetric matching

Let Ψ be an additive representation of $\oplus$,

$$\Psi(x,u) = \Psi(x,0) + \Psi(0,u),$$

and suppose (ϕ, W) is a *separable representation* of $\circ_{s,p}$ in the sense that

$$\phi(x \circ_{s,p} 0) = W(p)\phi(x). \tag{5.53}$$

Lemma 1. *The simple joint presentation decomposition*

$$(x \oplus_s y) \circ_{s,p} 0 = (x \circ_{s,p} 0) \oplus_s (y \circ_{s,p} 0) \tag{5.54}$$

holds if and only if there exist positive constants k, b such that

$$\phi = k\psi_s^b \tag{5.55}$$

and Ψ gives rise to constant bias, i.e.

$$\Psi(x,0) = \gamma\Psi(0,x). \tag{5.56}$$

Proof. Let f be the increasing homeomorphism connecting the two order-preserving functions ϕ and ψ_s,

$$\phi = f(\psi_s).$$

(5.54) $\Rightarrow$ (5.55) $\wedge$ (5.56): Making use of f and earlier results, we construct a chain of equivalent equations starting with (5.54).

$$(x \oplus_s y) \circ_{s,p} 0 = (x \circ_{s,p} 0) \oplus_s (y \circ_{s,p} 0) \Longleftrightarrow$$
$$f(\psi_s((x \oplus_s y) \circ_{s,p} 0)) = f(\psi_s((x \circ_{s,p} 0) \oplus_s (y \circ_{s,p} 0))) \Longleftrightarrow$$
$$\phi((x \oplus_s y) \circ_{s,p} 0) = f(\psi_s((x \circ_{s,p} 0) \oplus_s (y \circ_{s,p} 0))) \Longleftrightarrow$$
$$W(p)\phi(x \oplus_s y) = f(\Psi(x \circ_{s,p} 0, y \circ_{s,p} 0)) \Longleftrightarrow$$
$$W(p)f(\Psi(x,y)) = f(\Psi(x \circ_{s,p} 0, y \circ_{s,p} 0)) \Longleftrightarrow$$
$$W(p)f(\Psi(x,0) + \Psi(0,y)) = f(\Psi(x \circ_{s,p} 0,0) + \Psi(0,y \circ_{s,p} 0)) \Longleftrightarrow$$
$$f^{-1}(W(p)f(\Psi(x,0) + \Psi(0,y))) = \Psi(x \circ_{s,p} 0,0) + \Psi(0,y \circ_{s,p} 0). \quad (5.57)$$

We shall solve (5.57) using the uniqueness theorem, Theorem 2. We let $T(x,y) := \Psi(x,0) + \Psi(0,y)$, which is independent of p. The continuity of T, and its straight monotonicity in each variable over intervals, readily follow from the assumptions originally imposed on Ψ. Then $f_0(x) := \Psi(x,0)$, $g_0(y) := \Psi(0,y)$, and $h_0(z) := z$ give a particular solution of (5.32) with continuous and non-constant f_0, g_0.

Temporarily fixing p, let $f(x) = \Psi(x \circ_{s,p} 0,0)$, $g(y) = \Psi(0,y \circ_{s,p} 0)$ and $h = f^{-1}(W(p)f)$. From Theorem 2 we get that, for some constants a, c_1, c_2 (possibly depending on p),

$$f^{-1}(W(p)f(z)) = a(p)z + c_1(p) + c_2(p), \quad (5.58)$$
$$\Psi(x \circ_{s,p} 0,0) = a(p)\Psi(x,0) + c_1(p), \quad (5.59)$$
$$\Psi(0,y \circ_{s,p} 0) = a(p)\Psi(0,y) + c_2(p). \quad (5.60)$$

Boundary considerations (e.g., $\Psi(0^+,0) = \Psi(0,0^+) = 0$) imply that $c_1 = c_2 = 0$, which reduces (5.58) to

$$W(p)f(z) = f(a(p)z). \quad (5.61)$$

From the original assumptions on W, we can infer the continuity of the function $(p,z) \mapsto a(p)z$. Thus the uniqueness theorem, Theorem

3, is applicable to (5.61) (with p and z restricted to be strictly positive) and we obtain the relations

$$W(p) = c_3 a(p)^b, \tag{5.62}$$
$$f(z) = kz^b, \tag{5.63}$$
$$f(t) = c_3 kt^b. \tag{5.64}$$

However, these equations, once obtained, extend to include $z = 0$ and $t = 0$, using the continuity of f and the boundary condition $f(0) = 0$. The same considerations imply that $b > 0$. Comparing (5.63) and (5.64) we get $c_3 = 1$. Hence, $f(z) = kz^b$ for all $z \geq 0$, yielding (5.55).

Substituting $c_3 = 1$ into (5.62) yields $W(p) = a(p)^b$; substituting that into (5.59) and (5.60) yields

$$\Psi(x \circ_{s,p} 0, 0) = W(p)^{1/b} \Psi(x, 0), \tag{5.65}$$
$$\Psi(0, y \circ_{s,p} 0) = W(p)^{1/b} \Psi(0, y). \tag{5.66}$$

Comparing (5.65) and (5.66), and again applying Theorem 3, we get (5.56).

$(5.55) \wedge (5.56) \Rightarrow (5.54)$: It follows from (5.53) and (5.55) that ψ_s gives rise to a separable representation

$$\psi_s(x \circ_{s,p} 0) = S(p)\psi_s(x) \tag{5.67}$$

where $S(p) := W(p)^{1/b}$. Next, noting that

$$\psi_s(x) = \Psi(x, x) = \Psi(x, 0) + \Psi(0, x) = \psi_\ell(x) + \psi_r(x),$$

we get from (5.56) that the three functions ψ_s, ψ_ℓ, and ψ_r are scalar multiples of each other:

$$\psi_s = (\gamma + 1)\psi_\ell \text{ and } \psi_s - (1 + \gamma^{-1})\psi_r. \tag{5.68}$$

Thus (5.67) extends to all three functions:

$$\psi_i(x \circ_{s,p} 0) = S(p)\psi_i(x) \quad (i = \ell, s, r). \tag{5.69}$$

Finally,

$$
\begin{aligned}
\psi_s((x \oplus_s y) \circ_{s,p} 0) &= S(p)\psi_s(x \oplus_s y) && \text{[by (5.69)]}\\
&= S(p)\Psi(x,y)\\
&= S(p)[\Psi(x,0) + \Psi(0,y)]\\
&= S(p)\Psi(x,0) + S(p)\Psi(0,y)\\
&= \psi_\ell(x \circ_{s,p} 0) + \psi_r(y \circ_{s,p} 0) && \text{[by (5.69)]}\\
&= \Psi(x \circ_{s,p} 0, y \circ_{s,p} 0)\\
&= \psi_s((x \circ_{s,p} 0) \oplus_s (y \circ_{s,p} 0)).
\end{aligned}
$$

As ψ_s is strictly monotonic this proves (5.54).   □

Theorem 5. *If a p-additive representation exists for the joint presentation, then the following statements are equivalent.*

(i) $\circ_{s,p}$ has a separable representation (5.53) and simple joint presentation decomposition (5.54) holds.

(ii) Every additive representation Ψ for $\oplus$ satisfies

$$
\Psi(x,0) = \gamma \Psi(0,x) \tag{5.70}
$$

and renders separable representations

$$
\psi_i(x \circ_{s,p} 0) = S(p)\psi_i(x) \quad (i = \ell,s,r) \tag{5.71}
$$

for some distortion function S.

(iii) Every p-additive representation Ψ with $\delta > 0$ satisfies

$$
1 + \delta\Psi(x,0) = (1 + \delta\Psi(0,x))^\gamma
$$

and renders separable representations

$$
\log(1 + \delta\psi_i(x \circ_{s,p} 0)) = S(p)\log(1 + \delta\psi_i(x)) \quad (i = \ell,s,r) \tag{5.72}
$$

for some distortion function S.

Proof. (i) $\Rightarrow$ (ii): Assuming (i), we start with any fixed additive representation Ψ and separable representation (ϕ, W) for symmetric magnitude production (5.53). By Lemma 1, ψ_s and ϕ are related by (5.55), and constant bias (5.70) holds. Hence (ψ_s, S), where $S := W^{1/b}$, is a separable representation. This proves (5.71) for $i = s$. As (5.70) implies that the three functions ψ_s, ψ_ℓ, and ψ_r are scalar multiples of each other (*cf.* (5.68)), (5.71) holds for $i = r$ and $i = \ell$ as well (*cf.* (5.69)), proving (ii).

(ii) $\Rightarrow$ (i): This also follows from Lemma 1, upon taking $\phi := \psi_s$, $W := S$.

(ii) $\Longleftrightarrow$ (iii): This is immediate from the fact, noted earlier, that every p-additive Ψ gives rise to an additive $\tilde{\Psi} := \log(1 + \delta\Psi)$.   □

A theorem on asymmetric matching

I shall present my findings only in the case of left matching. The case of right matching is completely analogous.

Again, let Ψ be an additive representation of $\oplus$,

$$\Psi(x, u) = \Psi(x, 0) + \Psi(0, u).$$

Let (ϕ, W_ℓ) be a separable representation of $\circ_{\ell,p}$,

$$\phi(x \circ_{\ell,p} 0) = W_\ell(p)\phi(x). \qquad (5.73)$$

Lemma 2. *The simple joint presentation decomposition*

$$(x \oplus_\ell y) \circ_{\ell,p} 0 = (x \circ_{\ell,p} 0) \oplus_\ell (y \circ_{\ell,p} 0) \qquad (5.74)$$

holds if and only if there exist positive constants k, b such that

$$\phi = k\psi_\ell^b \qquad (5.75)$$

and Ψ gives rise to constant bias, i.e.

$$\Psi(x, 0) = \gamma\Psi(0, x). \qquad (5.76)$$

Proof. Let g be the homeomorphism of $[0, \infty[$ connecting the two representations,

$$\phi = g(\Psi_\ell).$$

(5.74) $\Rightarrow$ (5.75) $\wedge$ (5.76): Making use of g and earlier results, we construct a chain of equivalent equations starting with (5.74).

$$(x \oplus_\ell y) \circ_{\ell,p} 0 = (x \circ_{\ell,p} 0) \oplus_\ell (y \circ_{\ell,p} 0) \iff$$
$$g(\psi_\ell((x \oplus_\ell y) \circ_{\ell,p} 0)) = g(\psi_\ell((x \circ_{\ell,p} 0) \oplus_\ell (y \circ_{\ell,p} 0))) \iff$$
$$\phi((x \oplus_\ell y) \circ_{\ell,p} 0) = g(\psi_s((x \circ_{\ell,p} 0) \oplus_\ell (y \circ_{\ell,p} 0))) \iff$$
$$W_\ell(p)\phi(x \oplus_\ell y) = g(\Psi(x \circ_{\ell,p} 0, y \circ_{\ell,p} 0)) \iff$$
$$W_\ell(p)g(\Psi(x,y)) = g(\Psi(x \circ_{\ell,p} 0, y \circ_{\ell,p} 0)) \iff$$
$$W_\ell(p)g(\Psi(x,0)+\Psi(0,y)) = g(\Psi(x \circ_{\ell,p} 0,0)+\Psi(0, y \circ_{\ell,p} 0)) \iff$$
$$g^{-1}(W_\ell(p)g(\Psi(x,0)+\Psi(0,y))) = \Psi(x \circ_{\ell,p} 0,0)+\Psi(0, y \circ_{\ell,p} 0). \tag{5.77}$$

Subjecting (5.77) to the same steps that were applied to the analogous equation (5.57) in the proof of Lemma 1, we get

$$g(z) = kz^b, \tag{5.78}$$
$$\Psi(x \circ_{\ell,p} 0,0) = W_\ell(p)^{1/b}\Psi(x,0), \tag{5.79}$$
$$\Psi(0, y \circ_{\ell,p} 0) = W_\ell(p)^{1/b}\Psi(0,y), \tag{5.80}$$

where $b > 0$. (5.78) gives (5.75). Comparison of (5.79) and (5.80), using the uniqueness theorem, gives (5.76), constant bias. In a parallel with (5.65) and (5.66) we further obtain that ψ_ℓ, ψ_r and ψ_s are scalar multiples of each other, and that

$$\psi_i(x \circ_{\ell,p} 0) = S(p)\psi_i(x) \quad (i = \ell, r, s)$$

where $S := W_\ell^{1/b}$. $\qquad\square$

Lemma 2 can be used to prove the following theorem in the same way that Lemma 1 was used to prove Theorem 5.

Theorem 6. *If a p-additive representation exists for the joint presentation, then the following statements are equivalent.*

(i) *$\circ_{\ell,p}$ has a separable representation (5.29) and simple joint presentation decomposition (5.74) holds.*

(ii) *Every additive representation Ψ for $\oplus$ satisfies*

$$\Psi(x,0) = \gamma\Psi(0,x)$$

and renders separable representations

$$\psi_i(x \circ_{\ell,p} 0) = S(p)\psi_i(x) \quad (i = \ell, s, r)$$

for some distortion function S.

(iii) *Every p-additive representation Ψ with $\delta > 0$ satisfies*

$$1 + \delta\Psi(x, 0) = (1 + \delta\Psi(0, x))^{\gamma}$$

and renders separable representations

$$\log(1 + \delta\psi_i(x \circ_{\ell,p} 0)) = S(p)\log(1 + \delta\psi_i(x)) \quad (i = \ell, s, r)$$

for some distortion function S.   □

Acknowledgments

This work was supported in part by the NSERC of Canada Discovery Grant 8212.

References

Daróczy, Z. (2009). Mean values and functional equations. *Differential Equations and Dynamical Systems, 17*, 105-113.

Krantz, D. H., Luce, R. D., Suppes, P., & Tversky, A. (1971). *Foundations of Measurement, Vol. 1: Additive and Polynomial Representations.* New York, NY: Academic Press.

Luce, R. D. (2002). A psychophysical theory of intensity proportions, joint presentations, and matches. *Psychological Review, 109*, 520–532.

Luce, R. D. (2004). Symmetric and asymmetric matching of joint presentations. *Psychological Review, 111*, 446–454.

Luce, R. D. (2005). Measurement analogies: Comparisons of behavioral and physical measures. *Psychometrika, 70*(2), 227–251.

Narens, L. (1996). A theory of magnitude estimation. *Journal of Mathematical Psychology, 40*, 109–129.

Ng, C. T. (1973a). Local boundedness and continuity for a functional equation on topological spaces. *Proceedings of the American Mathematical Society, 39*, 525–529.

Ng, C. T. (1973b). On the functional equation $f(x) + \sum_{i=1}^{n} g_i(y_i) = h(T(x, y_1, y_2, \ldots, y_n))$. *Annales Polonici Mathematici, 27*, 329–336.

Steingrimsson, R., & Luce, R. D. (2005a). Evaluating a model of global psychophysical judgments—I: Behavioral properties of summations and productions. *Journal of Mathematical Psychology, 49*, 290–307.

Steingrimsson, R., & Luce, R. D. (2005b). Evaluating a model of global psychophysical judgments—II: Behavioral properties linking summations and productions. *Journal of Mathematical Psychology, 49*, 308–319.

Steingrimsson, R., & Luce, R. D. (2006). Evaluating a model of global psychophysical judgments—III: A form for the psychophysical function and intensity filtering. *Journal of Mathematical Psychology, 50*, 15–29.

Steingrimsson, R., & Luce, R. D. (2007). Evaluating a model of global psychophysical judgments—IV: Forms for the weighting function. *Journal of Mathematical Psychology, 51*, 29–44.

Stevens, S. S. (1973). *Psychophysics: Introduction to its Perceptual, Neural, and Social Prospects*. New York, NY: Wiley.

Model Structures

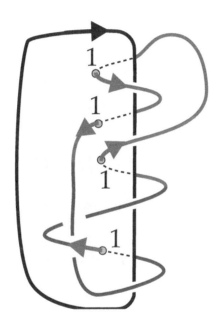

6

Functions of Structure in Mathematics and Modeling

Lee Rudolph

> The psychological is a good deal like the histological procedure in biology. You have to stop the dynamic life functioning and cut the specimen into thin sections, artificially distorted by hardening reagents and staining fluids, in order to analyze its structure. You have to kill it in order to state it. (Bawden, 1904, p. 304)

In our Introduction (p. 17) we quoted "three statements, by mathematicians, on mathematical modeling". Here is a fourth.

(D) Mathematics has its own structures; the world (as we perceive and cognize it) is, or appears to be, structured; mathematical modeling is a reciprocal process in which we *construct/discover/bring into awareness* correspondences between mathematical structures and structures 'in the world', as we *take actions that get meaning from, and give meaning to,* those structures and correspondences.

Later (p. 24 *ff.*) we briefly viewed modeling from the standpoint of "evolutionary epistemology" in the style of Konrad Lorenz (1941,

Keywords: cognition, evolutionary ontology, functional circle, *Funktionskreis*, hermeneutic circle, in-dwelt environment, mappings, morphisms, perception, projective plane, structure, symmetry, teleology, *Umwelt*.

1962). In this chapter, I view modeling from the standpoint informally staked out by **(D)**, which I propose to call "evolutionary ontology."[1] My discussion is sketchy (and not very highly structured), but may help make sense of this volume and perhaps even mathematical modeling in general.

Behind **(D)** is my conviction that there is no need to adopt any particular ontological attitude(s) towards "structures", in the world at large and/or in mathematics, in order to proceed with the project of modeling the former by the latter and drawing inspiration for the latter from the former. It is, I claim, possible for someone simultaneously to adhere to a rigorously 'realist' view of mathematics (say, naïve and unconsidered Platonism) and to take the world to be entirely insubstantial and illusory (say, by adopting a crass reduction of the Buddhist doctrine of Maya), *and still practice mathematical modeling in good faith* if not with guaranteed success. Other (likely or unlikely) combinations of attitudes are (I claim) just as possible, and equally compatible with the practice of modeling.

I have the impression that many practitioners, if polled (which I have not done), would declare themselves to be both mathematical 'formalists' and physical 'realists'. I also have the impression that a large, overlapping group of practitioners, observed in action (which I have done, in a small and unsystematic way), can reasonably be described to *behave* like thoroughgoing ontological agnostics. Mathematical modeling *as human behavior* is based, I am claiming, on acts of pattern-matching (or Gestalt-making)—which is to say, in other language, on creation/recognition/awareness of 'higher

---

[1]The phrase has, unfortunately, already been overloaded with several different meanings (which appear pairwise inconsistent, to me). None of those meanings applies obviously to mathematical modeling *per se*, so confusion should be minimal.

Two authors who haven't (as far as I know) used the phrase "evolutionary ontology", but who *have* written about ontology, mathematics, and mathematical modeling in ways that comport well with what I'm trying to get at with my informal definition, are Jody Azzouni (in Azzouni 1994, 2004—particularly Chap. 9, "Applied Mathematics and Ontology", and elsewhere) and Penelope Maddy (especially in Maddy, 2007). Both are professional philosophers with impeccable mathematical credentials, who (in those works) are addressing primarily fellow philosophers, especially of science and/or of mathematics. I am not even an amateur philosopher, nor is this chapter (or volume) particularly addressed to philosophers of any specialization, so it is entirely possible that my readings of Azzouni and/or Maddy are naïve, mistaken, or both; in any case, my arguments here are not based on either author's work, and certainly should not be read as claiming that either one agrees with any particular statement I may make.

order structures' relating some 'lower order structures'—that one performs (or that occur to one) independently of one's ontological stances. That is not all there is to it, as behavior; but that is its basis.

On the nature of mathematical structures

This section, the longest in this chapter, can be viewed as an expansion of my first assertion in **(D)**, that "Mathematics has its own structures".

Foundations of mathematics and the working mathematician[2]

In the second half of a long career, Morris Kline, an applied mathematician whose specialty was electromagnetism, wrote several books about mathematical history and mathematical education. In one, he summarized his view of "developments in [the 20th] century bearing on the foundations of mathematics" with a parable about

> a beautiful castle [that] had been standing for centuries. In the cellar of the castle, an intricate network of webbing had been constructed by industrious spiders who lived there. One day a strong wind sprang up and destroyed the web. Frantically, the spiders worked to repair the damage. They thought it was their webbing that was holding up the castle. (Kline, 1980, p. 277)

Of course, Kline was famously dismissive of whatever he didn't like (*cf.* Kline, 1973, 1977):

> The overall purpose of [Kline, 1980] is to advance as a philosophy of mathematics a mentalistic pragmatism which exalts "applied mathematics" and denigrates both "pure mathematics" and foundational studies. (Corcoran, 1982)

Still, I suspect many mathematicians ("applied" and "pure" alike), who may not share either Kline's general "mentalistic pragmatism" or more than a small and random selection from his (copious and

---

[2]This section's title is taken from Bourbaki (1949); its contents are unrelated.

well documented—see Kline, 1973, 1977, 1980, *passim*) specific likes and dislikes, have read Kline's parable with some enjoyment and sympathy, as I have. I suspect, in fact, that I am far from alone in *liking* the work of both spiders and researchers of "the foundations of mathematics", even as I maintain reservations on their structural indispensability to *my* work, as a working mathematician.

At least one fairly objective difference marks the distinction between Kline's position on "the foundations of mathematics" and mine (as I am trying to develop it for application in this chapter), namely, the difference between our uses of what happen to be overlapping quotations from a single text. Here is Kline (1980, pp. 316–317) quoting the last sentence of Lakatos (1962, p. 184),[3] with some of its preceding matter; the ellipsis points are Kline's.

> There are sound critics of the foundations who are even more impatient with the fine distinctions these mathematical foundationists make. If, they say, mathematics rests ultimately on intuitions, then to cite Imre Lakatos (1922–1974), why do we probe deeper and deeper?
>
> > Why then don't we stop earlier, why not say that "the ultimate test whether a method is admissible in arithmetic must of course be whether it is intuitively convincing." . . . Why not honestly admit mathematical fallibility, and try to defend the dignity of fallible knowledge from cynical scepticism, rather than delude ourselves that we can invisibly mend the latest tear in the fabric of our "ultimate" intuitions?

Now, in Chap. 2 (p. 43, note 5) I have also quoted that last sentence from Lakatos (1962), in my case prefacing it with the two

---

[3]The quotation is not entirely accurate. (i) "Why" is the second word of the first quoted sentence in the original (its first word is "But"). (ii) The word "arithmetic" is emphasized with italics in the original. (iii) The punctuation mark following "convincing" in the original is a comma. (iv) The word "fallible" is emphasized with italics in the original. Of course (i) and (iii) are trivial, and (iv) is not too serious, since it's fairly obvious that Lakatos's emphasis on "fallible" was purely rhetorical. But I think that (ii) seriously misstates Lakatos's intention: the emphasis on "arithmetic" really is germane to the *dialectic* of Lakatos's argument.

questions that Kline replaced with an ellipsis: "But why on earth have *'ultimate'* tests, 'final authority'? [...] Why foundations, if they are admittedly subjective?" (Lakatos, 1962, p. 184).[4] As I see it, Kline's use of his quotation and my use of mine differ just as would be predicted by the assessment of Kline (1980) by Corcoran (1982): Kline does not "make the distinction between (human) error and (mathematical) falsehood (p. 324)", nor does he "appreciate the conceptual distinction between (human) knowledge and (objective) fact", namely, that

> In order for a proposition to be true it must correspond to fact, but in order for it to be known to be true a human being must be aware of the correspondence. Insofar as a knower, or knowing subject, is essential to knowledge, all knowledge is subjective (in the strict philosophic sense) despite the fact that it must also be objective in that it involves correspondence with objective fact. "Purely objective knowledge", in the sense of knowledge not involving (fallible) human beings, is just as absurd as "purely subjective truth", in the sense of truth not involving (objective) fact. (Corcoran, 1982, fourth paragraph)

It is, I hope, substantially clear (see Chap. 2 and the Introduction of this volume) that I do make and appreciate those distinctions.[5]

With a final quotation from Lakatos (1962), I can explicate my position further and incidentally provide another, more subjective distinction between it and Kline's.

---

[4]The italics are Lakatos's; my ellipsis points replace a footnote number, only.

[5]Of course, as Corcoran (1982) puts it, those distinctions belong to "the complex of traditional philosophic distinctions which have been both exploited by and clarified by modern foundational work"; I appreciate them largely passively, as a fan of "foundational work". Lakatos certainly was intimately familiar with that complex, and would have appreciated the two distinctions in question— and the entire complex of distinctions—actively, through his own foundational work. I do not know whether (as Corcoran concluded) Kline was truly unaware of them, or if perhaps he were aware but (with or without due consideration) had dismissed them (*e.g.*, as being overly "fine distinctions"); whatever the case, I think he (accidentally or deliberately) misrepresented Lakatos by suggesting that his 'impatience' had been with "the fine distinctions" rather than with both 'infallibilism' *and* "cynical scepticism", and that his series of "Why not?" questions had been intended to have *only* the rhetorical effect of closing off discussion, and not *also* the rhetorico-dialectical effect of opening the discussion to alternative understandings and appreciations of the "fine distinctions".

> The logical theory of mathematics is an exciting, sophisticated speculation, like any scientific theory. It is an empiricist theory and thus can either be shown to be false or can remain conjectural for ever. [...] Pure-grained sceptics are rare: we find, however, that pessimistic dogmatists are virtually sceptics. These pessimistic dogmatists demand that we should abandon speculation and restrict our attention to some narrow field which they gracefully —but without any real justification—acknowledge to be safe. (Lakatos, 1962, pp. 178–179)

I think I do not overly misrepresent Kline by saying he is much more of a "pure-grained" skeptic and "pessimistic dogmatist" (save, perhaps, for the gracefulness) than I take myself to be.

More specifically, Kline's position seems to be, essentially, that research (by mathematicians and philosophers) on "the foundations of mathematics" is at best a distracting nuisance and at worst a positive evil, which has been a chief contributor to the "decline of the majesty of mathematics" in the 20th century (Kline, 1980, p. 3). My position is that it can be "an exciting, sophisticated speculation, like any scientific theory", and that—like any scientific theory—it can be appreciated by non-practitioners of the theory (including but not limited to "working mathematicians") without being taken with such deadly earnestness as either to throw them into full-on defense of their own practices in other theories (as though researchers on "foundations" had the real powers of a Grand Inquisitor to ban those other theories), or to paralyze them completely in their own practices (as though researchers on "foundations" had the imagined powers of a Grand Inquisitor to detect original sin, in the form of mathematical fraudulence, and thereby to make continued practice of un-, or wrongly, founded mathematics into mortal sin).[6]

Foundations and structures

It seems to me that the importance of "the foundations of mathematics" for working mathematicians in general, and mathematical modelers in particular, is more syntactic than semantic: that is,

---

[6]Whether some, many, none, or all full- or part-time researchers on "foundations of mathematics" view themselves (ever, sometimes, or always) as Grand Inquisitors is a different issue. The fact is, they're just not.

"foundations"—*any* foundations—provide working mathematicians with a single vocabulary in which to describe mathematical structures in a (more or less) coherent, (more or less) consistent[7] manner. That any particular such vocabulary comes with various ontological commitments is rarely if ever a concern of working mathematicians. Similarly, the 'grammar' that (necessarily, from the point of view of foundations) accompanies the vocabulary so as to make a 'language' (even, technically, a *formal language*) out of mathematics is usually of little concern to working mathematicians, at any level beyond its simplest constructions; except in unusual circumstances, they appear to act (rightly or wrongly) as though they had been "speaking prose all their lives", even if (to serious researchers in foundations) their prose is mostly baby-talk. In particular, they appear to find (individual and communal) meanings in the mathematical structures they study that in no way derive from any meanings inherent in (or absent from) the "foundations" (that they don't study).

It is surely true that almost all mathematicians working today, if asked on what "foundations of mathematics" their work ultimately relies, would answer "set theory".[8] If questioned further as to *which* "set theory" (there are several), most would probably give some answer (and most of those answers would probably be "ZFC", *i.e.*, "Zermelo–Frankel with Choice"). A smaller number (I am guessing, based on both introspection and years of observing other mathematicians) would even be able to enunciate all the axioms (or axiom-schemes) of ZFC. Progressively smaller numbers would be able to recall (if they had ever known) such definitions as that of "the cumulative hierarchy", the precise statements of various

---

[7]I use "consistent" purely in its common, non-technical sense; the *logical consistency* of the "foundations of mathematics" is something working mathematicians in general (with some very high-powered exceptions!) gladly leave to others.

[8]A small minority would answer "category theory". Although it originated essentially as an organizational tool for mathematicians, and—as a creation of its time—was at first very much described in set-theoretical language, it later was proposed as a better "foundations of mathematics", largely because of its (purportedly) better semantics. Its evident failure (to date) to win much support for that role (as contrasted with its continuing role as organizational tool, and more specifically a tool that helps mathematicians recognize and capitalize upon functional similarities between various mathematical structures that arise in very different contexts) may be just an accident of history, but does at least give *some* support for my claim that working mathematicians don't care much about the (internal) "semantics" of (any given candidate for) "foundations of mathematics".

theorems of ZFC that—logically—form the very foundations of the (ZFC version of the set-theoretical) foundations of mathematics, and the proofs of those theorems.

Yet nearly all working mathematicians, I am quite sure, could readily give definitions in the language of (some) set theory—often, set-theoretical definitions *from their own publications*—of the various mathematical structures they use, or have used, in their own work.[9] Although, as Michael Potter (2004, pp. 4–5) has made clear in his remarkably accessible treatment of set theory (combining its history and its philosophy with a complete mathematical development of considerably more than its rudiments),

> three roles for set theory—as a means of taming the infinite, as a supplier of the subject matter of mathematics, and as a source of its modes of reasoning— have all been important historically

it seems to me that for the "working mathematicians" I have been describing (a set which I can certify to have at least one element) the *contemporary* importance of set theory has been reduced to a combination of weak forms of Potter's second and third roles:[10] it is "a supplier of" a way to *talk about* "the subject matter of mathematics", and "a source of" mathematical modes of reasoning that have become so assimilated that their "source" never comes to mind.

Some examples of structures: projective planes

A set-theoretical definition of a mathematical structure typically looks like the following.

Definition. A *projective plane* is an ordered triple $(\mathcal{P}, \mathcal{L}, \iota)$, in which $\mathcal{P}$ and $\mathcal{L}$ are sets and ι is a binary relation (called *incidence*) between $\mathcal{P}$ and $\mathcal{L}$, satisfying the following axioms.

(PP-1) For all elements P and Q of $\mathcal{P}$, if $P \neq Q$ then there exists one and only one element L of $\mathcal{L}$ such that $P \iota L$ and $Q \iota L$.

(PP-2) For all elements L and M of $\mathcal{L}$, if $L \neq M$ then there exists one and only one element P of $\mathcal{P}$ such that $P \iota L$ and $P \iota M$.

---

[9] Analogues of the feats listed, with category theory replacing set theory, could doubtless be performed by (proportionately) fewer mathematicians at each step.

[10] Working mathematicians have long since taken the "taming" of "the infinite" as a matter of course—with, as always, notable exceptions.

(PP-3) There exist four elements P_0, P_1, P_2, and P_3 of $\mathcal{P}$ such that, for any assignment of three different values from the set of numerals $\{0,1,2,3\}$ to the set of letters $\{i,j,k\}$, there exists no element L of $\mathcal{L}$ such that $P_i \iota L$ and $P_j \iota L$ and $P_k \iota L$.

Conventionally, elements of $\mathcal{P}$ are called "points", elements of $\mathcal{L}$ are called "lines", and $P \iota L$ is read as "P lies on L", "L passes through P", or the like. Formally, nothing is lost by requiring each "line" $L \in \mathcal{L}$ to be identical (as a set) with $\{P \in \mathcal{P}: P \iota L\}$, but this is not mandatory and can be counterproductive.

Fig. 6.1 illustrates a projective plane with exactly seven points and seven lines,[11] and is (I hope) self-explanatory. This example could be called the *Boolean projective plane*, because it bears the same relation to the 2-element *Boolean field* $\{0,1\}$ used in Boolean algebra(which models propositional logic when 0 and 1 are taken to represent False and True respectively, the algebraic field operations $+$ and $\times$ are taken to represent the logical operations of exclusive or and and respectively, and $1+1$ is set equal to 0) as the *real projective plane* $\mathbb{RP}^2$ bears to the field $\mathbb{R}$ of real numbers. Despite this formal relation to logic, I do not know of—and have not managed to confect —a model in which the Boolean projective plane is used for logical purposes (or any other purposes relevant to this volume). However, larger finite projective planes, as well as their generalizations known as *balanced incomplete block designs* (BIBDs), are important in the

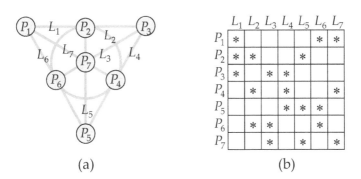

	L_1	L_2	L_3	L_4	L_5	L_6	L_7
P_1	*					*	*
P_2	*	*			*		
P_3	*		*	*			
P_4		*		*			*
P_5				*	*	*	
P_6		*	*			*	
P_7			*		*		*

(a)   (b)

Figure 6.1: The Boolean projective plane, depicted (a) 'geometrically' and (b) 'combinatorially' (by, in effect, its *incidence matrix*).

---

[11]There is a short, but not entirely obvious, proof from the axioms that no projective plane can have fewer than seven points or points or lines.

design of experiments in psychology (see, *e.g.*, Kirk, 2003) and other sciences.

Fig. 6.2 illustrates the just-mentioned real projective plane $\mathbb{RP}^2$; it needs explanation. First I explain (a).

(1) The gray disk depicts simultaneously
  (i) the *unit disk*

$$D := \{(x,y) \in \mathbb{R}^2 : x^2 + y^2 \leq 1\}$$

  in the Cartesian coordinate plane $\mathbb{R}^2$ (the most common intramathematical model of the Euclidean plane), and
  (ii) the upper hemisphere

$$U := \{(x,y,z) \in \mathbb{R}^3 : x^2 + y^2 + z^2 = 1 \text{ and } z \geq 0\}$$

  in Cartesian coordinate 3-space $\mathbb{R}^3$, seen from directly above (and so far away that parallax is negligible).

(2) The black line segment with both endpoints labeled P depicts simultaneously
  (i) a diameter of D, passing (as it must) through the center $O = (0,0)$, and
  (ii) a great semi-circle on U passing through the point $O = (0,0,1)$ (the "North pole").

(3) Each black curve that runs from one point P to the other depicts

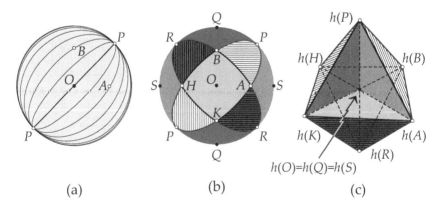

(a)   (b)   (c)

Figure 6.2: The real projective plane. (a) Four (not five) points and twelve lines in $\mathbb{RP}^2$. (b) A subdivision of $\mathbb{RP}^2$ into seven regions, three quadrilateral and four triangular. (c) A *heptahedron*, the image of a particular *topological immersion* h of $\mathbb{RP}^2$ into $\mathbb{R}^3$.

simultaneously
>(i) one of the *major semi-ellipses* of some ellipse in D that has the indicated diameter of D as its major axis (*e.g.*, the two black curves through the points A and B together form a single such ellipse), and
>(ii) a great semi-circle on U that does *not* pass through O.

These examples illustrate particular cases of the technical definitions of the sets $\mathcal{P}$ and $\mathcal{L}$ for $\mathbb{RP}^2$. Namely, $\mathcal{P}$ is to be taken to be either
(i) the union $\overset{\circ}{D} \cup \mathcal{A}$ of the *open unit disk*

$$\overset{\circ}{D} := \{(x,y) \in \mathbb{R}^2 : x^2 + y^2 < 1\}$$

and the set

$$\mathcal{A} := \{\{(x,y),(-x,-y)\} : (x,y) \in C\}$$

of *antipodal pairs* of (Cartesian) points on the *unit circle*

$$C := \{(x,y) \in \mathbb{R}^2 : x^2 + y^2 = 1\}$$

that bounds D, or

(ii) the union $\overset{\circ}{U} \cup \mathcal{A}'$ of the *open upper hemisphere*

$$\overset{\circ}{U} := \{(x,y,z) \in \mathbb{R}^3 : x^2 + y^2 + z^2 = 1 \text{ and } z > 0\}$$

and the set

$$\mathcal{A} := \{\{(x,y,0),(-x,-y,0)\} : (x,y) \in C\}$$

of antipodal pairs of (Cartesian) points on

$$C' := \{(x,y,0) \in \mathbb{R}^3 : x^2 + y^2 = 1\}$$

(*i.e.*, the unit circle in the (x,y)-plane, which bounds U) according as the depiction in (a) is interpreted by choosing (i) or (ii) consistently in (1)–(3). Similarly, $\mathcal{L}$ is to be taken to be either
(i) the union $\mathcal{D} \cup \mathcal{E} \cup \{C\}$ of the set $\mathcal{D}$ of all diameters of D, the set $\mathcal{E}$ of all major semi-ellipses in D like those described above *that are not semi-circles contained in C*, and the one-element set containing the unit circle C as its only member, or
(ii) the union $\mathcal{G} \cup \{C'\}$ of the set $\mathcal{G}$ of great semi-circles in U *that are not semi-circles contained in C'*, and the one-element set containing the unit circle C' as its only member.

Finally, the incidence relation ι between $\mathcal{P}$ and $\mathcal{L}$ must be defined case by case, depending on 'which kind' of point and 'which kind' of line are involved. I omit those details.[12]

Fig. 6.2(b) depicts (the reader's choice of) the same view of $\mathbb{RP}^2$ shown in (a), now subdivided into seven regions distinguished by shading. Four regions are triangular, namely, $\triangle ABP$, $\triangle BCR$, $\triangle CDP$, and $\triangle ADR$. The remaining three are quadrilateral, namely, $\lozenge ABHK$, $\lozenge APHR$, and $\lozenge BPKR$. Of these, only $\lozenge ABHK$ (light gray) appears at first glance to be connected. But, *e.g.*, although $\lozenge APHR$ (medium gray) appears to consist of two disconnected pieces, in fact what appear to be two distinct (antipodal) arcs $\overline{PSR}$ of C (or C') actually are a single arc (along which the two pieces are joined) seen twice—because the "points" of $\mathbb{RP}^2$ that are incident with C (or C') are in fact *antipodal pairs* of (Cartesian coordinate) points; similar considerations apply to $\lozenge BPKR$ (dark gray).

I did not specify any way to measure lengths or angles in $\mathbb{RP}^2$ (and if I had, $\mathbb{RP}^2$ would no longer have been *merely* a projective plane). However, the description in terms of U easily allows both to be taken over wholesale from spherical geometry.[13]

Once that is done, it is not hard to see that it is possible to choose all seven regions to be equilateral and equiangular (and in fact those in the figure were so chosen). Of course in spherical geometry an equilateral equiangular quadrilateral is *not* a "square", and in fact its four equal angles must all be *greater* than a right angle; likewise, an equilateral equiangular triangle has angles *greater* than $60°$. Nonetheless, one obtains a *topologically* (though not metrically)

---

[12]A 'cleaner' way to define $\mathbb{RP}^2$, which is however harder to illustrate, and which demands greater mathematical sophistication insofar as it uses the notions of an *equivalence relation* on a set and the corresponding set of *equivalence classes*, is as follows. Let $S := \{(x,y,z) \in \mathbb{R}^3 : x^2 + y^2 + z^2 = 1\}$ be the unit sphere. Define the binary relation α between S and itself by

$$(x,y,z)\,\alpha\,(u,v,w) \iff \text{either } (u,v,w) = (x,y,z) \text{ or } (u,v,w) = (-x,-y,-z).$$

Then α is an equivalence relation; each α-equivalence class is a pair of antipodal points of S. Let S/α denote the set of α-equivalence classes. Let Γ denote the set of all great circles $G \subset S$. Define the binary relation σ between S/α and Γ by $\{(x,y,z),(-x,-y,-z)\}\,\sigma\,G \iff \{(x,y,z),(-x,-y,-z)\} \subset G$. By using standard facts about S and $\mathbb{R}^3$ (and in particular the observation that a great circle G contains a point $(x,y,z) \in S$ if and only if G contains the antipodal point $(-x,-y,-z)$), it is easy to prove that $(S/\alpha, \Gamma, \sigma)$ is a projective plane. It is, in fact, $\mathbb{RP}^2$.

[13]This is virtually automatic if $\mathbb{RP}^2$ is defined as in note 12.

correct model of $\mathbb{RP}^2$ by attaching, in the abstract, three Euclidean squares and four Euclidean equilateral triangles to each other along pairs of edges according to the pattern in Fig. 6.2(b); and as indicated in Fig. 6.2(c), the result can even be realized in Cartesian coordinate 3-space $\mathbb{R}^3$ if one is willing to have 'apparent intersections' of each square with each of the other squares along their two diagonals (with a so-called 'triple point' at which all three squares intersect).

Shepard (1978) proposed that $\mathbb{RP}^2$ be used for models in the study of perception. On pp. 180–181 I sketch an application of $\mathbb{RP}^2$ to a new—and so far only imperfectly developed—"logic of ambivalence" (see Table 2.1, note 19, on p. 45).

On mathematical spaces[14]

The preceding discussion of projective planes provides not just explicit examples of how set-theoretical definitions are used mathematically, but also many examples (there not drawn explicitly to attention) of how (parts of) such definitions can acquire or give meaning in the course of their mathematical and para-mathematical interactions with other structures (some mathematical but having definitions that remain out of attention, others 'in the world' and defined —if at all—non-mathematically). In this section I pick up just one of those dropped threads.

Many mathematical structures have been called 'spaces' (usually with some modifier) since at least the discovery of non-Euclidean geometries—notably, the real projective plane $\mathbb{RP}^2$ and the *real hyperbolic plane*—early in the 19th century, and well before there were any set-theoretical "foundations of mathematics". By mid 20th century, mathematical 'spaces' were common not only in geometry but in algebra ('vector spaces'), mathematical analysis ('Hilbert spaces', 'Banach spaces', 'Hardy spaces', and many other kinds of vector spaces with extra structure, as well as 'metric spaces' and 'measure spaces'), abstract algebra ('representation spaces', 'prime ideal spaces'), probability theory ('probability spaces'), mathematical physics ('phase spaces'), and especially topology—the quintessential mathematics of the 20th century—in its many manifestations: point-set, combinatorial, algebraic, geometric, and differential. No single strictly mathematical property is shared by these many kinds

---

[14]Some of this section recapitulates a similarly titled section in Rudolph (2006b).

of 'spaces', but mathematicians in general seem content to agree that the metaphor is broadly appropriate.

Typically, when mathematicians call some mathematical structure S a *space* (here to be called a *mathematical space* in the hopes of averting confusion), they understand it to share, in some sense and to some degree, the following rather general pre-mathematical properties of the ordinary space of our daily experience.[15]

(1) A mathematical space is like a box that can 'contain' other sorts of 'things'.

(2) A mathematical space is like a stage on which various 'events' can 'happen' (*e.g.*, 'things' can 'move') in the course of time.[16]

Properties (1) and (2) are essentially extrinsic to a candidate S for 'spacehood': they depend almost entirely not on *what S is* but on *how S is used*.[17] In contrast, a third general property is chiefly intrinsic.

(3) A mathematical space 'has extent', and can (usually) be 'subdivided'; a 'piece' of a mathematical space, though (usually) of smaller 'extent', still has in its own right the quality of being a mathematical space.

Naturally, the interpretation of (3) depends on the meaning given to 'extent', 'piece', etc., and in that sense it is somewhat extrinsic.

What mathematicians typically try hard *not* to do, when calling a mathematical structure a 'space', is to attribute to that structure other properties of ordinary space that are not explicitly demanded by the context in which the structure is being used. Model-making scientists, be they physical, life, or social scientists, are often less fastidious when they adopt the metaphor of 'space' for mathematical models in their own disciplines: in contrast with mathematicians, they tend to incorporate into their models not only the general properties (1)–(3) of ordinary space, but also some or all of the following special properties.

---

[15]Many presuppositions are packed into the phrase "the ordinary space of our daily experience" and its variants, and most if not all of them are probably unjustifiably broad, particularly if "daily experience" is read so as to naïvely ignore or tendentiously suppress the considerable role of linguistic framings (cultural and sub-cultural, semi-permanent and evanescent) in that "experience". Still, the phrase and its variants have a reasonably well delimited denotation that is widely understood (until it is examined overly closely), so I take the risk of using it here.

[16]Somewhat confusingly, "time" is very often thought of as a mathematical space by mathematicians. See Chap. 10, p. 308, and Rudolph (2006a).

[17]In particular, one and the same mathematical structure S can be called a 'space' or not depending on the use to which it is being put.

(4) Ordinary space has (or can have imposed upon it) *metric* properties, including (but not limited to) numerical measures of distance, area, volume, and other forms of 'extent'; a mathematical space need not.

(5) Ordinary space has (or can have imposed upon it) *geometric* properties, such as notions of 'straightness' and 'curvature', 'convexity' and 'concavity', 'collinearity', 'congruence', and the like; a mathematical space need not.

(6) Ordinary space has properties of *continuity, homogeneity* (*i.e.*, indistinguishability among locations *per se*) and *isotropy* (*i.e.*, indistinguishability among directions *per se*);[18] a mathematical space need not.[19]

(7) In ordinary space, a 'point' is *atomic*, with no internal structure; in many important examples of mathematical spaces, each 'point' is a complex structure in its own right.

Although the policy of endowing mathematical spaces used as

---

[18]Or, at least, horizontal directions are (among themselves, ignoring their 'contents') indistinguishable in ordinary space; as Shepard (1992, p. 500) points out, gravity makes verticality salient for surface-dwellers (or rather, for those surface-dwellers that live above the "nanoscale" at which van der Waals forces, Brownian motion, etc., have effects much stronger than those of gravity).

[19]In this connection, it is almost incredible—to a mathematician educated in the second half of the 20th century—to read that, for instance, Bertrand Russell (1896)

> [i]n his first published paper [...] analyses the axioms of Euclidean geometry [...] and finds that some of the axioms are certainly true, and in particular *a priori* true, "for their denial would involve logical and philosophical absurdities" (p. 3). He classifies for instance the homogeneity of space as *a priori* true, the "want of homogeneity and passivity is . . . absurd: no philosopher has ever thrown doubt, so far as I know, on these two properties of empty space [...]." (Lakatos, 1962, p. 168; the unbracketed ellipsis points, and the italics, are Lakatos's.)

Moreover, Russell (1896, p. 1) purports to come to his conclusions *even though*

> we are not concerned with the correspondence of Geometry with fact; we are concerned with Geometry simply as a body of reasoning, the conditions of whose possibility we wish to examine [...] we have to do with the conception of space in its most finished and elaborated form, after thought has done its utmost in transforming the intuitional data.

Probably the best (though very difficult) course of action for the modern mathematician, incredulous in the face of what appears to be such an enormous blind spot, would be to take an appropriate modification of Stallings's advice (quoted on p. 64), and 'cultivate techniques leading to the abandonment' of one's own mechanisms for maintaining one's own (surely numerous) blind spots.

models with some or all of the special properties (4)–(7) has often been harmless, and occasionally useful, in the natural sciences, I see no evidence that it has often been useful (and some evidence that it has sometimes been harmful) in the human sciences. In any case, mathematicians typically see the denial, to a given mathematical space, of some or all of these special properties—especially (7)—as entirely normal, and frequently commendable.[20]

An application of $\mathbb{RP}^2$ plane to ambivalence logic

To conclude my discussion of the nature of mathematical structures —and incidentally to continue the segue, into a discussion of the mathematical structure of nature, that I have already begun with the cautionary distinctions (4)–(7) on the preceding page between properties of a typical "mathematical space" and those of "the ordinary space of our daily experience"—I briefly explore a certain fact about the Euclidean plane. Euclid neither assumed this fact as an axiom nor stated it as a theorem: and, actually, transplanted literally to the non-Euclidean geometry of $\mathbb{RP}^2$, the 'fact' becomes false. A generalization of this fact (in the Euclidean plane) has been implicitly relied on in various diagrammatic approaches to logic, in particular, the eponymous diagrams of John Venn (1866) and the "diagrammatic reasoning" of C. S. Peirce (1905). Using the failure of this 'fact' in $\mathbb{RP}^2$, I construct a diagrammatic method that may be a useful step towards the development of a "logic of ambivalence".

The fact is this: *any line L in the Euclidean plane* $\mathbb{E}$ *separates* $\mathbb{E}$ *into two connected pieces that are disconnected from each other.* Instead of formalizing this statement (which would involve, in the first place, making sense of "connected"), I leave it to the reader to check that it is 'obviously true' when $\mathbb{E}$ is modeled, as I have done consistently,

---

[20]Euclid's Definition 1, σημεῖον ἐστιν οὗ μέρος οὐδέν, states that "a point is that which has no parts". But, *e.g.*, in all the definitions of $\mathbb{RP}^2$, at least some points —being sets with two elements—have non-trivial, if meager, mereologies. A class of examples of mathematical spaces having far more radically "non-atomic" points than those of $\mathbb{RP}^2$ is that of *configuration spaces* of (mathematical or physical) systems of various sorts. In a configuration space, each point is a single configuration of the entire system. See Wehrle, Kaiser, Schmidt, and Scherer (2000) for an application to the dynamics of human affect that—in effect—constitutes a partial exploration of a mathematical space of "schematic facial expressions consisting entirely of theoretically postulated facial muscle configurations" (p. 105).

by the Cartesian coordinate plane $\mathbb{R}^2$.[21]

The generalization of the fact is this: *any simple closed curve*[22] K in $\mathbb{E}$ separates $\mathbb{E}$ into two connected pieces that are disconnected from each other. This generalization is the *Jordan Curve Theorem*, and requires non-trivial topological techniques for its proof. Of course for colloquially 'simple' simple closed curves like circles, the Jordan Curve Theorem does have simple proofs (but not until Euclid's axioms are supplemented by an "axiom of completeness" like that introduced by David Hilbert; again, that axiom is satisfied when $\mathbb{E}$ is modeled by $\mathbb{R}^2$).

The reliance that was put (unknowingly) on the Jordan Curve Theorem by both Venn and Peirce is that each assumed that, by drawing a simple closed curve $\mathcal{K}$, he had separated a planar region from a (single) complementary region; and that, therefore, he could assign an arbitrary proposition P (or logical formula) to one of the two regions, and its negation $\sim P$ to the other, without any possibility of the two contradictory propositions confronting each other anywhere except from opposite sides of the (barbed-wire?) fence $\mathcal{K}$. Of course, both of them were right (and provably so, by elementary methods) so long as they stuck to circles(or convex curves); but I don't believe either of them ever actually considered the potential danger lurking in a seemingly harmless extension to more general simple closed curves (which might, after all, have turned out to be practically necessary if their logical schemes had ever been applied to really complex logical arguments).

The failure of the Jordan Curve Theorem in the real projective plane $\mathbb{RP}^2$ is illustrated in Fig. 6.3. Certainly there are *some* Jordan curves in $\mathbb{RP}^2$; for instance, in the models described above in my discussion of Fig. 6.2(a), any simple closed curve entirely contained in the open unit disk $\mathring{D}$ divides $\mathbb{RP}^2$ into two pieces. But there are

---

[21]Hint: first consider the—apparently—special case in which L is the x-axis $\{(x,0): x \in \mathbb{R}\}$, and show that each of the two *open half-planes* $\mathbb{E}_+ := \{(x,0): x \in \mathbb{R}$ and $x > 0\}$ and $\mathbb{E}_- := \{(x,0): x \in \mathbb{R}$ and $x < 0\}$ is "connected", but that their union $\mathbb{E}_+ \cup \mathbb{E}_-$ is "disconnected". Then generalize....

[22]"Simple closed curve" is a topologist's term of art; again, I will not formally define it here. Circles, ellipses, and convex polygons (e.g., triangles, parallelograms, and all regular polygons) are simple closed curves, but so are many much less obviously "simple" curves, such as the one in Fig. 8.1(b). A curve is *simple* if it does not 'cross itself', and *closed* if it has no endpoints. Considered as curves, the Arabic numerals 0 and 8 are closed, 0, 2, 3, 5 and 7 are simple, and (in this font) 1, 4, 6, and 9 are neither simple nor closed.

also *non*-Jordan simple closed curves in $\mathbb{RP}^2$, most notably its lines, as shown in Fig. 6.3(a).

Next consider Fig. 6.4. In Fig. 6.4(a) I have drawn the most common example of an ordinary Euclidean Venn diagram, with (contrary to common usage) explanatory text on each circle. Each of three independent propositions (or classes) A, B, and C corresponds to the open disk inside one of three circles, and its denial (indicated by $\overline{A}$, $\overline{B}$, and $\overline{C}$ respectively) to the (infinite) open region outside that circle. Overlaps of regions correspond to conjunction of propositions. The text on each circle specifies that it separates one of the propositions from its denial.

Now, there is a standard method (known as *stereographic projection*) that allows the entire Euclidean plane to be modeled by the points of the unit sphere $S \subset \mathbb{R}^3$ (with one exception, a so-called 'point at infinity'). The symmetry of Fig. 6.4(a) allows an appropriate stereographic projection to carry it onto the highly symmetric arrangement of three great circles on S depicted in Fig. 6.4(b). There are still eight regions, but now they all look alike; I have omitted the labels on the back side of the sphere (and dashed the parts of the great circles on that side), as well as the labels on the circles themselves. Note that the region antipodal to ABC is $\overline{ABC}$, that antipodal to $\overline{A}BC$ is $A\overline{BC}$, and so on.

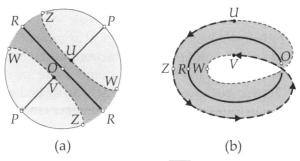

(a)   (b)

Figure 6.3: (a) The (heavy black) line $\overline{OR}$ through O and R does not separate $\mathbb{RP}^2$; for instance, the points U and V on the (light black) $\overline{OP}$, though "locally separated" by $\overline{OR}$, are the endpoints of an interval on $\overline{OP}$ with P as its midpoint. (b) It is also possible to connect U to V by a curved path that stays "close" to $\overline{OR}$ without crossing it; this alternative depiction of a (dark gray) *regular neighborhood* N of $\overline{OR}$ in $\mathbb{RP}^2$ makes it clear that N is a *Moebius band* with U and V on its boundary, joinable there as indicated.

Finally, Fig. 6.4(c) shows the result of passing from the sphere S to $\mathbb{RP}^2$. Antipodal regions have become identified. The great circles (which are Jordan curves on the sphere, separating it into two hemispheres) have become lines in $\mathbb{RP}^2$ (which do not separate $\mathbb{RP}^2$). It no longer makes sense to label a line as "the boundary between" one of the propositions and its denial; there may still be barbed wire on the fence, but now there is a way around it. Each polarity —as defined on p. 56: "a dimension equipped with a mapping to the dimension spanned by 'good' (or 'desirable', 'favored', etc.) and 'bad' (or 'undesirable', 'disfavored', etc.)", that mapping in this case being implemented by the choice of P rather than $\bar{P}$ as the unmarked form—collapses back into a (mere?) dimension, interpretable as "ambivalence between P and $\sim P$".

Whether anything can be made of this, I do not know; I offer it as an example of the use of one of many un- or under-utilized mathematical resources available for modeling in the human sciences.

On the mathematical structure of nature

In our Introduction (p. 9), we proposed that *structure* be defined recursively, starting from "*pattern* ... defined however you like— formally, informally ('I know it when I see it')" or "not at all". For

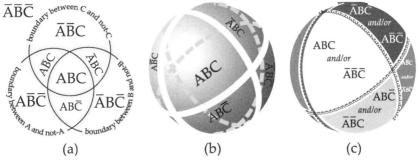

(a)   (b)   (c)

Figure 6.4: (a) An elaborated depiction of the most common Venn diagram. (b) The same diagram, transposed to spherical geometry as symmetrically as possible: the three circles are *great circles*, and all angles are (spherical) right angles. (c) By projecting the sphere in (b) onto a plane, we obtain a new kind of diagram consisting of three *projective lines* in the projective plane.

an evolutionary ontology, "I know it when I see it" is a natural choice.

The animal *Funktionskreis*

As Rosemarie Sokol Chang (2009a, p. xii) has noted,

> Nothing—except some charm of using another language term—can be gained from re-labeling environment by *Umwelt*—unless we look at the corresponding notion of the functional circle (*Funktionskreis*) that von Uexküll used to describe the dynamics of the ongoing processes within the *Umwelt*.

Following her editorial lead, Rudolph (2009) depicted a generic *Funktionskreis*, essentially duplicated in Fig. 6.5(a) (itself closely based on Uexküll, 1920b, Fig. 3, p. 155). The dotted pentagon has been added here; replacing its contents with those of the pentagon in Fig. 6.5(b) produces essentially Fig. 4 of Uexküll (1920b, p. 157).

It might seem that Fig. 6.5 can apply only to literal organisms, insofar as there are schematic elements labeled "organ", and only literal organisms have literal "organs". In fact, however, Uexküll —in different ways at different times—allowed himself sufficiently broad notions of "organ" and "organism" to make it possible for at least some kinds of intra-specific animal organism-groups to have

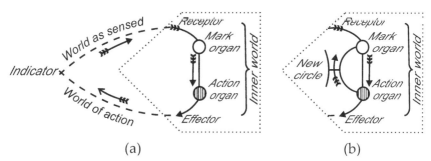

(a)   (b)

Figure 6.5: Animal *Funktionskreis*es (adapted from figures in Uexküll, 1920b). (a) Generic. (b) More structured, as "[i]n the highest animals" (Uexküll, 1920b, p. 157).

"organs" and so (potentially) *Funktionskreise*s.[23] I will allow myself the same liberty, exercised rather differently: the organism-groups to which I extend the notion of "functional circle"/*Funktionskreis* and the related notion of in-dwelt "environment"/*Umwelt* are certain (intentional or contingent) groups of human beings, existing at various temporospatial scales that need not be (and often are not) coextensive with the life-space/lifetime of any particular individual human-being-as-organism.

My justification for the extension is purely formal: it is the observation that an animal *Funktionskreis* as depicted in Fig. 6.5 is *teleologically*, *topologically*, and *logically* an hermeneutical circle. This observation is somewhat problematic—not only has the term "hermeneutical circle" not been defined in this book, but even a cursory survey of the literature shows that it has no single, universally accepted definition, and certainly no single, universally accepted *formal definition*.[24] That being the case, the meaningfulness of asserting that an animal *Funktionskreis* "is … an hermeneutical circle", in the three ways listed, is thrown into doubt. Yet the situation is not actually that bad, as will be shown in detail by Rudolph (in preparation).

---

[23]See, in addition to Uexküll (1920b) and its English translation (Uexküll, 1920/1926), the second German edition (Uexküll, 1928, heavily revised, and never published in English translation); Uexküll (1920a), also never published in English translation, and the apparent source of much of the new matter in Uexküll (1928), which Stjernfelt (2007, p. 459, note 269) has described—with entire justice, as near as I can judge—as "no less than a fascist biologist doctrine of state"; and Uexküll (1933), a revised reissue of Uexküll (1920a) published by the Hanseatic League, which concludes "with the hopeful observation that Adolf Hitler's rise to power was bringing an end to the forms of pathological decay Germany had been displaying for years" (Burkhardt, 2005, p. 247).

I find it curious that never, apparently, did Uexküll—who otherwise had rich and extensive, if in some respects terribly flawed, foresight and imagination—consider the usefulness (or necessity) of extending the notions of *Funktionskreis* and *Umlaut* either to *inter*-specific animal organism-groups or to non-animal organisms and organism-groups. To retain biological (not to mention ecological) usefulness, both restrictions should be lifted; see, *e.g.*, recent work on animal microbiomes (the human case is described by Robinson, Bohannan, & Young, 2010 and in resources available through the Human Microbiome Project) and on "neuronal aspects of plant life" (Baluška, Mancuso, & Volkmann, 2006).

[24]In particular, it has no *mathematical* definition that is in any sense "universally accepted". Of course it has various definitions that are partly or fully mathematized; see, *e.g.*, Arecchi (2007) and Klüver and Klüver (2010) for proposals from physics and from mathematical economics/artificial intelligence, respectively.

Human and human-group *Funktionskreis* and *Umwelt*

Humans, as social beings living in cultures, surely have—even at the raw biological level—peculiarities in their *Funktionskreis*es and *Umwelt*s that non-social animals do not have (a point that Uexküll made much of, at various times). It is not novel to suggest that human groups, too, can usefully be said to have *Funktionskreis*es and *Umwelt*s (this theme recurs throughout Sokol Chang, 2009b). What may be novel is to suggest evolutionary ontology in general, and mathematical modeling in particular, as particularly striking and important instances of the coevolution of *Funktionskreis*es and *Umwelt*s of human groups of many sorts. Reading this volume (particularly the three chapters that follow) with this perspective in mind may suggest some interesting working hypotheses.

References

Arecchi, F. T. (2007). Physics of cognition: Complexity and creativity. *The European Physical Journal – Special Topics, 146,* 205–216. (From the issue entitled New Trends, Dynamics and Scales in Pattern Formation)

Azzouni, J. (1994). *Metaphysical Myths, Mathematical Practice: The Ontology and Epistemology of the Exact Sciences.* Cambridge, GB: Cambridge University Press.

Azzouni, J. (2004). *Deflating Existential Consequence: A Case for Nominalism.* New York, NY: Oxford University Press.

Baluška, F., Mancuso, S., & Volkmann, D. (2006). *Communication in Plants: Neuronal Aspects of Plant Life.* Heidelberg, DE: Springer.

Bawden, H. H. (1904). The meaning of the psychical from the point of view of the functional psychology. *The Philosophical Review, 13*(3), 298–319.

Bourbaki, N. (1949). Foundations of mathematics for the working mathematician. *Journal of Symbolic Logic, 14*(1), 1–8.

Burkhardt, R. W. (2005). *Patterns of Behavior: Konrad Lorenz, Niko Tinbergen, and the Founding of Ethology.* Chicago, IL: University of Chicago Press.

Corcoran, J. (1982). *Untitled [Review of the book* Mathematics: The Loss of Certainty*].* Retrieved from http://www.ams.org/mathscinet-getitem?mr=584068 (On-line resource, not paginated)

Human Microbiome Project. (2011). *Human Microbiome Project DACC—Home.* Retrieved November 2, 2011, from http://www.hmpdacc.org/

Kirk, R. E. (2003). Experimental design. In J. A. Schinka & W. F. Velicer (Eds.), *Handbook of Psychology: Research Methods in Psychology. Vol. 2* (pp. 3–32). Hoboken, NJ: John Wiley & Sons. (Series editor: Irving B. Weiner)

Kline, M. (1973). *Why Johnny Can't Add: The Failure of the New Mathematics.* New York, NY: St. Martin's Press.

Kline, M. (1977). *Why the Professor Can't Teach: Mathematics and The Dilemma of University Education.* New York, NY: St. Martin's Press.

Kline, M. (1980). *Mathematics: The Loss of Certainty.* New York, NY: Oxford University Press.

Klüver, J., & Klüver, C. (2010). *Social Understanding: On Hermeneutics, Geometrical Models and Artificial Intelligence.* Heidelberg, DE: Springer.

Lakatos, I. (1962). Infinite regress and foundations of mathematics. *Proceedings of the Aristotelian Society, Supplementary Volumes, 36,* 155–184. (Reprinted with editorial corrections in *Mathematics, Science, and Epistemology: Vol. 2 of Philosophical Papers of Imre Lakatos,* J. Worrall and G. Currie, Eds., Cambridge University Press, 1980, Chap. I, pp. 3–23)

Lorenz, K. (1941). Kants Lehre vom Apriorischen in Lichte gegenwärtiger Biologie [Kant's doctrine of the *a priori* in the light of contemporary biology]. *Blätter für Deutsche Philosophie, 15,* 94–125. (Reprinted in Lorenz & Wuketits, 1983; translated as Lorenz, 1962)

Lorenz, K. (1962). Kant's doctrine of the *a priori* in the light of contemporary biology (C. Ghurye, Trans.). In L. V. Bertalanffy & A. Rapoport (Eds.), *General Systems: Yearbook of the Society for General Systems Research* (Vol. VII, pp. 23–35). Ann Arbor, MI: Society for General Systems Research.

Lorenz, K., & Wuketits, F. M. (1983). *Die Evolution des Denkens [The Evolution of Thought]* (2nd ed.). München, DE: R. Piper.

Maddy, P. (2007). *Second Philosophy: A Naturalistic Method.* New York, NY: Oxford University Press.

Peirce, C. S. (1905). Prolegomena to an apology for pragmaticism. *The Monist, 16,* 492–546.

Potter, M. (2004). *Set Theory and its Philosophy.* Oxford, GB: Oxford

University Press.

Robinson, C. J., Bohannan, B. J. M., & Young, V. B. (2010). From structure to function: The ecology of host-associated microbial communities. *Microbiology and Molecular Biology Reviews*, 74(3), 453–476.

Rudolph, L. (2006a). The fullness of time. *Culture & Psychology*, 12, 157–186.

Rudolph, L. (2006b). Spaces of ambivalence: Qualitative mathematics in the modeling of complex, fluid phenomena. *Estudios Psicologías*, 27, 67–83.

Rudolph, L. (2009). A unified topological approach to *Umwelt*s and life spaces, Part I: *Umwelt*s and finite topological spaces. In R. I. S. Chang (Ed.), *Relating to Environments: A New Look at* Umwelt (pp. 185–206). Charlotte, NC: Information Age Publishing.

Rudolph, L. (in preparation). Funktionskreis as *hermeneutische Zirkel.*

Russell, B. (1896). The logic of geometry. *Mind*, 5(17), 1–23.

Shepard, R. N. (1978). The circumplex and related topological manifolds in the study of perception. In S. Shye (Ed.), *Theory Construction and Data Analysis in the Behavioral Sciences* (pp. 29–80). San Francisco, CA: Jossey-Bass.

Shepard, R. N. (1992). The perceptual organization of colors: An adaptation to regularities of the terrestrial world? In J. H. Barkow, L. Cosmides, & J. Tooby (Eds.), *The Adapted Mind* (pp. 495–532). New York, NY: Oxford University Press.

Sokol Chang, R. I. (2009a). Introduction. In R. I. Sokol Chang (Ed.), *Relating to Environments: A New Look at Umwelt* (pp. vii–xii). Charlotte, NC: Information Age Publishing.

Sokol Chang, R. I (Ed.). (2009b). *Relating to Environments: A New Look at Umwelt*. Charlotte, NC: Information Age Publishing.

Stjernfelt, F. (2007). *Diagrammatology: An Investigation on the Borderlines of Phenomenology, Ontology, and Semiotics*. Dordrecht, NL: Springer.

Uexküll, J. von. (1920a). *Staatsbiologie [Biology of the State]*. Berlin, DE: G. Paetel. Retrieved from http://babel.hathitrust.org/cgi/pt?id=njp.32101068558020

Uexküll, J. von. (1920b). *Theoretische Biologie [Theoretical Biology]*. Frankfurt, DE: Suhrkamp.

Uexküll, J. von. (1926). *Theoretical Biology* (D. L. MacKinnon, Trans.).

New York, NY: Harcourt, Brace & Company, Inc. (Original work published 1920)

Uexküll, J. von. (1928). *Theoretische Biologie [Theoretical Biology]*. Berlin, DE: J. Springer. (Revised ed.)

Uexküll, J. von. (1933). *Staatsbiologie [Biology of the State]* (2nd ed.). Hamburg, DE: Hanseatische Verlagsanstalt.

Venn, J. (1866). *The Logic of Chance: An Essay on the Foundations and Province of the Theory of Probability, with Especial Reference to its Logical Bearings and its Application to Moral and Social Science*. London, GB: Macmillan.

Wehrle, T., Kaiser, S., Schmidt, S., & Scherer, K. R. (2000). Studying the dynamics of emotional expression using synthesized facial muscle movements. *Journal of Personality & Social Psychology*, *78*(1), 105–119.

7

Structure and Hierarchies in *Ganzheitspsychologie*

Rainer Diriwächter

Renewed discussions of holistic approaches to psychology in an effort to create theoretical syntheses (see Diriwächter & Valsiner, 2008) should help contribute to the ongoing endeavors to incorporate qualitative mathematical models for the social sciences. In this chapter, I let one such holistic approach—*Ganzheitspsychologie*—guide my discussion about the importance of considering underlying structures for theory construction.[1]

I start by outlining a model (Lothar Kleine-Horst's Acht Welten Model,[2] or AWM) that captures the move from basic configurations to higher level existence. This model is broken down via a selective analytical approach in order to highlight what I consider to be important structural and hierarchical components vital to psychological analyses. Particular attention is given to person in environment, with an emphasis on the relevance of cultural psychology. The chapter ends with a discussion of one example taken from Herbart's mathematical model of the psyche.

My hope is that through these discussions, readers with an interest in qualitative mathematics will be able to draw new ideas and important considerations for working on formulæ that should

---

Keywords: development, "Eight-World Model", feelings, *Ganzheitspsychologie*, Gestalt, hierarchical nature of the whole, holistic psychology, structure.

[1]Literally, *Ganzheitspsychologie* is *holistic psychology*; the original term is kept to emphasize its theoretical origins in the second Leipzig School of Psychology.

[2]That is, "Eight-World Model".

eventually help capture the complex psychological processes of the person in his/her world.

The nature of the Whole

When speaking of holistic approaches, one pertinent question that consistently arises is, of what does the 'Whole' actually consist? Indeed this has been a difficult concept to answer. That is, how does one define a whole? Can we define the whole as the sum of its parts, and if so, how do we define these parts? Or is the whole to be defined precisely through the lack of any concrete parts and boundaries (it is diffuse; Krueger, 1928/1953); and if yes, how should we study such a diffuse reality?

The discipline of *Ganzheitspsychologie* has proposed four central tenets along which investigators can orient themselves: holism — structure — development — feelings (see Diriwächter, 2008). Briefly put, the tenet of holism emphasizes the primacy of the whole over its parts. We must remember that elements or parts, which are predominantly the focus of investigations, receive their properties in relation to the whole. Yet not keeping in mind the genetic and functional primacy of the whole over its parts may lead the investigator to lose vital and necessary properties in his/her analysis. In short, the tenet of holism (1st tenet of *Ganzheitspsychologie*) provides the underlying meaning attributed to the focus of investigation. Without a look towards the whole, its elements become artificial. In the realm of human life, the whole encompasses a living structure (2nd tenet) that is open to development (3rd tenet). Life means continuous development and to neglect this fact leads to inaccurate results in research practices. Lastly, the experienced qualities of the whole are manifested in our feeling states (4th tenet). Feelings are ever-present and their given states are bound to our biology and relation to the environment.

Although in this chapter my focus is on the structural tenet of *Ganzheitspsychologie*, it should be obvious that structure ought never to be seen as independent of the other three tenets proposed.

The importance of structure

Structure is to be seen as the conditions out of which human experiences emerge. These structural configurations are not to be seen as closed systems, which would negate the tenet of development; rather life's structure is open to development, therefore not static. As we progress along the irreversible time continuum, the nature of the structure transforms. As such, structure is the carrier of development: it is through an analysis of the structure that we can see how development occurs. Yet somewhat perplexingly, in order to see structural configurations, most researchers are forced to stop the time continuum by means of data collection to gain a 'snap-shot' of an approximation of what the given structure presently looks like.

When examining structure, it needs to be reiterated that the observer himself/herself is embedded in a structural reality out of which he/she investigates particular occurrences. This little truth becomes apparent when we speak of the *Weltanschauungen* of particular persons. For science, such *world-views* manifest themselves in the underlying axioms of research methodologies that are adopted in order to study nature in all its multiplicity. We may be reminded of Pepper's (1942) four *world-hypotheses* (formism—mechanism —organicism—contextualism), which have competed throughout the ages yet which have proven fairly inadequate for thorough explanatory power if they are not combined.

The importance of hierarchies

Just as development proceeds on several levels simultaneously (*e.g.*, phylo-, onto-, and microgenetically), structure equally is not one-dimensional. Fundamental to understanding the whole is its hierarchical nature. That is, the whole can be layered, with each layer representing a sub-whole that is nevertheless dependent on the larger whole to which it belongs. For example, we can say that the cosmos represents the infinite nature of the whole but for social sciences there is little utility in trying to incorporate the entirety of the universe in theory construction. Our planet's ecosystem is relatively self-contained yet dependent on the outside influences of larger cosmic events (*e.g.*, our sun, asteroids, etc.). As we move down the hierarchical ladder, we eventually arrive at the domain to which the social sciences have devoted themselves: human life and society. What we should not forget, however, is that the layer of human life or society is not independent of nature in which it is embedded. Human civilization is dependent on nature (the larger whole) and while humans have learned to cultivate the natural state (see Valsiner, 2000), a natural event can devastate the environment in which humans live (as shown in August 2005 by Hurricane Katrina, when it destroyed large parts of New Orleans and other areas in the United States). Thus, the whole (the earth ecosystem) dominates over its parts (in our case, human society). Similarly, the human organism represents a system that is not independent of the larger whole to which it belongs. For example, the structural configurations of society and culture—the values, norms, customs, and traditions that are transmitted from one generation to the next—impact each human life on a daily basis. Of course, the boundaries of each layer (*e.g.*, see Fig. 7.1) of the whole are only as concrete as the investigator defines them to be. The whole, in its entirety, is in fact much more diffusing. It is also important to note that the relationship between the different layers of the whole is not unidirectional. While the whole dominates and gives the lower level hierarchies their properties, sub-wholes equally have the ability to change the structural configurations of the whole through the axiom of development.

To reiterate, the layers just described can be seen as sub-wholes which are not, however, independent of a greater whole. A question that may emerge pertains to how these layers are 'linked' to each other. One proposition is that we can study these layers and their corresponding hierarchy in terms of functionality—that is, how they are functionally linked. From a meta-perspective, the four-

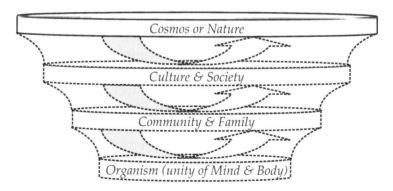

Figure 7.1: Hierarchical nature of the Whole

level, four-"manner of being" model proposed by Kleine-Horst (2001, 2004, 2008) provides a good framework for conceptualizing the structural and hierarchical relationship of whole-parts. The aspects necessary for understanding this model include not only four *evolutionary levels* but also four *manners of being*: (a) order, (b) matter, (c) function, and (d) phenomenon/consciousness. As Fig. 7.2 shows, the evolutionary level and the manners of being interact to form a total of eight well-definable *domains* (or "worlds" —the *Acht Welten Model* [AWM]). It is important to emphasize that the hierarchies Kleine-Horst proposes are developmental, that is, evolutionary in nature. The model attempts to highlight the move from basic configurations to higher abstract existence.

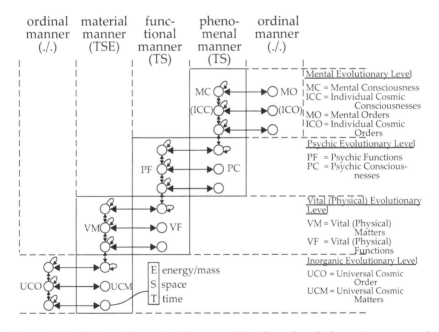

Figure 7.2: Kleine-Horst's schema of the four-level, four-"manner of being" model (adapted with the author's permission from Kleine-Horst, 2008, p. 25, Fig. 1.11). Vertical axes indicate the quadrialism of the ordinal—material—functional—phenomenal manners of being, and horizontal axes indicate the quadrialism of the inorganic—physical—psychic —mental evolutionary levels. The interfaces of the two groups of entities form eight "worlds" (AWM), of which the branched hierarchies are symbolized through three circles.

In Kleine-Horst's model, the lowest world of matter is inorganic or Universal Cosmic Matter (UCM). We can say that out of a particular Universal Cosmic Order (UCO)—consisting of the conditions for energy/mass, space, and time, matter comes into existence. Matter is an actualization of the hierarchy of the conditions for matter (UCO) and represents the lowest evolutionary level. This follows the theorem of isomorphism, according to which the relationships between the left and the right world of every evolutionary level are the same (see Fig. 7.2). Hence, both worlds— UCO and UCM—represent the lowest of four evolutionary levels. The lowest entity of inorganic matter is the TSE hierarchy: the hierarchy of the dimensions (from bottom up) "time" (T), "space" (S), and "energy/mass" (E). The TSE hierarchy comes to existence through the actualization of the hierarchy of the TSE-conditions for matter (UCO). Thus, non-matter (the UCO) represents the conditions for matter, which seems to agree with the modern speculations of quantum theorists. The higher levels of UCM develop together with their correlates of the UCO, thereby forming a broadly branched-out hierarchy of entities which are combinations of the basic TSE hierarchies. Again, the TSE hierarchy is the basic unity of matter: Each elementary particle and each quantum field demonstrates the hierarchical TSE-structure.

Furthermore, one could say that—evolutionarily speaking—the four-manner four-level system can be interpreted as an actualization and a subsequent progressive deactualization of the UCO hierarchy whose actualization began with the Big Bang (thereby producing the basic TSE hierarchy). Such a progressive deactualization of the UCO-condition of the TSE hierarchy implies that with a cessation of the hierarchically ordered dimensions (time-space-energy/mass), from the top down (first energy, then space, then time), frees the restrictiveness of the respective entities and leads to the formation of new entities (*e.g.*, more content is contained in each further step— see below).

The material entities of the uppermost level of the universal cosmic matter (or UCM) have the capability to reproduce themselves (see the reflexive arrows in Fig. 7.2), thus allowing material entities to exist at a 'higher level' (*i.e.*, in the world of Vital Matter—VM). For Eigen and Schuster (1979), the circular processes between nucleic acids and proteins (the "hypercycles") entail the foundation for life. This allows for a new primary hierarchy, namely that of life, which is manifested through the body of an organism, which develops both phylogenetically as well as ontogenetically. We, thus, can say that inorganic matter (UCM) can take on a specific configuration that eventually develops into organic matter (vital matter—VM). That the two are connected can be exemplified through the large quantity of H_2O that is contained within our own bodies (one can say that these are the traces of the universal cosmos within us).

At the second evolutionary level, the Vital Matters hierarchy of time-space-energy/mass creates *function*. That is, we can say this new hierarchy (*e.g.*, the organism) can be studied in terms of its functionality (Vital Functions—VF). For example, the heart (VM) has the function to pump blood in order to keep the circulation going (VF). Yet at the same time, the functional manner differs from the material manner in that it is bound on matter but does not itself possess mass or energy. While the energetic (energy/mass) dimension is absent in function, function still refers to space and time. Kleine-Horst highlights that this is the first stage in the "progressive deactualization" of the TSE-conditions.

But how can something function without the energy/mass dimension? We can conceive of evolution to include a process of learning in which the left entities produce the right entities (see arrow pointing towards the right in Fig. 7.2), yet of which the latter provide a feedback loop (arrow towards the left) which helps direct the learning process. In this way, a phylogenetic learning process takes place in the VM of the physical evolutionary level whose results are stored in the genome (VM). When a genome actualizes (gene-expression), enzymes (VM) are created which serve (*i.e.*, have the function) as catalysts (VF). As catalyst, enzymes cause specific reactions in metabolic pathways (sequence of reactions). Catalysts increase the rates of likelihoods of reactions but don't put their own energy into the reaction. In other words, the enzyme initiates at a particular place (S) and at a particular time (T) an energetic process, without using its own energy. This means that while the enzyme-matter is TSE-dimensional, its function is only TS-dimensional.

Life-forms are 'things' which have reached at least the second evolutionary level; that is, which consist of the worlds UCO to VF. The 'most evolved' matter in our bodies (VM + VF), the neuron (VM), has the function of *excitation* (VF). Through this function, further functional circuits (the neuronal circuits) are formed which are the basis for developing the next higher world: that of Psychic Functions (PF) at the third evolutionary level. Thus, an organism with a psyche (PF + PC) consists of the worlds UCO to PC (or Psychic Consciousness). The hierarchy of psychic functions evolves on the basis of individual implicit (unconscious) learning processes. For example, in the case of visual perception of "figure in its outfield" (Kleine-Horst, 2001, *passim*) the learning material consists of the relationships and relation-relationships between the excitations (VF) of the retinal receptors (VM) that are stimulated by photons (UCM). After a PF hierarchy of visual Gestalt functions has evolved, it will be actualized by specific *Gestalt stimuli* that follow retinal stimulation. By this process, the hierarchy of psychic functions (PF) produces a hierarchy of psychic consciousnesses (PC)—the *Gestalt qualities* such as conscious visual experiences. In other words, the actualization of psychic functions (PF) causes Psychic (perceptual) Consciousness (PC) to exist. With the existence of PC develops a fourth level—the phenomenal manner of being— whose contents are subjective experiences. Thus the structure of the external world transforms itself—so to speak—via our sense organs into an *experience*. Note that the outer world is not apperceived as a series of single random stimulations, but rather as a *whole* (*Ganzheit*) —as a *Gestalt*. This is a fundamental characteristic of higher beings and belongs to the realm of the *psyche* (PF + PC).

At the third evolutionary level, when PF of the functional manner has been actualized in the phenomenal manner of being (here: PC), not only is energy/mass (E) lacking (as is the case in the functional manner), but also space. For Kleine-Horst, perceptual consciousness such as visual experiences—the central focus of Kleine-Horst's (2001) main body of work—is neither space nor located in space. Rather space belongs only to the contents of perceptual consciousness that are often spatially organized, such as when we visually perceive a certain spatial configuration. The phenomena themselves (PC)—not their perceptual contents—are spatially neither one-, two-, or three-dimensional. In that regard, we can see a further (a second) step of "progressive deactualization" (Kleine-Horst, 2001) of the TSE condition via elimination of the space (S) dimension. However, since phenomena require a certain duration to exist, we can say that they happen in time (*cf.* Lewin, 1936, pp. 35–35) and thus remain temporally organized.

Psychic Consciousness (PC) is pure perception, pure awareness, pure experience of material things, or of feelings. The experience of the color blue, or of Gestalt qualities like line or field, are PC. However, pure experience (awareness) of color has nothing to do with reflection on color experience. To be conscious of having a color experience is not a thing of PC, but of an even higher order, that of Mental Consciousness (MC). PC refers immediately to the outer or inner world; but MC refers to PC (and other MC). To 'see' a red spot of a certain form is a content of PC, but to think: 'What I see is a red square' is a content of MC. One 'explicitly' (consciously) learns (*e.g.*, in school) that a spot of such a form is called a square. Yet, to be able to just simply see (to be aware of) the specific form of a square does not necessitate formal schooling, rather it is 'implicitly' (*i.e.*, unconsciously) learned during the first few weeks of life. This is the difference between the two sorts of consciousness, the lower-level world (PC) and the higher-level world (MC).

Our capacity to reflect upon Gestalt qualities (PC) is an important characteristic of mental consciousness (MC) (compare with Carini, 1983, p. 109, "awareness of our own awareness"). The new MC hierarchy is grounded on the highest entity of psychic consciousness (*e.g.*, the awareness PC—see the reflexive arrow in Fig. 7.2). Mental consciousness takes the actuality of our environment (*Umwelt*) and allows us to manipulate it in the abstract realm alone. The actualization of mental consciousness (MC) expresses itself in the world of Mental Order (MO) of the ordinal manner of being. For example, thinking is a process of MC; but the 'products' of thinking (the thoughts themselves) are contents of the world of Mental Order (MO), which belongs to the ordinal manner of being.

On the one hand, Mental Consciousnesses (MC) produce Mental Orders (MO); on the other hand MC receives feedback from MO (see the left-pointing arrow in Fig. 7.2). Thus Mental Order (MO) can be seen as governing rules of thought, such as those internalized through the culture into which we are born. Abstract thought systems, such as our laws, systems of logic, arithmetic, and geometry are all configured through mental consciousness (MC). It is precisely in this world that the reality of our lives is processed and this reality is always holistically oriented. This fourth (mental) evolutionary level (the mind, MC + MO) brings with it the final (and third) step of deactualization of the three-level conditions of the TSE-dimensions since the contents of MO (as with UCO) lack not only the energy and space dimensions but also the time dimensions.

In that sense, their withdrawal increases the relative 'freedom' of reality, which was previously constricted by the limitations of time, space, and energy/mass.

The implications of this four-level four-"manner of being" model are manifold. For one, the model is structural and developmental in nature, thus complying with the tenets of *Ganzheitspsychologie* and thereby accounting for both the hierarchical nature of life as well as its continuous changes—including those of the phylogenetic, ontogenetic, and/or microgenetic kind. The model also provides a renewed look at the dilemma of the mind-body dualism that has plagued philosophy for centuries. That is, how are mind and body linked? What the "Eight-world-model of being" (AWM) proposes is that we no longer look at this linkage in terms of dualism (of mind/body, or matter/consciousness), but rather in terms of double quadrialism, where matter and consciousness are linked by a third manner of being (the functional manner). These three manners of being are embedded in a fourth, the "ordinal manner of being", which represents the most fundamental part of "being" or existence. We thus conceive of four evolutionary levels—inorganic matter, body, psyche, and mind—instead of merely two, mind and body, as traditional philosophy suggests. The "interfaces" of the four evolutionary levels with the four manners of being form the eight "worlds" of being so that this model (AWM) as a whole represents a system of impressive symmetry.

While this model is a work in progress (as all models are)—contingent upon further evidence from empirical work—it nevertheless provides a holistic framework (a 'super-theory') along which researchers (and interested mathematicians) can orient themselves for the larger perspective. This implies that partial theories will either contribute towards a more thorough elaboration of this model or suggest alterations to account for possible discrepancies in its present form.[3] Subsequent research will need to further investigate the structural processes on each evolutionary level within the aspects of the manner of being model. For example, through his Empiristic Theory of Visual Gestalt Perception (ETVG), Kleine-Horst (2001) has proposed a ten-level (functional) hierarchy of 25 psychic visual factors (the Gestalt factors) that govern visual perception. While it is not possible to provide the elaborate details of this model

---

[3]The AWM model as presented here expands on Kleine-Horst (2004).

within this book chapter, it is worth mentioning that ETVG presents a testable theory that takes place predominantly within the psychic evolutionary level (the psyche), containing both the PF hierarchy of the functional manner of being and the correlated PC hierarchy of the phenomenal manner of being. The importance of the functional hierarchy approach is that it helps overcome the limitations of the past Elementarists' proposed constancy hypothesis, which to this day faces difficulties in explaining deviations from perceptual experience and physical occurrences (*e.g.*, why the experience of brightness does not always correspond to the luminance of light impinging on the retina, or why color does not always correspond to its wavelength). ETVG expands upon the Gestalt approach of the early 20th century by breaking down the Gestalt law of *Prägnanz*, which stated that "psychological organization will always be as 'good' as the prevailing conditions allow" (Koffka, 1935, p. 110). According to Koffka (*ibid.*), "in this definition the term 'good' is undefined". What ETVG does is not only expand on basic Gestalt factors (like the laws of proximity, good continuation, closure, good shape, etc.), but also provide explanations of why a Gestalt perception occurs in the first place, after objects have been optically projected upon the retina. All Gestalt laws rest upon the associations between the Gestalt factors or Gestalt functions.

While one can say that a primary effect of a Gestalt factor is qualitatively and quantitatively informational, a secondary effect is "formative" and ends in a normative (*e.g.*, *Prägnanz*) experience (see Kleine-Horst, 2001). This "formative" (and thus developmental) effect does not just take place in the here-and-now context (as microgenetic studies can show—*e.g.*, Wohlfahrt, 1932) but originates in early childhood experiences, thus making memory (especially implicit memory) a key factor to understanding visual perception and thereby including an ontogenetic component in the study of perception. In that regard, we may be reminded of Sander's (1962b) studies involving the *parallelogram illusion*, in which he demonstrated that the *coefficient of illusory effect* drops with increasing age: if we present a child with a parallelogram as in Fig. 7.3, and ask him/her to rate the two objectively equal lengths ($\overline{AF}$ and $\overline{DF}$), the younger the child is the more he/she is likely to base his/her perception according to larger wholes. The seeming inequality of the lengths of $\overline{AF}$ and $\overline{DF}$ is due to their belonging to two parallelograms of different size. The parts receive their psychological properties in relation to the whole; consequently, the 'apparent' size of the lines is dominated by the 'actual' size of the whole. However, when the isosceles triangle $\triangle$AFD is lifted out of the complex, the two (actually equal) sides $\overline{AF}$, $\overline{DF}$ tend towards equivalence.

The structure of situations

The ontogenetic component contributes to the psychological properties of situations. The terminus of *Prägnanz*-tendency need not apply only to visual configurations but conceivably is also evident within situational constellations of the field-theoretic kind. Kurt Lewin's use of Gestalt terminology is translated to a systemic model of social-psychological components. Lewin's (1936) field theory of topological psychology is captured in mathematical notation via the often cited formula $B = f(L = P + E)$ expressing that behavior B is seen as a function of Life Space (or *Lebensraum*) L, which in turn is seen as comprised of the person P and his/her environment E. Based on certain tendencies (in Lewin's terms, "quasi-needs" or *Quasibedürfnisse*), objects in our environment receive their valences (invitation-characters or *Aufforderungscharakter*). That is, based on a person's quasi-needs, an object receives its *Aufforderungscharakter*, which can be expressed either positively or negatively (*i.e.*, with positive or negative valence). These states are translated into forces that act upon the person's behavior. Hence, the person is seen in the functional relationship between various *field forces*.

Lewin's approach predominantly captures the dynamic relationships that take place within the worlds of MC/MO and to a large extent drops the time (T) dimension for making sense of peo-

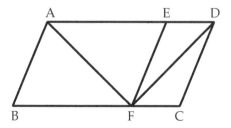

Figure 7.3: Sander's Parallelogram Illusion.

ple's actions. Lewin's position is one of "contemporaneity" where influences of the past (as having happened in the past) are not necessary to be taken into account (as Freudians would maintain). Nor is Wilhelm Wundt's (1922) belief—*causa finalis* (teleology)— that future events influence the present events a non-dispensable form of analysis. As Lewin (1936, pp. 34–35) put it:

> In opposition to this assumption we shall here strongly defend the thesis that neither past nor future psychological facts but only the present situation can influence present events. This thesis is a direct consequence of the principle that only what exists concretely can have effects. Since neither the past nor the future exists at the present moment it cannot have effects at the present. In representing the life space therefore we take into account only what is contemporary.

Both past and future are present field conditions given the person's consciousness. This is not to deny that the person's current state (*e.g.*, quasi-needs) is a partial result of his or her past but rather that the past is only considered insofar that it is a part of the present situation. Particulars of the past which are not part of the present field conditions (*e.g.*, not within a person's consciousness) have little bearing upon a person's actions. It is this line of reasoning that has allowed researchers to create (and vary) situational conditions in order to see the transformations and structure of activities within a given field (for an excellent review of the contributions of Lewin's Berlin group, see de Rivera, 1976).

In an effort for greater generalizability, psychologists have now begun to examine the properties of situations through the use of a descriptive Q-sort tool (the Riverside Situational Q-sort, or RSQ) that forces people to categorize characteristics of a situation being described (Wagerman & Funder, 2009). The result is a nine-step forced distribution of 81 RSQ items (*e.g.*, "someone is trying to impress someone or convince someone of something") which are believed to be the properties of situations (personal experiences) that are described by research participants. These quantified properties are then co-related, as well as related to other factors (such as personality or negative/positive affect dimensions) in order to establish associations between these variables. For example, when participants gave a high rating to the RSQ item "P(articipant) is being insulted", it directly related to ratings of having fantasized in response (r = .25), having been sarcastic (r = .25), or having blamed others (r = .24). In other analysis of the same data, Wagerman and Funder found simple (positive/negative) affect relationships with particular ratings of situational properties, such as positive affect being related to the potential for being complimented or praised (r = .41). The promise stemming from this line of research is to refine an instrument that would capture the properties of situations and in turn relate these to behaviors, emotional experiences, and personality. The underlying effort would be to allow for predicting in which specific situations people are likely to act a certain way.

The underlying premise in research of this kind is that the situation can be viewed independently from the person's perception of that situation. Yet at the same time, the unit of analysis seems to be not the independent situation *per se* but rather the person's (rater's) judgments of these 'independent' situational properties and their co-relating factors (*e.g.*, a person's affect or personality). Furthermore, research of this kind follows Lewin's closed system (monadic) approach (see Valsiner & Diriwächter, 2008), in which the unit of analysis and greater whole become separated. By separating quality (considered determinate) from quantity (indeterminate), Lewin and traditional psychometrics share a common approach. However, unlike Lewin, traditional psychometrics tends to eradicate the dynamic variability of person and field-whole by positing the existence of a 'true score', then seeking ways for researchers to bypass 'noise' ('error') that stands in the way of their 'measuring' the 'true score'. The introduction of the RSQ overcomes this latter 'problem' to the extent that the RSQ prioritizes situational properties. But it still relates these properties to 'true scores' of personality or affect ratings that follow traditional true/false or Likert scales (for discussion of the shortcomings of such standardized approaches, see Diriwächter & Valsiner, 2005; Diriwächter, Valsiner, & Sauck, 2004; Valsiner, Diriwächter, & Sauck, 2004).

Culture: an underlying structure of psychological function

In order to fully understand human experiences (and actions), it may not be sufficient to examine the immediately given as isolated from the past. This does not conflict with Lewin's assertions (see above) but rather traces the immediately given to the structural configurations from which it emerges. Instead of examining the present as a closed system (a 'monad', researchers could draw from developmentally oriented cultural psychology traditions. The origins of cultural psychology can be traced to its predecessor, *Völkerpsychologie*, which under Lazarus and Steinthal (and later Wundt) made up a significant part of German theorizing but never really achieved a coherent synthesis (Diriwächter, 2004). However, more recent advances—largely through the incorporation of theorists including James Mark Baldwin, Georg Simmel, George Herbert Mead, Lev Vygotsky, and Mikhail Bakhtin (several of whom were familiar with the *Völkerpsychologie* tradition)—have led to new momentum in establishing a developmentally focused cultural psychology (see Valsiner, 2000). This truly genetic approach is based on systemic models that are non-linear and open to development.

In this regard, modern day *Ganzheitspsychologie* incorporates cultural psychology, whereby implicit emphasis is given to Wilhelm Dilthey's (1894/1961) structural views insofar as they dictated an understanding of phenomena not in and of themselves but rather as part of a larger whole outlasting any given phenomena under investigation. A phenomenon progresses in the direction of an encompassing whole whose structure has traditionally been examined through analytically or synthetically oriented psychology. Early holistic approaches gave precedence to analytic interpretations, by creating an opposition to Wundt's (1894) principle of *creative synthesis* (*Schöpferische Synthese*), which emphasized that the melding of unrelated elements would create properties that contain characteristics not found within the elements themselves. Hans Volkelt (1922) would advocate for an abandonment of meaningless elements in psychological analysis by stating that psychic synthesis is never created entirely new —rather it merely represents transformed relationships. The concept of creative synthesis would thus be replaced by the concept of *synthesis transformations* (old syntheses transform into new ones). Kurt Koffka (1915) would refute Wundt's principle by showing that perceptions cannot be derived purely from sensations— thereby advocating a top-down approach he called *creative analysis* (*Schöpferische Analyse*).

Similarly, the developmental psychologist Heinz Werner (1959) stated, in regards to a holistically and genetically oriented psychology (and cultural psychology in particular), that efforts to obtain a structure of higher units through the aggregation or synthesis of elements highlight the difficulties in formulating the problems of meaning-making. It is easy to show that a totality can be founded in various ways and that the so-called elements which make up this totality can be changed without losing the character of the whole. For this reason it is not possible to say it is because of the dots[4] that we see a circle (see Fig. 7.4(a)), and also not because of a summation through synthesis. A particular (different) configuration (*e.g.*, the oval in Fig. 7.4(b)) can exist through the same elementary building blocks (the dots); or the 'same' configuration can exist through different elementary building blocks (in Fig. 7.4, compare the 'dot' circle (a) with the 'square' circle (c)).

In a similar vein, a synthesis through a number of individuals will never lead to an "over-individual", *i.e.* collective, totality (*Über-individuum*; *cf.* Münsterberg, 1900). The totality is not derivable through a melding (or mixing)—a synthesis—of the elements. Thus, Werner (1959, p. 8) demanded a reformulation of the old approaches that attempt to build the whole by means of elements: "If the to-tality can not be derived from the elements, then this totality can only be explained through itself." This, in return, shifts the way we approach the problem:

> [...] not the elements are the precondition of the whole,
> but rather the whole as point of origin is the precondition

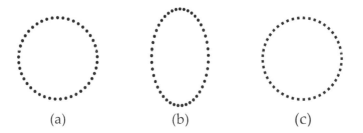

(a)   (b)   (c)

Figure 7.4: (a) Circle consisting of dots; (b) oval consisting of dots; (c) circle consisting of squares.

---

[4]In that regard, a dot can never be experienced on its own, rather only as part of a holistic partial visual complex; as a point that differs in brightness and color tone from an underlying field (for a psychological and epistemological investigation on the axioms of Euclidean geometry see Bergfeld, 1928).

> for its parts. Psychology, including cultural psychology
> (*Völkerpsychologie*) needs to start with the living units
> and through analysis derive the lower level units. Not
> the term creative synthesis, but the term creative analysis
> will lead to fruitful results. (Werner, 1959, p. 8)

According to Werner, humans are members of an "over-individual"
unit through which they receive characteristics that can be understood only through this higher unit. For example, expressions of
the "over-individual" unit are seen in language, religion, law, and
customs that can never be rooted within the individual alone. What
these hierarchically higher placed units have in common is that they
are the products of and originate from *collective mental life*.

Mathematical models that attempt to answer the question of
whole-part relationships through addition and subtraction theory
alone seem limited. If we subtract from the individual those characteristics (or factors) that are not common within other individuals
of a given culture, and consequently believe that what is left are
the common characteristics, then we find ourselves working with
characteristics that are inferior precisely because they are the result
of subtraction (or addition) efforts. Werner notes that this approach
is certainly one of convenience but in the end, such a mathematical
psychology in which the mental life becomes mathematized and
physicalized seems somewhat absurd. For is it really the same for
ten strangers (or one hundred, for that matter) to do the same thing
as for the identical actions to be the result of shared inner intentions
that stem from belonging to a group with a positive, shared sense
of community?

Can all characteristics of the masses really be explained through
individual traits, or does the person receive an array of characteristics that without any sociological ('macro') interpretation can
never be explained? According to Werner, concepts like ambition,
love for fatherland, or national pride are understandable only through
an examination of the real purpose and structure of the masses
themselves. Such purposes as the sense of living or experienced
belonging (*lebendig erlebtes Zusammenstehen*) and co-functioning (*Mit-
sammenwirken*) are the reasons that highlighted characteristics of the
masses are much more affect-, rather than intellect-, prone.

Hence, any mathematical approach needs to be guided through
descriptive analysis. Cultural psychology highlights the structural

unity of both person (individual culture) and his/her belonging to a group of people that share similar psychological distancing or semiotic devices that mediate their dependence upon a 'here-and-now' context. As Valsiner (2000, p. 50) mentions, within our immediate life contexts "the person becomes simultaneously an actor, who is immersed in the given 'situated activity context', and a reflexive agent who is distanced from the very setting in which he or she is immersed." This fact often goes unnoticed by traditional standardized approaches that at best measure the *outcome* of reflective activities in which the agent has distanced himself/herself from the rated activity, but ironically do not themselves capture *how this outcome was achieved*, nor take into account the actor as being immersed in a particular life context (*e.g.*, filling out a questionnaire) at the time of the rating.

Such semiotic devices as distancing are not purely intellectual, but rather focus on personal relations that are often highly emotionally colored (*e.g.*, in aesthetic experiences; see Bullough, 1912). According to Valsiner (2000),

> Distancing is possible thanks to the construction of hierarchically organized self- (and other-) regulation mechanisms—through meanings. Any labeling of an object or situation entails distancing: the person who makes the statement 'THIS IS X' actually performs the operation 'I (from my position) IN RELATION TO THIS (object that is indicated) CREATE THE MEANING OF THAT OBJECT, WHICH I CLAIM IS X'. Two moments of distancing are involved here: first the SUBJECT (I) is distanced from the OBJECT ('this object out there'); then, secondly, the OBJECT *is distanced from itself* by way of meaning (X). (Valsiner, 2000, p. 51; original emphasis)

As more general meanings replace the original act of assigning meaning to something, the process of meaning-making becomes increasingly complex and abstract (as in the case of a statement along the lines of "Life is dangerous"). Semiotic mediation allows for the construction (as well as destruction) of flexible meaning hierarchies. This open-systemic developmental process (in this case of microgenetic origin), and the person's immediate dependence upon the here-and-now context, are what allow a sense of autonomy.

Meaning-constructions that occur in any given here-and-now context transpire from a larger socio-cultural setting. To understand a person's meaning-making process within the topological structure of situations requires an understanding of culture-bound character- istics of the process that outlast any given situation. In this regard, *Ganzheitspsychologie* has early on included an appeal for an organic (in contrast to mechanistic) approach. Krueger (1915) noted the importance of both cultural and developmental sensitivity in em- pirical investigations. This kind of sensitivity reveals the structural configurations and developmental trajectories that represent the conditions out of which human experiences emerge.

This relevancy becomes clear upon examination of our experi- ences within the structure of everyday human environments. For example, if we take the case of the street (as Valsiner, 2004, has done), we may first maintain that from a formal standpoint, it is an empty space (an aperture) among human-made structures (build- ings, yards, parking lots . . .). However, on a closer look we can see that the street as a life space is much more complex. As part of a culturally textured environment, the street contains a dynamic structure of socially fluid realities that can be viewed from multiple perspectives. As Valsiner notes (2004, p. 9),

> the study of streets goes through people—and the tem- porary appearance of people in the street makes such study possible. The street is a stage—and all people who participate in its spectacle are actors on that stage. Psy- chologically, encountering that stage entails the transi- tion to settings of public exposure of the self, uncertainty of experiences, use of cognitive tools to anticipate these experiences, dialogical negotiation of the personal self with public social order, and seeking of new thrills. The street is simultaneously dangerous and alluring, uniting and separating, free and un-free.

The cultural meaning we give to the street sets up dichotomies (along the lines of "A" vs. "non-A"—see Josephs, Valsiner, & Surgan, 1999), which in return allow for relativity of those meanings. In the case of the street, which may stand in opposition to one's home, the relativity between street/home may be changed (or even reversed) when the street simultaneously serves *as* one's home. The person under these changed circumstances draws upon different semiotics during the process of meaning construction. The people we commonly classify as 'homeless' (because they lack the distinct locale we have come to designate 'a home') or 'streetchildren' (who often have a distinct locale to return to, but whose main activities take place within the streets) develop their own meaning system that is focused on a particular notion of survival (*e.g.*, from raids by police or local gangs). This survival is linked sociopsychologically to the economic situation of these street dwellers. In economic terms, the case of streetchildren could be functionally equivalent to those children who worked in textile mills or in mines some 100 years earlier. However, as Valsiner (2000, p. 138) points out, there is a major psychological difference: by working in factories or mines, children would undergo a strict behavioral routine which was determined by the production process. Today, streetchildren are much more prone to activities that require independence through personal initiative and flexible adaptability.

An understanding of these children's actions and personal experiences requires an examination of socio-cultural and political structures that reach beyond the immediate context in which these children find themselves (*e.g.*, on the street). These socio-cultural structures reach beyond their own lives, which are often marked by great varieties of single-case scenarios (*e.g.*, the child may live on the street in order to escape the psychological horrors of home— or to help his/her parents gain economic resources). Instead, an understanding of the immediate life context may necessitate an investigation of the person's *exosystem*, *i.e.*, the settings that do not directly involve a person as an active participant but ultimately do impact the settings that involve that person (Bronfenbrenner, 1979). The rules and laws we follow are chiefly a result of the exosystem, yet a critical examination of this system (as a structural and dynamic component of our experiences) is often neglected.

Feelings—the experienced qualities of our world

From a psychological vantage point, the street is culturally structured to allow our moving—both into places already known (*e.g.*, home), and out of known places into the increasingly unknown (unpredictable). Certain dwellings are not just places of living— they are *homes*. At the core of such person-location relationships stand our feelings—the experienced qualities of a structure. Precisely in the case of feelings, we can have multidimensional occurrences that often lack a concrete structure (Krueger, 1928/1953). Human beings can create over-generalized feeling-fields to capture the holistic meaning of a place (see also Valsiner, 2005). Saying that a nicely decorated street in the month of December gives me the 'Christmas spirit' indicates this level of generalization that is mediated by a particular *Gefühlston* ("feeling-tone"). These levels of human semiotic mediation of affective experiencing provide the guidance for new experiences in a given environment. However, at the level of over-generalized feeling-field, it is difficult for the individual to make clear sense of what he/she is experiencing. Hence, it requires the levels of specific categories of emotions in order to articulate the felt phenomena (a process we can call *entification*). At the same time, it should be clear that all levels of feelings are integrative. In terms of the AWM model described earlier, they span evolutionary levels from the vital to the mental; that is, from the material manner of being (where we can say that, on the physiological level, feelings consist of excitation and inhibition—VM/VF) to the phenomenal and ordinal manner of being (where we make sense of our experiences—MC/MO).

Feelings have colored our lives since the dawn of our time. They are components of control systems and ultimately the perception of Self (Laird, 2007). This perception matches the level of organization in the world (its structure) and the levels of organization by our stacked (or hierarchically ordered) control systems. According to Krueger's (1928/1953) first law of feelings, "any change in a holistic (*ganzheitlicher*) complex will be more readily noticed than changes in its parts" (p. 211). Every life space sets boundaries. In the case of a person's home, the boundaries pertain to the private and public domain of the social world. These boundaries support tensions between the known and the unknown, between "inside" and "outside", or the near and the afar. It is through such cultural psychological processes that emotion-laden concepts like *Fernweh* (the longing for the afar and the unknown) or *Heimweh* (also known as homesickness: the longing for the known, the familiar, the safe) become understandable.

At the heart of *Heimweh* (homesickness) stands *Sehnsucht* (longing/addiction), which promises an alternative to an "is-condition" (Boesch, 1998, p. 16). The immigrant in a new country may have his

/her house but does not feel 'at home'. The goal of *Sehnsucht* is luck and happiness (*Glück*). Yet we also know that luck and happiness are fleeting conditions, and thus regret too belongs to *Sehnsucht* and thereby adds to the felt tension. An analysis of the structure and process of *Heimweh* reveals how affected humans structure their environments to cope with the particular shift from one holistic complex (*e.g.*, the home experience) to another (*e.g.*, the experience in the new environment; Diriwächter, 2009). Music, pictures, and other objects associated with the home environment serve a dualistic purpose during semiotic construction. For one, they help bridge the distance between the old and new environment. A foreign student may display the Swedish national flag in his dorm room to remind himself that the "old world is still there." The flag stands for everything he has left behind and is an important focus point while remembering the past. Yet at the same time, such objects (flags, pictures, etc.) also serve to increase the distance between the self and the new environment: the lingering in the past necessitates a separation from the present and, furthermore, serves as a reminder of the physical distance between the current locale and one's place of origin. Thus, the semiotic *demand setting* extends the here-and-now context through inclusion of conditions that were central to past (and often pleasurable) experiences.

These processes are not of pure intellectual origin, but are tainted by the feelings which help define our experiences. This applies equally to the researcher investigating human experiences. As Volkelt (1962b, p. 51) once noted, the less complete our descriptions of human experiences are, the more intellectualized the qualities become. Feelings are much richer than simplistically expressed negative/positive affect and are never secondary in the sense of simple accessories. Feelings capture the experiential qualities of a holistic complex and, as such, become the center of our experiences (Krueger, 1928/1953, p. 204). Taking into account the historicity of a person gives rise to entirely new dimensions of feeling states, beyond the traditional one (or three—see Diriwächter, 2008) dimensions commonly discussed. For example, when the person speaks of deep feelings (such as during a religious ceremony or when loving someone), he/she is highlighting an experience that points towards a general direction, namely that of his/her values. This in turn can only be understood through an examination of the structure— of cultural origin—that allows such values to exist.

The person in context

Albert Wellek's (1950, p. 14) claim—"In the beginning was the complex quality (*Komplexqualität*)"—characterizes the fundamental conviction by the second school of Leipzig (so-called Genetic *Ganzheitspsychologie*) that the person is an emotional creature who experiences through the structural qualities of his/her world. Not *logos*, but feeling; not the concrete, but the diffuse; not the person, but the world (environment) came first. An examination of the person in the environment requires considering him/her as a moving-living organism who progresses along a particular developmental trajectory. Furthermore, this person is embedded in a cultural structure that (aside from the biological constraints) provides laws for the manner in which this individual develops.

The person as a living-moving organism who acts upon his/her environment implies a developmental approach that is purposeful from the perspective of that person's self concept. Valsiner (2004) coined this characteristic as the *Strivingly Moving Self* (*SMS*). This notion of self is dialogical in its nature and built upon the axiom of doubleness of the singular events (*i.e.*, that events include their opposition—*A* includes non-*A*).[5] Hence, we can say that any action is the opposition between tendencies towards acting and non-acting (for further discussions on theories of dialogical self, see Hermans, 2002). This in return implies that the Self is not static (the unmoved mover)[6] but rather itself consists of a multiplicity of dynamic processes.

The notion of multiple forms of Self is by no means a new concept. William James (1890) proposed this over a century ago in the forms of the material Self, the social Self, and the spiritual Self. To account for the process (movement) of self-consciousness, James (p. 400) draws upon his concept of "stream of thought", which highlights the dialogical and constructive nature of agency (the *I*) and creation (the *me*). More recently, Valsiner (2007) has proposed that we should therefore examine the Self as a frame for viewing unity in the processes of creating personal experiences. Traditional psychometric- or personality-type questions along the lines of "who am I?" are actually outcomes of a self-as-semiotic-

---

[5]This can be contrasted with note 4.

[6]Compare with the discussion on Herbart below.

process. The axiomatic stand to any question that inquires "who am I?" pertains to a locally constructed fiction which needs to be theoretically treated as such. Taking the notion of Self as something that 'is' could have led the psychology of self on an unproductive theoretical road for the past century which generated an avalanche of consensually accepted data about the 'self' (or personality) without actually understanding the dynamic complexity of the system that generates it all. Valsiner suggests that the only possibility of making sense of the self lies in observing the self-system in action —that is, by examining the Self as a frame for viewing unity in the processes of creating personal experiences (*e.g.*, via the trialogic self-system discussed by Valsiner).

We may therefore want to consider studying the process of 'Selfing' in action. One such microgenetic study (Pereira & Diriwächter, 2008) was conducted by investigating participants' reported daydreams. Since participants are always a component of a larger holistic complex (the complex qualities), microgenetic development during daydreaming involves a complex interplay between a series of apperceptive shifts (both analytically and/or synthetically oriented) with frequent referencing to the general impression of the phenomena. There is a push towards establishing levels of concreteness out of the diffuse situation (*e.g.*, while observing a boring screen matrix during the study). Participants often drew upon their immediate environment (*e.g.*, scribbles on the desk) as a launching point for subsequent daydreams. In each instance where novelty (a new daydream theme) became visible on the narrative level, it was simultaneously embedded in the layers of the general impression previously expressed. Hence, the findings suggested "daydream monads" consisting of fluid *Simultangestalten* (simultaneous Gestalts) whereby one (the general impression) incorporates the other (a sub-whole or *Unterganzes*).

In addition, daydream monads during synthesis formations became noticeably colored by a particular feeling-tone (*Gefühlston*). Of course, the developmental trajectories themselves bring possible shifts of feelings expressed as the person "feels into" (*Einfühlen*) the new (hypothetical) situation that was set up by that person. That a person's thoughts during synthesis formation of a microgenetic event need not be confined to the context of the immediately given (*e.g.*, what can be evidently heard or seen in the physical surroundings by the person), but can venture into a hypothetical realm of future directives, has been proposed by Volkelt (1962a) some time ago. He saw that the process of *Einfühlung* into an immediate given also implies a certain degree of anticipation of future developmental directives, so that our feeling into the situation is also guided by our expectation of what the future holds for us. The result of these microgenetic events is the *Endgestalt*— or, in terms of the above discussion on Self, a transformed Self.

The Self as a reflexive agent (and not as a fixed entity) draws upon the immediate environment, past experiences, and future directives in order to negotiate its position in the here-and-now context. It is under such conditions that we may begin to make the processes of human thought and experiences (like daydreaming) theoretically understandable.

Mathematizing the mind

The process of daydreaming incorporates a cognitive component. The history of psychology has seen many attempts to mathematize the functions of the human mind. One of the first comprehensive efforts to express the activities of the mind mathematically can be attributed to Johann Friedrich Herbart (1776–1841). Probably his most important work for psychology was his long, rather difficult *Psychologie als Wissenschaft: Neu gegründet auf Erfahrung, Metaphysik und Mathematik [Psychology as a Science: Newly Founded on Experience, Metaphysics and Mathematics]* (1824–1825), in two volumes.[7] The first, synthetic part constructs a psychological theory on the basis of certain abstract principles. The second, analytic part uses the results of the first part to describe and analyze concrete phenomena of the mind (for an overview of Herbart's psychology, without the display of his elaborate formulæ, see Stout, 1888a,b, 1889).

Herbart was critical of psychology as an experimental science (in the traditional sense) since experimentation necessitated dividing up its subject matter.[8] In his *Psychologische Untersuchungen* (1839), he reminds the reader that any approach to psychology may not forget that over the particular rests the whole (*"über dem Einzelnen darf das Ganze nicht aus den Augen verloren werden"*,

---

[7]An earlier and shorter book—*Lehrbuch zur Psychologie [Textbook in Psychology]* (1813/1834)—provides a systematic account of his psychological doctrines.

[8][*Cf.* the epigraph from Bawden (1904) on p. 163. (Ed.)]

pp. v–vi). Herbart's approach followed that of Gottfried Wilhelm von Leibniz (1646–1716) in assuming that our *Vorstellungen* can be likened to monads (such as the daydream monads mentioned above). The concept of *Vorstellung* (the singular form of *Vorstellungen*) roughly means 'idea' or 'presentation', but literally translates into 'placing before': we can place objects before our mind (represent them), even when these objects are not in our immediate environment (*e.g.*, the homesick student who is thinking of his/her home, family, or friends). Yet *Vorstellung* in Herbart's theory is more than just pure thought, internal dialogue, or visual images. For Herbart, all subjective phenomenal experiences—including feelings —were part of a person's *Vorstellung* (for further details on Herbart's dealings with feeling states, see Stout, 1888b).

For Herbart, *Vorstellungen* are derived from experiences, but he also believed that once these *Vorstellungen* existed, they then contained a force or energy of their own and struggled to gain expression in consciousness. *Compatible* (rather than incompatible) *Vorstellungen* are more likely to group, thereby creating an *apperceptive mass* (either *outer* or *inner*—see Stout, 1888b). We can also compare the apperceptive mass to all that we are attending to. If a new *Vorstellung* outside the apperceptive mass (*i.e.*, one of which we are not conscious) is not compatible with the other *Vorstellungen* already present, tension builds and a certain magnitude of force is required for the new *Vorstellung* to be permitted its expression (this can also occur when enough repressed *Vorstellungen* group together and thereby gain access to consciousness). The term *repression* was used to describe the force that would hold incompatible ideas out of consciousness, and *limen* (or *threshold*) to describe the boundary between what lies in and out of consciousness.

One of Herbart's (1850/1968) tasks was to show (mathematically) how new *Vorstellungen* would be kept out of consciousness. It is important to note that *Vorstellungen* are never completely destroyed, rather they are *gehemmt* (repressed or restrained). Thus, Herbart goes on to define the *Hemmungssumme* (inhibition-sum) associated with two *Vorstellungen*: "The inhibition-sum (*die Summe der Hemmung*) is the quantity in consciousness (*das Quantum des Vorstellens*) which will inhibit the competing *Vorstellungen* taken together"[9]

[9]*Die Summe der Hemmung ist das Quantum des Vorstellens, welches von den einander entgegenwirkenden Vorstellungen zusammengenommen, muss gehemmt werden.*

(p. 140). As Boudewijnse, Murray, and Bandomir (1999) observe, Herbart believed that the inhibition-sum was determined predominantly by the force of the weaker of two *Vorstellungen*. Similar to Newton's attempts to describe the physical world mathematically, Herbart focused his efforts on the psychological world. Thus, given two opposing *Vorstellungen*, Herbart's approach sees these two exerting a force (like gravity in the physical world) on each other in order to achieve a state of equilibrium.

Herbart chose to measure the magnitude of a *Vorstellung* by introducing numerical values representing its strength (and therefore its clarity) in consciousness. The following example follows closely the discussion by Boudewijnse et al. (1999). We could assume that a person has two *Vorstellungen*, having magnitudes of a and b respectively, with a greater than b. It can be said that to keep the smaller *Vorstellung* (b) repressed, the larger *Vorstellung* (a) need not exert any more strength than the maximum magnitude of b, since the smaller *Vorstellung* can muster no more strength than that, which will always be less than a. Thus, Herbart could make the assumption that the total inhibition force would be a function of b, and consequently the inhibition-sum s—given through the formula

$$s = mb$$

in which m represents the *degree* (*Grad*) to which the two *Vorstellungen* are conflicting—would always lie between 0 and 1 inclusive. If $m = 0$, then no inhibition is exerted by either a or b; whereas when $m = 1$ the total inhibition-sum exerted would be equal to the original maximum magnitude of b.

Herbart assumed that a, the stronger *Vorstellung*, would receive the lesser proportion of the total inhibition-sum, while the weaker *Vorstellung*, b, would receive the larger proportion. Consequently the inhibition-sum would divide itself over a and b so as to reduce a as follows:

$$mb \cdot \frac{b}{a + b}$$

This formula states that m stands in relation to b as well as to $a+b$, which is b's portion of the total initial magnitude of the two *Vorstellungen*, together giving the inhibition exerted by b on a. In other words, the amount of resistance left to a after being inhibited by b is:

$$a - mb \cdot \frac{b}{a+b}$$

while the amount of resistance left to b after being inhibited by a is:

$$b - mb \cdot \frac{b}{a+b}.$$

If we set the total inhibition-sum equal to the original magnitude of b by declaring $m = 1$, then the remainder of a after being inhibited by b from $a - [mb][b/(a+b)]$ will equal

$$a - \frac{b^2}{a+b};$$

similarly, the remainder of b after being inhibited by a will equal

$$b - \frac{ab}{a+b} = \frac{b^2}{a+b}.$$

What these mathematical expressions ultimately show is that b is always larger than its remainder, even when $m = 1$. Similarly, if a is greater than b from the onset (as defined above), then even when $m = 1$, a will always be greater than its remainder. As Boudewijnse et al. (1999) further noted, the consequence is that even when the inhibition-sum is set to its maximum (*i.e.*, mb is set with $m = 1$), a cannot eliminate b out of consciousness, nor can b push a completely out of the mind.[10]

It is relatively easy to notice similarities between Herbart's psychic mechanics and the approach taken by English Associationists. What 'saves' Herbart from the atomistic standpoint for which the English Associationists have been so severely criticized (*e.g.*, see Wundt, 1922) is his metaphysical doctrine on what handles our *Vorstellungen*. The Self (in Herbart's term, *die Seele*; *i.e.*, soul) is intrinsically a simple, unchanging being that originally is without any plurality of states, activities, or powers (*cf.* Valsiner, 2007). As Stout (1888a, p. 324) summarizes, "the variety of mental phenomena,

---

[10]Only the simplest case is discussed here, but the reader is encouraged to examine the above cited works by Herbart (or for continuation of specific cases, Boudewijnse et al., 1999) for further development of these mathematical examples, such as Herbart's notion of the *fusion* of several *Vorstellungen*, and his dealings with the time continuum.

as they actually exist, is ultimately referred by him to the reactions of the soul, whereby it resists a diversity of disturbances *ab extra* due to its relations with other simple beings." This *doctrine of simplicity* implies that the 'thing' that does the processing (the soul or 'that which has agency') resists outside disturbances in the form of simple acts. It can only be considered in the form of multiples insofar as that there are multiplicities of disturbing conditions. For Herbart the question was not how isolated atoms form a whole, but rather how demarcation and partition grow up within an original distinctionless unity.

Herbart's approach may in fact provide new ideas for existing mathematical approaches towards ambivalence and attentiveness (*e.g.*, those proposed by Rudolph, 2006a, 2006b, 2006c). While Herbart's theory has endured some heavy criticism since the second half of the 19th century, much of this criticism may be unwarranted (Boudewijnse, Murray, & Bandomir, 2001). For one, Herbart's approach could reduce our reliance on probability distributions (such as during recognition or recall tasks) that have found their way into today's mainstream psychology yet shed little light on the actual processes underlying *yes/no* responses or obtained response times. In fact, although psychology often seems to suffer from "physics-envy" (Valsiner, 2000), distributions of probabilities of events were not discussed by Newton, nor are they used in many contemporary courses on physics for applied scientists. As Boudewijnse et al. further mention,

> Analytic equations, such as the one expressing the fact that the force exerted by a moving object is equal to its mass times its acceleration, are adequate for most practical application. Herbart believed that analytic equations sufficed to describe mental events; even though these equations were sometimes quite elaborate, Herbart did not need to mention distributions of probabilities in his system. (Boudewijnse et al., 2001, p. 126)

Boudewijnse et al. also note that Herbart's theory shows compatibility with today's dynamic-systems theory. Though mental events cannot be localized in terms of spatial coordinates (see Fig. 7.2 of the AWM model discussed earlier), they could be located on a temporal scale (PC/MC). Since dynamic-systems approaches are non-linear and emphasize the process in real time (microgenesis), the potential exists to describe them analytically (just as the Newtonian concept of momentum can be introduced to the mathematical treatment of operant conditioning).

Herbart's approach was never really considered by Felix Krueger or any of the Leipzig *Ganzheitspsychologen*; nor did the Berlin Gestalt psychologist Kurt Koffka mention Herbart in his voluminous *Principles of Gestalt Psychology* (1935) (see Boudewijnse et al., 2001, for further discussion on the fate of Herbart's psychology). While Herbart's philosophy was still a part of early *Völkerpsychologie* (under Lazarus and Steinthal), it also began to disappear from early cultural psychology, largely under the authority of Wundt's *Völkerpsychologie*. Granted, Herbart's views may seem to be too mechanistic to be considered for a developmentally focused cultural psychology, and at first glance incompatible with the doctrine of *Ganzheitspsychologie* in general. One of Wundt's (1922) strongest points of criticism pertains to Herbart's metaphysical doctrine of the soul (see above). Wundt believed that if we remove the axiom of soul, the entire metaphysical system that rests upon this assumption (and with it all its mathematical proofs) collapses like a house of cards. However, if we were to replace 'soul' with the concept of 'agency', and from there re-examine the feasibility of Herbart's proposed formulæ, it may well help contribute towards establishing and formalizing new concrete mathematical relationships of mental content within the microgenetic domain.

Yet what needs to be remembered is that the psychic holistic complex in fact knows no boundaries (Buchholz, 1928a). It is a unit based on a particular way of examining psychic complexes, such as the experience of space or of time. When we examine these partial wholes (or elements as some may want to call them), we stop continuity and furthermore we alter the quality of the diffuse whole. A mathematical approach towards understanding the human mind requires creating differentiations within this holistic complex, thereby establishing units that can lead to quantitative explanations (including those that may lead to causal inferences). Of course, the 'appropriate' unit of analysis in all areas of psychology is a difficult concept to tackle (for socio-cultural research see Matusov, 2007). Yet I believe that this chapter has provided several examples of suitable units relevant for psychological analyses. In addition, a mathematical social science faces the difficult task of taking into account the quality of quantity. Formalism by itself may not help overcome this difficulty. Providing concrete operational terms may in fact lead to the illusion of objectivity masking the reality that they serve merely to specify a concept being discussed (Buchholz, 1928a).

Some final thoughts

In this chapter I have outlined a general theory (AWM) that is hierarchical and developmental in nature and helps situate the particulars (or *Unterganze*) of a given investigation. I have further broken down the holistic complex by highlighting structural components (and conditions) that demonstrate the complexity of person in environment, with an inclusion of Herbart's all-too-often forgotten approach from which mathematicians could draw new ideas to deal with the most specific cases of mental content and processes.

But any form of psychological analysis needs to account for our past and our belonging to a greater whole, and thus will need to rely on qualitative descriptions of both components and conditions. It is not that *Ganzheitspsychologie* denies any form of quantifications for psychological analyses. In fact, it was Krueger, who together with his English friend Charles Spearman (1863–1945), introduced correlations into German psychology (Krueger & Spearman, 1907), where for the first time a "General Factor", to which all correlating achievements can be traced back, was established (a concept that Spearman would later continue to develop). Furthermore, nearly an entire volume (Vol. 3, 1928) of Krueger's journal *Neue Psychologische Studien* was devoted to the psychological understanding of theories in mathematics and geometry.

However, Sander (1962a, p. 374) reminds us of an important proclamation by Goethe: "Measurement and numbers in their nakedness dissolve and ban the living spirit of observation." The nature of scales, such as those that measure constructs like 'personality' (*e.g.*, the dimensions introversion/extroversion), allows for nothing more than that which is measurable, that is, one cannot imagine anything beyond that which goes through the co-variation. According to such a quantitative approach (*e.g.*, factor analysis), a personality is a finite whole, which can be determined through an exact number of parts. Individual people differ in the number of these parts of which they consist. The more people possess the same parts, the more similar these people are. This portrays a mosaic of elements, independent of what lies outside the person, through a reflex-like system that constructs a finite whole. This whole takes on the nature of a closed system, making it non-developmental and thus fictional.

New (qualitative) mathematical approaches need to allow not only for open systemic views, but also for taking into account the historicity and structure in which a person is embedded. This requires mathematicians to deal with the problem of continuity (Bergfeld, 1928; Buchholz, 1928a) as well as the problem of trying to apply metric for measurements (Buchholz, 1928b). Yet in the end, it is obvious that the purely descriptive analysis (beyond reductionism) for psychology can never be entirely eliminated. In this regard, we may be reminded of the prominent *Gestalt qualities* or *complex qualities* of all degrees for that matter. We may think of the rich variety of feeling states that color our lives. They are the *qualities* of the whole that develop and are most profound when the experience has not been reduced in its nature (*e.g.,* to numbers). To capture the reality of our lives—our experiences as we live through them —requires that we tread carefully when dealing with methods of convenience (such as statistics) and remember that underlying all those numbers still stand greater philosophical questions. It is now left to the research community to develop methods that are not just pragmatically, but also theoretically, integrative.

Acknowledgments

The author wishes to thank Mindy Puopolo, Steven Kissinger, and Lothar Kleine-Horst for their helpful comments on earlier versions of this manuscript, and to Victor Dorff for his help on clarifying some mathematical concepts.

References

Bawden, H. H. (1904). The meaning of the psychical from the point of view of the functional psychology. *The Philosophical Review*, *13*(3), 298–319.

Bergfeld, E. (1928). Die Axiome der Euklidischen Geometrie psychologisch und erkenntnis-theoretisch untersucht [The axioms of Euclidean geometry investigated psychologically and epistemologically]. *Neue Psychologische Studien, 3*(2), 135–202.

Boudewijnse, G. J. A., Murray, D. J., & Bandomir, C. A. (1999). Herbart's mathematical psychology. *History of Psychology*, 2(3), 163–193.

Boudewijnse, G. J. A., Murray, D. J., & Bandomir, C. A. (2001). The fate of Herbart's mathematical psychology. *History of Psychology*, 4(2), 107–132.

Bronfenbrenner, U. (1979). *The Ecology of Human Development.* Cambridge, MA: Harvard University Press.

Buchholz, H. (1928a). Das Problem der Kontinuität [The problem of continuity]. *Neue Psychologische Studien*, 3(2), 5–110.

Buchholz, H. (1928b). Die Unmöglichkeit absoluter metrischer Präzision und die erkenntnistheoretischen Konsequenzen dieser Unmöglichkeit [The impossibility of absolute metric precision and the epistemological consequences of this impossibility]. *Neue Psychologische Studien*, 3(2), 111–134.

Bullough, E. (1912). "Psychical distance" as a factor in art and an aesthetic principle. *Journal of Psychology*, 5(2), 87–118.

Carini, L. (1983). *The Theory of Symbolic Transformations: A Humanistic Scientific Psychology.* Lanham, MD: University Press of America.

de Rivera, J. (Ed.). (1976). *Field Theory as Human-Science: Contributions of Lewin's Berlin Group.* New York, NY: Gardner Press, Inc.

Dilthey, W. (1961). Ideen über eine beschreibende und zergliedernde Psychologie [Thoughts about a descriptive and analytic psychology]. In *Gesammelte Schriften, Bd. V* (pp. 139–240). Stuttgart, DE: B. G. Teubner Verlagsgesellschaft. (Original work published 1894)

Diriwächter, R. (2004). *Völkerpsychologie:* The synthesis that never was. *Culture & Psychology*, 10(1), 85–109.

Diriwächter, R. (2008). Genetic *Ganzheitspsychologie.* In R. Diriwächter & J. Valsiner (Eds.), *Striving for the Whole: Creating Theoretical Syntheses* (pp. 21–46). New Brunswick, NJ: Transaction Publishers.

Diriwächter, R. (2009). *Heimweh* or homesickness: A nostalgic look at the Umwelt that no longer is. In R. I. S. Chang (Ed.), *Relating to Environments: A New Look at Umwelt* (pp. 185–206). Charlotte, NC: Information Age Publishing.

Diriwächter, R., & Valsiner, J. (2005). Qualitative developmental research methods in their historical and epistemological contexts.

Forum Qualitative Sozialforschung/Forum: Qualitative Social Research [On-Line Journal], 7(1), 53 paragraphs. Retrieved December 30, 2005, from `http://www.qualitative-research.net/fqs-texte/1-06/06-1-8-e.htm`

Diriwächter, R., & Valsiner, J. (2008). *Striving for the Whole: Creating Theoretical Syntheses.* New Brunswick, NJ: Transaction Publishers.

Diriwächter, R., Valsiner, J., & Sauck, C. (2004). Microgenesis in making sense of oneself: Constructive recycling of personality inventory items. *Forum Qualitative Sozialforschung/Forum: Qualitative Social Research* [On-Line Journal], 6(1), 49 paragraphs. Retrieved December 1, 2004, from `http://www.qualitative-research.net/fqs-texte/1-06/06-1-8-e.htm`

Eigen, M., & Schuster, P. (1979). *The Hypercycle: A Principle of Natural Self-Organization.* Heidelberg, DE: Springer.

Herbart, J. F. (1839). *Psychologische Untersuchungen [Psychological Investigations].* Göttingen, DE: Dietrich.

Herbart, J. F. (1968). *Psychologie als Wissenschaft, neu gegründet auf Erfahrung, Metaphysik und Mathematik. (Erster, synthetischer Theil) [Psychology as a Science, Newly Founded on Experience, Metaphysics and Mathematics. (First, Synthetic Part)].* Amsterdam, NL: E. J. Bonset. (Original work published 1850)

Hermans, H. (Ed.). (2002). *The Dialogical Self* (Vol. 12, No. 2 of *Theory & Psychology*). (Special issue)

James, W. (1890). *Principles of Psychology* (Vol. 1). New York, NY: Henry Holt & Company. (In two volumes)

Josephs, I. E., Valsiner, J., & Surgan, S. E. (1999). The process of meaning construction: Dissecting the flow of semiotic activity. In J. Brandtstädter & R. Lerner (Eds.), *Action & Self-Development: Theory and Research Through the Life Span* (pp. 227–282). Thousand Oaks, CA: Sage.

Kleine-Horst, L. (2001). *Empiristic Theory of Visual Gestalt Perception. Hierarchy and Interactions of Visual Functions.* Köln, DE: Enane-Verlag. Retrieved November 11, 2006, from `http://www.enane.de/cont.htm`

Kleine-Horst, L. (2004). *Der Anfang des nach- naturwissenschaftlichen Zeitalters: Gedanken und Experimente jenseits der Lehrmeinungen [The Beginning of the Post-Scientific Age: Thoughts and Experiments Beyond Doctrines].* Köln, DE: Enane-Verlag.

Kleine-Horst, L. (2008). From visual actual genesis and ontogenesis toward a theory of man. In E. Abbey & R. Diriwächter (Eds.), *Innovating Genesis: Microgenesis and the Constructive Mind in Action* (pp. 3–40). Greenwich, CT: Information Age Publishing. (Book Series: Advances in Cultural Psychology: Constructing Human Development, Vol. 5)

Koffka, K. (1915). Beiträge zur Psychologie der Gestalt- und Bewegungserlebnisse. III. Zur Grundlegung der Wahrnehmungspsychologie. Eine Auseinandersetzung mit V. Benussi [Contributions to the psychology of form and movement experiences. III. Foundations of cognitive psychology. A confrontation with V. Benussi]. *Zeitschrift für Psychologie und Physiologie des Sinnesorgane (1. Abt.)*, *73*, 11–90.

Koffka, K. (1935). *Principles of Gestalt Psychology*. New York, NY: Harcourt, Brace & Company, Inc.

Krueger, F. (1915). *Über Entwicklungspsychologie: Ihre sachliche und geschichtliche Notwendigkeit [On Developmental Psychology: Its Factual and Historical Necessity]*. Leipzig, DE: Verlag von Wilhelm Engelmann.

Krueger, F. (1953). Das Wesen der Gefühle [The nature of feelings]. In E. Heuss (Ed.), *Zur Philosophie und Psychologie der Ganzheit: Schriften aus den Jahren 1918–1940* (pp. 195–221). Berlin, DE: Springer. (Original work published 1928; reprinted from Archiv für die gesamte Psychologie, **65**, pp. 91–128, 1928)

Krueger, F., & Spearman, C. (1907). Die Korrelation zwischen verschiedenen geistigen Leistungsfähigkeiten [The correlation between various mental abilities]. *Zeitschrift für Psychologie und Physiologie der Sinnesorgane*, *44*, 50–114.

Laird, J. D. (2007). *Feelings: The Perception of Self*. New York, NY: Oxford University Press.

Lewin, K. (1936). *Principles of Topological Psychology* (F. Heider & G. M. Heider, Trans.). New York, NY, and London, GB: McGraw-Hill.

Matusov, E. (2007). In search of 'the appropriate' unit of analysis for sociocultural research. *Culture & Psychology*, *13*(3), 307–333.

Münsterberg, H. (1900). *Grundzüge der Psychologie [Handbook of Psychology]*. Leipzig, DE: Barth.

Pepper, S. C. (1942). *World Hypotheses: A Study in Evidence*. Berkeley, CA: University of California Press.

Pereira, S., & Diriwächter, R. (2008). Morpheus awakened: Microgenesis in daydreams. In E. Abbey & R. Diriwächter (Eds.), *Inno-*

vating Genesis: Microgenesis and the Constructive Mind in Action (pp. 159–186). Greenwich, CT: Information Age Publishing. (Book Series: Advances in Cultural Psychology: Constructing Human Development, Vol. 5)

Rudolph, L. (2006a). The fullness of time. *Culture & Psychology, 12,* 157–186.

Rudolph, L. (2006b). Mathematics, models and metaphors. *Culture & Psychology, 12,* 245–265.

Rudolph, L. (2006c). Spaces of ambivalence: Qualitative mathematics in the modeling of complex, fluid phenomena. *Estudios Psicologías, 27,* 67–83.

Sander, F. (1962a). Das Menschenbild in der neueren Psychologie [The Image of Man in Modern Psychology]. In F. Sander & H. Volkelt (Eds.), *Ganzheitspsychologie* (pp. 369–382). München, DE: C. H. Beck'sche Verlagsbuchhandlung.

Sander, F. (1962b). Experimentelle Ergebnisse der Gestaltpsychologie [Experimental results of Gestalt psychology]. In F. Sander & H. Volkelt (Eds.), *Ganzheitspsychologie* (pp. 73–124). München, DE: C. H. Beck'sche Verlagsbuchhandlung. (Originally presented in the *Bericht über den 10. Kongress der Deutschen Gesellschaft für Psychologie, Bonn 1927,* Jena, 1928)

Stout, G. F. (1888a). The Herbartian Psychology (i). *Mind, 13*(51), 321–338.

Stout, G. F. (1888b). The Herbartian Psychology (II). *Mind, 13*(52), 473–498.

Stout, G. F. (1889). Herbart compared with English psychologists and with Beneke. *Mind, 14*(53), 1–26.

Valsiner, J. (2000). *Culture and Human Development.* London, GB: Sage.

Valsiner, J. (2004). The street. *Khôra II, 5* (Ment, Territori i Societat), 69–84.

Valsiner, J. (2005). Affektive Entwicklung im kulturellen Kontext [Emotional development in a cultural context]. In J. B. Asendorpf (Ed.), *Enzyklopädie der Psychologie, Vol. 3: Soziale, emotionale und Persönlichkeitsentwicklung* (pp. 677–728). Göttingen, DE: Hogrefe.

Valsiner, J. (2007). *Locating the Self: Looking for the Impossible? Or Maybe the Impossible is the Only Possibility.* Retrieved from http://www.tu-chemnitz.de/phil/ifgk/ikk/cs/secure/Valsiner.pdf (Paper presented at the conference "Culturalization of the Self", Chemnitz, DE, December 1, 2007)

Valsiner, J., & Diriwächter, R. (2008). Returning to the whole: A new theoretical synthesis in the social sciences. In R. Diriwächter & J. Valsiner (Eds.), *Striving for the Whole: Creating Theoretical Syntheses* (pp. 211–238). New Brunswick, NJ: Transaction Publishers.

Valsiner, J., Diriwächter, R., & Sauck, C. (2004). Diversity in unity: Standard questions and non-standard interpretations. In R. Bibace, J. Laird, K. Noller, & J. Valsiner (Eds.), *Science and Medicine in Dialogue* (pp. 385–406). Stamford, CT: Greenwood.

Volkelt, H. (1922). Die Völkerpsychologie in Wundts Entwicklungsgang [Cultural psychology in Wundt's development]. In A. Hoffmann (Ed.), *Wilhelm Wundt — Eine Würdigung* (pp. 74–105). Erfurt, DE: Verlag der Keyserschen Buchhandlung.

Volkelt, H. (1962a). Simultangestalten, Verlaufsgestalten, und „Einfühlung" [Simultaneous Gestalts, historical Gestalts, and "feeling-in"]. In F. Sander & H. Volkelt (Eds.), *Ganzheitspsychologie* (pp. 147–158). München, DE: C. H. Beck'sche Verlagsbuchhandlung. (Originally presented in 1959 for the Festschrift für Friedrich Sander in Göttingen)

Volkelt, H. (1962b). Grundbegriffe der Ganzheitspsychologie [Basic concepts of holistic psychology]. In F. Sander & H. Volkelt (Eds.), *Ganzheitspsychologie* (pp. 31–65). München, DE: C. H. Beck'sche Verlagsbuchhandlung.

Wagerman, S. A., & Funder, D. C. (2009). Personality psychology of situations. In P. J. Corr & G. Matthews (Eds.), *The Cambridge Handbook of Personality Psychology* (pp. 27–42). Cambridge, GB: Cambridge University Press.

Wellek, A. (1950). *Die Wiederherstellung der Seelenwissenschaft im Lebenswerk Felix Kruegers [The Restoration of the Soul of Science in Felix Krueger's Lifework]*. Hamburg, DE: Richard Meiner Verlag.

Werner, H. (1959). *Einführung in die Entwicklungspsychologie [Introduction to Developmental Psychology]* (4th ed.). München, DE: Johann Ambrosius Barth.

Wohlfahrt, E. (1932). Der Auffassungsvorgang an kleinen Gestalten. Ein Beitrag zur Psychologie des Vorgestalterlebnisses [The perception of small Gestalts. A contribution to the psychology of the pre-Gestalt experience]. *Neue Psychologische Studien, 4,*

347–414. (Dissertation, Jena, 1925)

Wundt, W. (1894). Ueber psychische Causalität und das Princip des psychophysischen Parallelismus [About mental causality and the principle of psychophysical parallelism]. *Philosophische Studien, 10*(1), 1–125.

Wundt, W. (1922). *Grundriß der Psychologie [Foundations of Psychology]* (15th ed.). Leipzig, DE: Alfred.

8

Mind-Knots and Mind-Relations: Knot Theory Applied to Psychology[1]

Akio Kawauchi

A mathematical *knot* is a loop embedded in ordinary 3-dimensional Euclidean space $\mathbb{R}^3$. Less formally, a knot is a geometric idealization of a loop made by tying a physical knot in a physical rope, then splicing the ends of the rope to each other. The 2-*bridge knots* are particularly well understood; they include the trivial knot, figure-8 knot, trefoil knot and others illustrated below. In this chapter, we model minds by special 2-bridge knots that we call *mind-knots*.

Two knots K_1 and K_2 have the same *knot type* if K_1 can be continuously deformed into K_2 (allowing stretching and shrinking, but not cutting and re-splicing). The knot type of a mind-knot models *personality* as evaluated in the Five-Factor Model established by Eysenck (1947/1997, 1967/2006), Eysenck and Eysenck (1969, 1976), and Costa and McCrae (1988). Relationships among two or more minds —briefly, *mind-relations*—are modeled as *mind-links* composed of two or more mind-knots. In our Mind-Knot Model, "twistedness" of personalities and relationships is represented by knottedness and linkedness, respectively: a mind with "untwisted" personality is modeled by a *trivial knot*, and a group of individuals with mutually "unentangled" personalities is modeled by a *split link*.

We describe knot-theoretic operations on mind-knots and mind-links, which we call *mind-changes*, that model changes of person-

---

Keywords: Five-Factor Model, human mind, knot, knot diagram, link, mind-relations, personality, psychology, self-releasability relation.

[1]This chapter extends and improves on Kawauchi (2007).

Table 8.1: Knots, links, tangles, strings, etc., and states of mind.

JAPANESE (ROMANIZED)	ENGLISH
(sunaona seikaku)	tame/untangled personality/character
(hinekureta seikaku)	twisted personality/character
(omoi no ito)	string connecting human feelings
(kokoro no kotosen)	heartstring
(ningen kankai no motsure)	emotional entanglement of human relationships

alities and relationships in time. A mind-change of a mind-knot is defined "locally", but typically has the "global" effect of altering the knot type, *i.e.*, the personality represented by the mind-knot. Similarly, a mind-change of a mind-link is defined "locally", but typically has "global" effects, among them, changes in the entanglements of the component mind-knots. We define the *self-releasability relationship type* of a mind-link with $n \geq 2$ components, and classify these equivalence classes for $n \leq 3$.

Introduction: topology in psychology

Several years ago (*cf.* Kawauchi, 1998) the author concluded that there is no contradiction in considering the human mind as a knot, with its *knot type* representing the personality at a given time, and a *crossing change* operation being regarded as a 'mind-change'. One motivation is that in both English and Japanese various expressions refer to mental states using imagery from daily life involving strings (see Table 8.1).[2] Also, the author knew from Rucker (1984, p. 83) that B. Stewart and P. G. Tait (1894)[3] "seem[ed] to say that the *soul* exists as a knotted vortex ring in the æther", although the 'æther hypothesis' is known to be wrong, and the author does not clearly understand what Stewart and Tait meant.

For our purpose, the following points are difficulties we must overcome to construct our model (*cf.* Enomoto & Kuwabara, 2004):

---

[2][Other such expressions in European languages are to have a "knotted stomach" (French "estomac noué") or "frayed/frazzled nerves, nerves in tatters" (Portuguese "nervos em frangalhos"). (Ed.)]

[3]Tait is known as a pioneer of knot theory.

(I) **Birth-Time Mind Situation.** Since 'personality' may have a genetic component,[4] we cannot assume that the mind at birth is necessarily untwisted.

(II) **Mind-Changes and Environment Changes.** Mind-changes come from many sources, including (i) age, (ii) history, and (iii) non-standard events.[5]

Our knot model of a mind takes account of conditions (I) and (II).

We mention here some other applications of topology to psychology. E. C. Zeeman's work (1962) on the topology of the brain and visual perception, and the application of general topological spaces to psychology by Lewin (1938), were pointed out by R. Fenn and by J. Simon, respectively, with further remarks by S. Kinoshita, during the International Workshop on Knot Theory for Scientific Objects, held in Osaka, Japan, March 8–10, 2006. L. Rudolph reported to the author several applications of topology to psychology (Rudolph, 2006a, 2006b, 2006c), including a rehabilitation of Lewin's topological psychology (Rudolph, 2008, 2009). Incidentally, the author also learned of J. Valsiner's work on personality (1998) from Rudolph.

Mathematical knot theory

Knots and *links* are the main research objects in *knot theory*, an area within the mathematical field of topology. Our aim is to use knot theory as an aid to visualization of mind situations. The

---

[4][Some models of personality explicitly distinguish heritable (genetic) factors that make up 'temperament' from non-heritable (epigenetic, environmental) factors that make up 'character'. For instance, in the model created by Cloninger (1987; see also Cloninger, Syrakic, & Przybeck, 1993),

> The four temperament dimensions consist of Novelty Seeking, Harm Avoidance, Reward Dependence, and Persistence, which are theoretically assumed to be independently heritable and manifested early in life. [...] The three dimensions of character consist of Self-directedness, Cooperativeness, and Self-transcendence, which mature in adulthood and influence personal and social effectiveness by insight learning about self concepts. (Tomotake, Harada, Ishimoto, Tanioka, & Ohmori, 2003, p. 1163)

For the Five-Factor Model, heritability of all 'factors' has been supported by research on twins separated at birth (Jang & Livesley, 1996). (Ed.)]

[5][Compare (iii) to the "unexpected, one-time events" with "major impact on the ontogenetic level" mentioned on p. 15 and discussed in Chap. 7, *passim*. (Ed.)]

introduction here is quite condensed and mainly pictorial. For further details, the reader is referred to Kawauchi (1996).

Knots and knot types

Throughout this chapter, we mean by a *knot* an unoriented loop embedded in ordinary (Euclidean) 3-space $\mathbb{R}^3$, and by a *link* the union of one or more pairwise disjoint knots. Knots and links will be illustrated by planar *diagrams* with some *crossings* drawn in a standard manner to indicate places where one 'strand' passes over another.[6] Fig. 8.1 shows some examples of knot and link diagrams.

As previously mentioned, two knots K_1 and K_2 have the same *knot type* if K_1 can be continuously deformed in $\mathbb{R}^3$ until it coincides with K_2. Having the same knot type is an equivalence relation on knots. (Link type, defined similarly, is also an equivalence relation.) Different diagrams may depict knots or links of the same type (*e.g.,* in the top row of Fig. 8.1). However, a fundamental theorem of knot theory (Alexander & Briggs, 1926/1927; Reidemeister, 1926) asserts that two diagrams depict the same type of knot (or link) if and only if one can be changed to the other by a sequence consisting only of (a) continuous deformations of a diagram in the plane where it is drawn—technically, *isotopies* of the diagram; and (b) local applications, somewhere in a diagram, of the three types $RM1$, $RM2$, and $RM3$ of *Reidemeister moves* illustrated in Fig. 8.2. (for

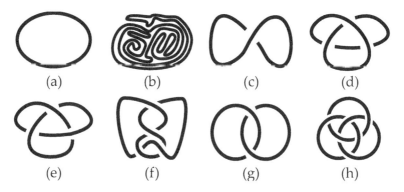

(a)   (b)   (c)   (d)

(e)   (f)   (g)   (h)

Figure 8.1: (a)–(d) Trivial knots. (e) Trefoil knot. (f) Figure-8 knot. (g) Non-trivial link of two trivial components. (h) Borromean rings.

---

[6][See Chap. 13 for some remarks on knot theorists' use of diagrams. (Ed.)]

proofs, see Crowell & Fox, 1963; Kawauchi, 1998).[7] Fig. 8.1(b), (c),
and (d) differ Fig. 8.1(a) by isotopy, *RM*1, and *RM*2, respectively.
In Fig. 8.3, Reidemeister moves are used to simplify a 6-crossing
diagram of a trefoil knot. (Shaded regions in diagrams show where
various moves are being locally applied, the unshaded part staying
untouched.)

A *crossing change CC* is, like the Reidemeister moves *RM*1,
*RM*2, and *RM*3, a "local" operation on a knot or link diagram
(Fig. 8.4(a)). Unlike *RM*1–*RM*3, which preserve knot and link type,
CC can (but need not) change them; see Fig. 8.4(b), which shows
how to turn a trefoil knot into a trivial knot with one crossing
change. A link diagram is called *split* if none of the crossings in
the diagram involve strands from two different component knots;
a link is *split* if it has some split diagram. More generally, two
disjoint sub-links L_1, L_2 of a link L are *splittable from each other* if
L has some diagram in which no crossing involves a strand from

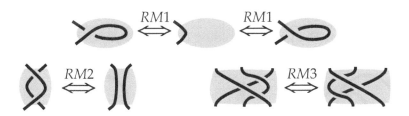

Figure 8.2: The three types of Reidemeister moves.

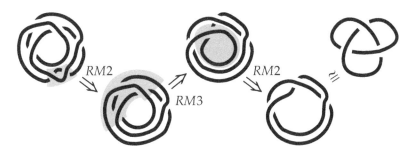

Figure 8.3: Simplifying a knot diagram with Reidemeister moves.

---

[7][The diagrammatic definition is more practical: it is much easier to manipulate
and modify knot and link diagrams drawn in two dimensions than to manipulate
or modify models of knots and links in three dimensions. (Ed.)]

L_1 and a strand from L_2.[8] Fig. 8.5 illustrates these concepts with the Borromean rings, which are obviously (and provably!) not split, although any two components are splittable.[9]

2-bridge knots and links

The 2-*bridge knots*, introduced by Bankwitz and Schumann (1934) and studied deeply by Schubert (1956), Conway (1970), and others, are now very well understood. In this section we reprise from

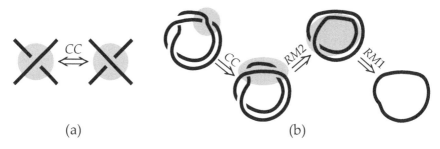

(a)   (b)

Figure 8.4: (a) The generic crossing change. (b) Trivializing a trefoil.

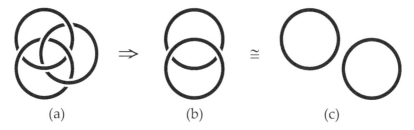

(a)   (b)   (c)

Figure 8.5: Splitting a 2-component sublink of the Borromean rings.

---

[8][For any given knot, some sequence of crossing changes turns it into a trivial knot, and for any given link some sequence of crossing changes turns it into a split link of trivial knots (see Kawauchi, 1996, Chap. 11). It is noteworthy that crossing changes occur in nature: enzymes called *topoisomerases* allow strands of DNA to pass through each other by performing physical crossing changes, without which DNA could not replicate within the confines of a cell (see Menasco & Rudolph, 1995, and references cited there and by Kawauchi, 1996). (Ed.)]

[9][Historically, this fact has often been felt to have powerful symbolic significance (*cf.* Thomson, Bruckner, & Bruckner, 2011, pp. 181–182). It is interesting to consider that symbolism in the light of the author's investigations of "self-releasability relations" below, p. 243 *ff.* (Ed.)]

Kawauchi (1996, Section 2.1) several standard definitions and results that will be used in the study of our Mind-Knot Model.

There are different characterizations of 2-bridge knots and links. Several are diagrammatic, including the following, which we use as our definition. First we define a *Conway normal form* (CNF) diagram $C(p_1, q_1, \ldots, p_n, q_n)$ by Fig. 8.6(b). Here n is a positive integer and $p_1, \ldots, p_n$ and $q_1, \ldots, q_n$ are *non-zero* integers,[10] and the $|p_1| + |q_1| + \cdots + |p_n| + |q_n|$ crossings of $C(p_1, q_1, \ldots, p_n, q_n)$ are grouped into *tangles* as in the figure. We then define a 2-*bridge knot* (or link) to be a knot (or link) that has some CNF diagram.[11]

This diagrammatic characterization requires a lot of data—all of the integers $p_1, \ldots, q_n$. Furthermore, CNF diagrams with different data can represent the same knot type—e.g., $C(p_1, q_1, \ldots, p_n, q_n)$ and $C(p_1, q_1, \ldots, p_n, q_n \pm 1, \mp 2, \mp 1)$ (so long as $q_n \neq \mp 1$); see Fig. 8.7.

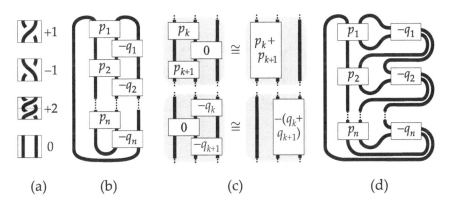

(a)   (b)   (c)   (d)

Figure 8.6: (a) Four *tangles*: the basic ± 1-tangles, a $+2$-tangle (composed of two copies of the $+1$-tangle) and a 0-tangle. (b) A CNF diagram $C(p_1, q_1, \ldots, p_n, q_n)$; a box labeled m contains an m-tangle comprising $|m|$ copies of the $m/|m|$-tangle ((a) shows this for $m = 2$). (c) Eliminating a 0-tangle reduces n to $n - 1$. (d) This diagram is isotopic to (b) but is not a CNF diagram.

---

[10] Allowing some p_i or q_j to be 0 does not expand the class of CNF diagrams (see Fig. 8.6(c)), and disallowing this has some formal advantages.

[11] A diagram like Fig. 8.6(b) but without any crossings would arise if we formally allowed either $n = 0$ or $n > 0$ with *all* p_i and q_j equal to 0. Such a diagram would be a 2-bridge presentation of a trivial knot. Trivial knots are actually 1-*bridge knots* (conversely, every 1-bridge knot is trivial), but here it is convenient to give them honorary 2-bridge status.

However, by imposing restrictions almost all of this ambiguity can be eliminated, as follows. Call a CNF diagram $C(p_1, q_1, \ldots, p_n, q_n)$ *special* all p_i and q_j have the same sign (positive or negative), and both $|p_1|$ and $|q_n|$ are greater than 1. It can be proved (*cf.* Kawauchi, 1996, Theorem 2.1.11, p. 25) that two distinct special CNF diagrams $C(p_1, q_1, \ldots, p_n, q_n)$ and $C(p'_1, q'_1, \ldots, p'_{n'}, q'_{n'})$ represent the same knot type of 2-bridge knot if and only if $n' = n$ and

$$(p'_1, q'_1, \ldots, p'_{n'}, q'_{n'}) = (-q_n, -p_n, \ldots, -q_1, -p_1)$$

(see Fig. 8.8 for an example).

Another characterization of 2-bridge knots and links, this one numerical, was given by Schubert (1956). It condenses the two sequences $\{p_1, \ldots, p_n\}$, $\{q_1, \ldots, q_n\}$ into *a single rational number* without losing track of the 2-bridge knot or link that they determine. We form the *continued fraction*

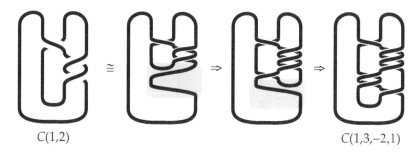

Figure 8.7: $C(-1, -2)$ is transformed into $C(-1, -3, 2, -1)$ by an isotopy and two Reidemeister moves of type *RM2*.

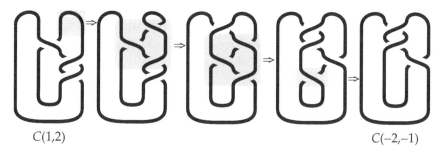

Figure 8.8: $C(-1, -2)$ is transformed into $C(2, 1)$ by a sequence of four Reidemeister moves of types *RM1*, *RM3*, *RM3*, and *RM1*.

Table 8.2: Rational invariants of some 2-bridge knots.

CONTINUED FRACTION & COMMON FRACTION	KNOT DIAGRAM	RATIONAL INVARIANT $\{P/Q, P/Q'\}$
$[-2,-1] = -2 + \dfrac{1}{-1}$ $= -3/1$	 negative trefoil	$\{1/(-1), 3/2\}$ [Check: $(-1) \times 2 - 1$ $= -3 = (-1) \times 3$.]
$[2,2] = 2 + \dfrac{1}{2}$ $= 5/2$	 figure-8	$\{5/2, 5/(-2)\}$ [Check: $2 \times (-2) - 1$ $= -5 = (-1) \times 5$.]
$[2,4,2,4] = 2 + \dfrac{1}{4 + \dfrac{1}{2 + \dfrac{1}{4}}}$ $= 2 + \dfrac{1}{4 + \dfrac{4}{9}}$ $= 2 + \dfrac{9}{40} = 89/40$	 $12a_{690}$	$\{89/40, 89/(-20)\}$ [Check: $40 \times (-20) - 1$ $= -801 = -9 \times 89$.]

$$[p_1, q_1, p_2, \ldots, q_n] := p_1 + \cfrac{1}{q_1 + \cfrac{1}{p_2 + \cdots + \cfrac{1}{q_n}}}$$

and then reduce it to a common fraction P/Q in lowest terms: *i.e.*, $P > 0$ and Q are integers with no common factor greater than 1.

It turns out that $C(p_1, \ldots, q_n)$ represents a knot if and only if P is odd. The *rational invariant* of $C(p_1, \ldots, q_n)$ is P/Q. Schubert's *classification of 2-bridge knots* says two CNF diagrams represent the same 2-bridge knot type if and only if their rational invariants P/Q, P/Q' either are identical or have $P' - P$ and $QQ' - 1$ an integer multiple of P. The *rational invariant* of a 2-bridge knot type is the set $\{P/Q, P'/Q'\}$ of all (*i.e.*, either one or two) distinct rational invariants of CNF diagrams of that type. (Number theory shows that there is exactly one Q' for which $-P < Q < P$ and $QQ' - 1$ is an integer multiple of P; but that Q' may be Q itself.) We let the rational invariant of the trivial knot type be $\{0, 0\}$. Some examples of 2-bridge knots are given in Table 8.2.

Schubert's classification is vital to our Mind-Knot Model.

Constructing a knot model of a mind

To clarify our viewpoint of a mind, in this section we first set up several axioms. We use evaluations of personality factors to determine several integers, which we interpret as measuring several manners in which a mind can be "twisted". In our initial Mind-Knot Model, some linear combinations of these measures, for a mind at age n, are used to create a 2-bridge knot model $M(n; a, b)$ of that mind. We classify the possible untwisted and twisted knots $M(n; a, b)$, and suggest further constructions of more refined mind-knot models.

Mind-knot hypotheses

(A) A *mind* is understood as a knot. An *untwisted mind* is a trivial knot. A *twisted mind* is a non-trivial knot.[12] (B) A *personality* is understood as a knot type; an *untwisted personality* is the knot type of a trivial knot, and a *twisted personality* is the knot type of a non-trivial knot. (C) A *mind-change* is understood as a *crossing change*.

To construct a concrete knot model of a mind, we consider several basic personality factors described by Eysenck (1947/1997, 1967/2006) and Eysenck and Eysenck (1969, 1976) as:

   (1)   Introversion-Extroversion,
   (2)   Neuroticism,
   (3)   Psychoticism,

and refined by Costa and McCrae (1988) in the Five-Factor Model (or Big-Five Model) as:

---

[12][N.B.: The author's use of "twisted" and "untwisted" is *topological*, not geometric—"twistedness" is independent of the extrinsic point of view imposed on a knot by a particular knot diagram; more generally, it is independent of the extrinsic point of view imposed on a knot type by any one of the knots it represents. As already seen in the top row of Fig. 8.1, even the trivial knot type has many diagrams with geometrically "twisted" appearance. The subtle interplay between intrinsic and extrinsic, and the challenge of discovering intrinsic properties of a knot or knot type behind its infinitely varied extrinsic manifestations, are among the attractions of knot theory to mathematicians. (Ed.)]

(1) Introversion-Extroversion,
(2) Neuroticism,
(3_1) Openness to Experience,
(3_2) Agreeableness, and
(3_3) Conscientiousness.

We evaluate the degrees of these five factors at age n years as follows (see *Big Five Personality Traits*, n.d., for a concrete test).[13]

Personality factor evaluations

(1) The *introversion-extroversion degree* at age n, denoted IE_n, takes the value -1 or 0, according to whether the mind at age n is introverted or not. (2) The *neuroticism degree* at age n, denoted N_n, takes the value -1 or 0, according to whether the mind at age n is neurotic or not. (3) The *psychoticism degree* at age n, denoted OAC_n, takes the value $O_n \cdot A_n \cdot C_n \in \{-1, 0\}$, where O_n, A_n, and C_n denote the *openness-to-experience degree*, the *agreeableness degree*, and the *conscientiousness degree*, respectively, at age n, and

(3_1) O_n takes the value 0 or -1,
(3_2) A_n takes the value 0 or -1,
(3_3) C_n takes the value 0 or -1,

according to whether or not the mind at age n is open to experience, agreeable, and conscientious, respectively.

We also need similar data representing the personality factors of Parents at the birth-time of the mind being modeled. Denoted IE_F, N_F, and OAC_F for Father and IE_M, N_M, OAC_M for Mother, they are defined analogously; *e.g.*, IE_F takes the value -1 or 0 according to whether or not Father's mind is introverted at birth-time, and OAC_M takes the value $O_M \cdot A_M \cdot C_M \in \{-1, 0\}$, where O_M is 0 or -1 according to whether or not Mother's mind is open to experience at birth-time, etc. Further, we define *Parents' Introversion-Extroversion Degree*, *Parents' Neuroticism Degree* and *Parents' Psychoticism Degree*, denoted IE_P, N_P, and OAC_P, respectively, by

$$IE_P := \ell_F \, IE_F + \ell_M \, IE_M$$
$$N_P := m_F \, N_F + m_M \, N_M$$
$$OAC_P := n_F \, KOAC_F + n_M \, OAC_M$$

---

[13]To simplify the exposition, we restrict n to be an integer.

in which ℓ_F, ℓ_M, m_F, m_M, n_F, and n_M are suitable non-negative integral constants. Here we take $\ell_F = \ell_M = m_F = m_M = n_F = n_M = 1$; in general, values should be chosen by careful estimation of the Birth-Time Mind Situation (*cf.* p. 229).

We define *total introversion-extroversion, neuroticism,* and *psychoticism degrees* of a mind at age n years, respectively, as follows.

$$IE[n] := IE_P + \sum_{i=1}^{n} IE_i$$

$$N[n] := N_P + \sum_{i=1}^{n} N_i$$

$$OAC[n] := OAC_P + \sum_{i=1}^{n} OAC_i$$

Let $a := IE[n] + N[n]$ and $b := OAC[n]$. We have

$$-2n - 4 \le a \le 0 \text{ and } -n - 2 \le b \le 0.$$

Our Mind-Knot Model of a mind at age n is the knot $M(n; a, b)$ represented by the knot diagram with $|a| + 2|b|$ crossings in Fig. 8.9(a). Consulting Fig. 8.6, we see that if either a or b is 0, then $M(n; a, b)$ is a trivial knot, and if a and b are both strictly negative, then $M(n; a, b)$ is a 2-bridge knot with CNF diagram $C(2b, a)$. For example, $M(n; -1, -1)$ is the negative trefoil and $M(n; -2, -1)$ is the figure-8 knot. (Compare Fig. 8.9(b) and (c) to Table 8.2, rows 1 and 2. N.B.: the figure-8 knot is its own mirror image.)

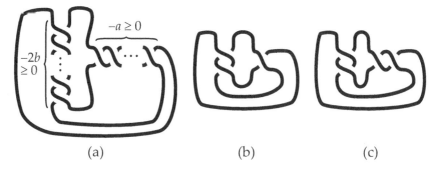

(a)   (b)   (c)

Figure 8.9: (a) A general mind-knot $M(n; a, b)$. (b) $M(n; -1, -1)$, the negative trefoil knot.(c) $M(n; -2, -1)$, the figure-8 knot.

Classification of mind-knots

Lemma. *If a and b are both strictly negative, then the rational invariant of $C(2b, a)$ is $\{(2ab + 1)/a, (2ab + 1)/(-2b)\}$.*

Proof. We calculate

$$[2b, a] = 2b + \frac{1}{a} = (2ab + 1)/a.$$

Now we must verify that (i) the numerator $2ab + 1$ is positive, (ii) the denominator a has no factor greater than 1 in common with $2ab + 1$, (iii) $-2b$ is in the range $-(2ab + 1) < -2b < 2ab + 1$, and (iv) $a \times (-2b) - 1$ is divisible by $2ab + 1$.

As to (i), we assumed $a < 0$ and $b < 0$, so $2ab + 1 > 2ab > 0$.

As to (ii), suppose $k > 0$ is a common factor of $2ab + 1$ and a, say $2ab + 1 = ks$ and $a = kt$ for some integers s and t; then k must be 1, for k divides $1 = (2ab + 1) - (2b)a = ks - (2b)kt = k(s - 2bt)$.

As to (iii), $-(2ab + 1) < -2b$ because $-(2ab + 1) < 0$ (by (i)) and $0 < -2b$ (by hypothesis); and $-2b < 2ab + 1$ because $0 < 2b + 2ab + 1 = 2(a + 1)b + 1$ by the same reasoning used for (i).

Lastly, as to (iv), $a \times (-2b) - 1 = -2ab - 1 = -1 \times (2ab + 1)$.   □

Proposition (Mind-Knot Classification). (i) *A mind-knot $M(n; a, b)$ is untwisted if and only if $a = 0$ or $b = 0$.* (ii) *Two twisted mind-knots $M(n; a, b)$, $M(n'; a', b')$ have the same personality if and only if $(a', b') = (a, b)$.*

Proof. Both statements follow from Schubert's classification of 2-bridge knots.

As to (i), we have already remarked that, if one or both of a and b equals 0, then $M(n; a, b)$ is a trivial knot, *i.e.*, untwisted, and has rational invariant $\{0, 0\}$. Conversely, if both a and b are non-zero, then $M(n; a, b)$ is a 2-bridge knot with rational invariant

$$\{(2ab + 1)/(-a), (2ab + 1)/(-2b)\},$$

and this is not $\{0, 0\}$, since $2ab + 1$ is an odd integer.

As to (ii), we know by (i) that $M(n; a, b)$ and $M(n'; a', b')$ are both genuine 2-bridge knots. By the classification theorem, they have the same knot type if and only if their rational invariants are equal. By the preceding lemma, these invariants are

$$\{(2ab + 1)/a, (2ab + 1)/(-2b)\}$$
$$= \{(2a'b' + 1)/a', (2a'b' + 1)/(-2b')\}.$$

In each of these pairs of rational numbers, one is strictly negative and the other is strictly positive: so $(2ab + 1)/a = (2a'b' + 1)/a'$ and $(2ab + 1)/(-2b) = (2a'b' + 1)/(-2b')$. Since all these fractions are in lowest terms, $a' = a$ and $b' = b$.   □

Mind-changes on mind-knots

When the knot diagram of $M(n; a, b)$ differs from the knot diagram of $M(n - 1; a', b')$, we consider that some *mind-change* has occurred between the $(n - 1)$st and nth birthdays. Thus, in our model, the total picture of a mind-knot from birth to the nth birthday ($n \geq 1$) can be considered as *a cylinder properly immersed in* $(3 + 1)$-*dimensional space-time* $\mathbb{R}^3 \times [0, n]$.

Fig. 8.10(d) illustrates the simplest mind-change that changes knot type; Fig. 8.10(a), (b), and (c) work up to that point. In Fig. 8.10(a), a cylindrical surface in $(2 + 1)$-dimensional space-time $\mathbb{R}^2 \times [0, 4]$ has been sliced at five levels ($t = 0, 1, \ldots, 4$). Each slice is a perfect circle depicted in its own (x, y)-plane labeled t.

In Fig. 8.10(b), a "dented" cylinder in $\mathbb{R}^2 \times [0, 4]$ has been subjected to the same treatment. The dent in the cylinder is reflected by dents in the slices at $t = 1, 2, 3$. Each of these slices is a loop, embedded in its own (x, y)-plane, but is not a perfect circle.

The cylinders depicted in Fig. 8.10(a) and (b) are *embedded* in $\mathbb{R}^2 \times [0, 4]$. Fig. 8.10(c) depicts a cylinder that is not embedded, but

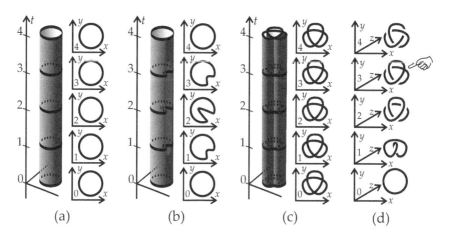

Figure 8.10: Four cylinders, sliced at five evenly-spaced levels.

only *immersed*, in $\mathbb{R}^2 \times [0,4]$. Again, the cylinder is sliced at five levels. Each slice is itself immersed, but not embedded, in its own (x,y)-plane. No crossings are depicted on these immersed planar loops, because there are no crossings to depict: these diagrams *are not knot diagrams*.

Finally, Fig. 8.10(d) depicts five slices of a cylinder immersed in space-time of $(3+1)$ dimensions. Unlike the preceding cases, in this case only the slices are shown; the cylinder in its totality is left to the imagination.[14] Each slice is a loop in its own (x,y,z)-space (depicted on the 2-dimensional page by a diagram with crossings). These loops, except the one at $t = 3$, are actually embedded, *i.e.*, they are knots, and their diagrams, which include crossings, are knot diagrams. In fact, they are mind-knots. At $t = 3$, however, the slice is a *non-embedded loop*; it has a *double-point* where indicated in the depicted *singular knot diagram*. For $0 \le t < 3$, the time-t slice of the immersed cylinder is a trivial mind-knot $M(t;0,0)$. For $3 < t \le 4$, the time-t slice is a negative trefoil mind-knot $M(t;-2,-1)$. The mind-change at $t = 3$ changes the knot type from trivial to trefoil.

Further constructions of human mind models

With a finer evaluation of the degrees of the five factors, we can re-fine our construction. By taking $a_P = IE_P + N_P$, $b_P = OAC_P$, $\tilde{a} = a - a_P$, and $\tilde{b} = b - b_P$, we obtain the knot model $M(n;a_P,b_P,\tilde{a},\tilde{b})$ of a mind illustrated in Fig. 8.11(a). If $b_P = 0$, then $M(n;a_P,b_P,\tilde{a},\tilde{b}) = M(n;0,0,\tilde{a},\tilde{b})$, and the proposition can be used for the classifica-tion. However, $M(n;a_P,b_P,a,b)$ with $b_P \ne 0$ generally has different behavior than $M(n;a,b)$. A further refinement can be obtained by decomposing $\tilde{a}$ and $\tilde{b}$ into numbers $a_i = IE_i + N_i$ and $b_i = OAC_i$ $(i = 1,2,\ldots,n)$ and considering the knot $M(n;a_P,b_P,a_1,b_1,\ldots,a_n,b_n)$ illustrated in Fig. 8.11(b). This refinement encodes all mind-changes from birth to age n very directly in the mind-knot.

If we allow arbitrary integers $c,c_P,\tilde{c},c_i$ instead of even integers $2b,2b_P,2\tilde{b},2b_i$ (*i.e.*, if we permit the use of arbitrary 2-bridge links), then our knot model becomes a link model, which represents a mind either by a mind-knot (twisted or untwisted) or by a *mind-link* that comprises two individually untwisted mind-knots. In the latter case, all the "twistedness" of the mind so represented must

---

[14]For further guidance in 4-dimensional visualization see Rucker (1984).

lie in the linking of the two components. Fig. 8.12(a) and (b) show by example that this linking can be quite complicated geometrically.

For the same example, Fig. 8.12(c) indicates the calculation of the *linking number* of the two components. In general, linking number is the first stage in algebraicizing the geometric linking of two (disjoint) knots. It is calculated by orienting both components (indicated in the figure by arrowheads), spanning one component by a two-sided surface (the light gray disk in the figure), and summing up how many times the other component passes through that surface, where a passage through the surface from its "top" side to its "bottom" side is reckoned as 1 or −1, respectively. The resulting integer (−4 in the figure), changes sign if the orientation of either

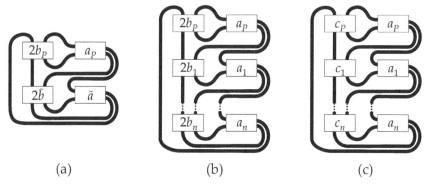

(a)   (b)   (c)

Figure 8.11: Refinements of the mind-knot construction.

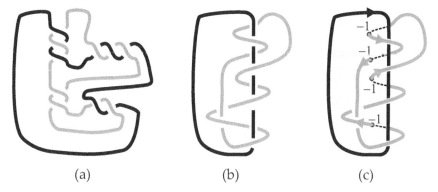

(a)   (b)   (c)

Figure 8.12: (a) A mind-link $M(n;3,3,2,2)$. (b) A link isotopic to $M(n;3,3,2,2)$. (c) The components of $M(n;3,3,2,2)$ have linking number −4 (shown) or 4; $M(n;3,3,2,2)$ is not a split link.

component, alone, is reversed. If a link is split, the linking number of any two components is 0 (the converse is false). More generally, if L is a 2-component link, and the linking number of its components (with some orientation) is ℓ, then L cannot be transformed to a split link by any series of fewer than $|\ell|$ crossing changes.

Mind-links and self-releasability

A natural extension of modeling one mind as a knot is to model several minds in relationship together as a link of several mind-knots, which we call a *mind-link*. In this context, we extend our definition of *mind-changes* to include also crossing changes between two different mind-knot components of a mind-link. Then every mind-link is generated by mind-changes on a split mind-link.[15] In this section, we describe the concept of *self-releasability relationship types* for links of $n(\geq 2)$ knots, in particular for mind-links. We classify these relationship types for $n = 2$ and $n = 3$.

Self-releasability relationship types

A crossing change CC of a link L is called *homogeneous* if the two strands involved in the change belong to the same component knot of L. Let L_1 and L_2 be disjoint sublinks of a link L. We say that L_1 *is self-releasable from* L_2 if L_1 can be transformed, using only homogeneous crossing changes on L_1, to a link L_1' that is splittable from L_2 as defined on p. 231. The special case of this in which L_1' is identical to L_1 (*i.e.*, no crossing changes are needed) is of course the same as L_1 and L_2 being splittable, and is therefore symmetric in L_1 and L_2. However, in the general case, there exist asymmetric examples in which L_1 is self-releasable from L_2 but L_2 is not self-releasable from L_1 (for instance, example 2-3 in Fig. 8.13, p. 245).

Given a link L with $\ell \geq 2$ component knots numbered $K_1, K_2, \ldots,$ K_ℓ, we define the *self-releasability relationship of L* to be the set of all pairs (I, J) of disjoint subsets of $\{1, \ldots, \ell\}$, say

$$I =: \{i_1, \ldots, i_p\}, \quad J =: \{j_1, \ldots, j_q\},$$

for which the disjoint sublinks

$$L_I =: K_{i_1} \cup \cdots \cup K_{i_p}, \quad L_J =: K_{j_1} \cup \cdots \cup K_{j_q}$$

---

[15][Compare this to note *8 on p. 232. (Ed.)*]

of L are such that L_I is self-releasable from L_J. Given two links L and L' with the same number $\ell \geq 2$ of component knots, we say that L and L' have *the same self-releasability type* if the components of L and L' can be numbered $K_1, K_2, \ldots, K_\ell$ and $K'_1, K'_2, \ldots, K'_\ell$, respectively, in such a way that the self-releasability relationships of L and of L' are identical. "Having the same self-releasability type" is an equivalence relation on links.

Let L be a mind-link, with component mind-knots representing individual minds mutually "entangled" in L. A sublink of L is also a mind-link. Let L_I and L_J be disjoint sublinks of L. We can understand the self-releasability of L_I from L_J as expressing that there are some homogeneous mind-changes on L_I (*i.e.*, mind changes of those individual personalities that are entangled in L_I) that entirely disentangle it from L_J. An interesting particular case is that in which $I = \{i\}$ is a 1-element set, so that $L_I = K_i$ is a single mind-knot (one individual's personality).

Self-releasability types of 2-component mind-links

Let L be a mind-link with two component mind-knots, A and B. Clearly the following is an exhaustive and exclusive enumeration of the logically possible self-releasability types L might have.

(1) Neither mind-knot component is self-releasable from the other. This relation is denoted by $A - B$.

(2) Each mind-knot component is self-releasable from the other (this is, for instance, trivially the case if L is split). This relation is denoted by $A \leftrightarrow B$.

(3) One mind-knot component is self-releasable from the other, but not conversely. This relation is denoted by $A \rightarrow B$ if A is self-releasable from B, and by $B \rightarrow A$ in the opposite case.

Proposition. (a) *Each of the types (1), (2), and (3) is realized by some mind-link $L = A \cup B$.* (b) *For type (2), L need not be split.* (c) *For type (3), the self-releasable mind-knot is necessarily twisted.*

Proof. We claim that Fig. 8.13 proves (a) and (b). To justify this claim and prove (c) we must give some technical discussion.

As to (a), if the linking number of A and B (see p. 242 *ff.*) is non-zero, then L falls into case (1). This is so because, as may be deduced from the earlier discussion, a homogeneous crossing change of one component leaves its linking number with the other component unchanged. The linking number of A and B in the example of type (a) is easily seen to be ± 1.

As to (b), the reader is invited to verify (with knot-diagrams and Reidemeister moves, or with physical models) that in the proposed example, a single homogeneous crossing change of the crossing adjacent to either of the labels A and B suffices to transforms $A \cup B$ into a split link (of two trivial components). Thus A is self-releasable from B, and B from A. There are various ways to show that $A \cup B$ is not itself split. One is to observe that it has the same link type as the 2-bridge link with CNF diagram $C(2,1,3,-1)$, which has continued fraction $[2,1,3,-1] = 8/3$ and rational invariant $\{8/3\}$; the rational invariant of the trivial 2-bridge 2-component link is $\{\infty, \infty\}$ (where the symbol ∞ stands for the non-numerical result of dividing an integer by 0).

As to (c), here we must move into algebraic topology. To any knot K in $\mathbb{R}^3$ there is associated a *knot group* G_K,[16] and to any knot K' disjoint from K a subset $[K'] \subset G_K$, called a *conjugacy class* (unique up to at worst a two-fold ambiguity dependent on choices of orientation for K and K'). Furthermore, (i) every conjugacy class G_K arises as $[K']$ for some knot K' disjoint from K, (ii) $[K']$ is the (1-element) conjugacy class of the identity element of G_K if and only if K' is self-releasable from K, (iii) $[K'] = [K'']$ (up to the mentioned ambiguity) if and only if some series of homogeneous crossing changes on K' transforms it into K'', and (iv) if $[K'] = [K'']$, then the linking number of K' and K has the same absolute value as the linking number of K'' and K. Finally, but fundamentally, it is a theorem that K is a trivial knot if and only if its knot group is isomorphic to the group $\mathbb{Z}$ of integers.

1   $A \;-\; B$ 2 $A \leftrightarrow B$ 3 $A \rightarrow B$

Figure 8.13: The three self-releasability types for two mind-knots.

---

[16][G_K is a group in the sense of Chap. 9. (Ed.)]

On the one hand, from these technicalities it follows at once that if mind-knots A and B have linking number 0, and both are untwisted, then $A \leftrightarrow B$. This proves (c). On the other hand, suppose that A and B have linking number 0, A is twisted, B is untwisted, and $[B]$ *is not the conjugacy class of the identity of* G_A. Then the technicalities show that $A \to B$. This is true, in particular, in the example of type (3) presented in Fig. 8.13, in which A is a trefoil knot. It is known (see, *e.g.*, Crowell & Fox, 1963) that G_A has a *group presentation* $\langle x, y \mid xyx = yxy \rangle$. The obviously trivial knot B can be shown to represent the conjugacy class of the element $x^{-1}y$ of that group, which is not the conjugacy class of the identity.   □

Self-releasability types of 3-component mind-links

We present our results on self-releasability of 3-component mind-links in Figs. 8.14–8.17 (adapted from Kawauchi, 2007, Figs. 6 and 7, with a correction of diagram 3-11) and Table 8.3.

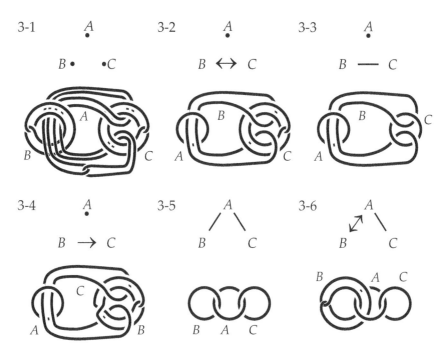

Figure 8.14: Self-releasability relationship types for three mind-knots.

In general, the self-releasability relationship of a non-splittable link of three components A, B and C consists (up to permutations of A, B, and C) of one *triangle relationship* on A, B and C, and a triplet of *1:2 relationships* each having one of four types.

1:2(1) Neither A nor $B \cup C$ is self-releasable from the other. This is denoted $A - BC$.

1:2(2) Each of A and $B \cup C$ is self-releasable from the other. This is denoted $A \leftrightarrow BC$.

1:2(3) A is self-releasable from $B \cup C$, but $B \cup C$ is not self-releasable from A. This is denoted $A \to BC$.

1:2(4) A is not self-releasable from $B \cup C$, but $B \cup C$ is self-releasable from A. This is denoted $A \leftarrow BC$.

The figures display examples of mind-links $A \cup B \cup C$ realizing all possible types of triangle relationships. For each example, the table gives the triplet of 1:2 relationship types.

We omit proofs of correctness for the figures and table, which are similar to the 2-component case but sometimes more complicated.

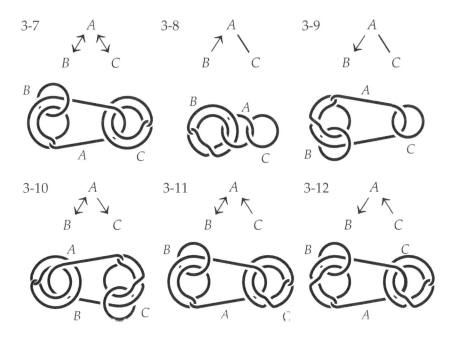

Figure 8.15: Self-releasability relationship types for three mind-knots (continued).

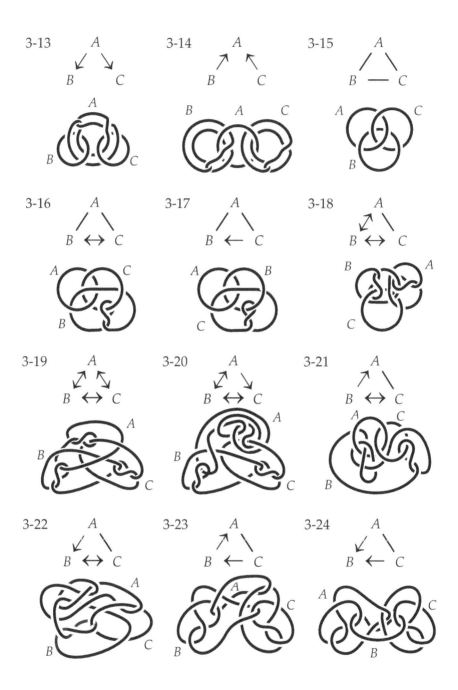

Figure 8.16: Self-releasability relationship types for three mind-knots (continued).

An interesting example is the Borromean rings, which has tri-angle relationship of type 3-1 (since, as shown in Fig. 8.5, any 2-component sublink is split), but, unlike the example for that type of triangle relationship in Fig. 8.14, has the triplet $A - BC$, $B - AC$ and $C - AB$. This "higher-order linking" property of the Borromean rings is a well-known consequence of the *link-homotopy classification* theorem of J. W. Milnor (1954). In terms of our mind-link model, a Borromean situation is one in which the minds of persons A, B, and C (whose individual personalities may or may not be twisted) are mutually entangled in such a way that any two of them are disentangled, but no individual mind-changes of A can release A from entanglement with the dyad $B \cup C$, nor can any individual mind-changes of B and C release the dyad $B \cup C$ from entanglement with A (and similarly with A, B, C permuted).

A complete classification of self-releasability relationship types for mind-links of 3 or more components appears possible but com-plicated, and remains as an open problem.

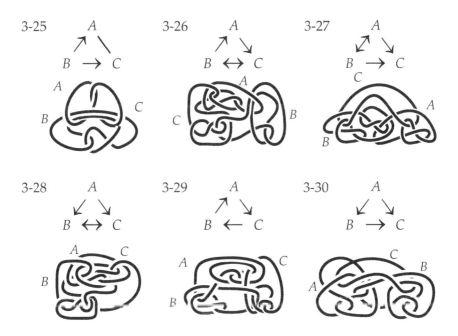

Figure 8.17: Self-releasability relationship types for three mind-knots (concluded).

Table 8.3: Triplets of 1:2 relationship types for the 30 types of triangle relationships.

Triangle type of $A \cup B \cup C$	1:2 relationship types of:		
	A and $B \cup C$	B and $A \cup C$	C and $A \cup B$
3-1	$A \leftrightarrow BC$	$B \leftrightarrow AC$	$C \leftrightarrow AB$
3-2	$A \leftarrow BC$	$B \leftrightarrow AC$	$C \leftrightarrow AB$
3-3	$A \rightarrow BC$	$B - AC$	$C - AB$
3-4	$A \leftarrow BC$	$B \rightarrow AC$	$C \leftarrow AB$
3-5	$A - BC$	$B - AC$	$C - AB$
3-6	$A - BC$	$B \leftrightarrow AC$	$C - AB$
3-7	$A \leftrightarrow BC$	$B \leftrightarrow AC$	$C \leftrightarrow AB$
3-8	$A - BC$	$B \rightarrow AC$	$C - AB$
3-9	$A - BC$	$B \leftarrow AC$	$C - AB$
3-10	$A \rightarrow BC$	$B \leftrightarrow AC$	$C \leftarrow AB$
3-11	$A \leftarrow BC$	$B \leftrightarrow AC$	$C \rightarrow AB$
3-12	$A - BC$	$B \leftarrow AC$	$C \rightarrow AB$
3-13	$A \rightarrow BC$	$B \leftarrow AC$	$C \leftarrow AB$
3-14	$A \leftarrow BC$	$B \rightarrow AC$	$C \rightarrow AB$
3-15	$A - BC$	$B - AC$	$C - AB$
3-16	$A - BC$	$B - AC$	$C - AB$
3-17	$A - BC$	$B - AC$	$C - AB$
3-18	$A - BC$	$B \leftrightarrow AC$	$C - AB$
3-19	$A \leftrightarrow BC$	$B \leftrightarrow AC$	$C \leftrightarrow AB$
3-20	$A \rightarrow BC$	$B \leftrightarrow AC$	$C \leftarrow AB$
3-21	$A - BC$	$B \rightarrow AC$	$C - AB$
3-22	$A - BC$	$B \leftarrow AC$	$C - AB$
3-23	$A - BC$	$B - AC$	$C - AB$
3-24	$A - BC$	$B \leftarrow AC$	$C - AB$
3-25	$A - BC$	$B \rightarrow AC$	$C - AB$
3-26	$A - BC$	$B \rightarrow AC$	$C \leftarrow AB$
3-27	$A \rightarrow BC$	$B \rightarrow AC$	$C \leftarrow AB$
3-28	$A \rightarrow BC$	$B \leftarrow AC$	$C \leftarrow AB$
3-29	$A - BC$	$B - AC$	$C - AB$
3-30	$A \rightarrow BC$	$B - AC$	$C \leftarrow AB$

Acknowledgments

This work was done while the author served as the program leader of the 21st century COE program "Constitution of wide-angle mathematical basis focused on knots" of the Ministry of Education, Culture, Sports, Science and Technology of Japan.

References

Alexander, J. W., & Briggs, G. B. (1926/1927). On types of knotted curves. *Annals of Mathematics, 28*(1–4), 562–586.

Bankwitz, C., & Schumann, H. G. (1934). Über Viergeflechte [On four-plats]. *Abhandlungen aus dem Mathematischen Seminar der Hamburgischen Universität, 70*, 263–284.

Big Five Personality Traits. (n.d.). Wikipedia, The Free Encyclopedia. Retrieved December 29, 2009, from http://en.wikipedia.org/wiki/Big_five_personality_traits

Cloninger, C. R. (1987). A systematic method for clinical description and classification of personality variants: A proposal. *Archives of General Psychiatry, 44*, 573–588.

Cloninger, C. R., Syrakic, D. M., & Przybeck, T. R. (1993). A model of temperament and character. *Archives of General Psychiatry, 50*, 975–990.

Conway, J. H. (1970). An enumeration of knots and links, and some of their algebraic properties. In J. Leech (Ed.), *Computational Problems in Abstract Algebra* (pp. 329–358). Oxford, GB: Pergamon.

Costa, P. T., & McCrae, R. R. (1988). From catalog to classification: Murray's needs and the five-factor model. *Journal of Personality and Social Psychology, 55*, 258–265.

Crowell, R. H., & Fox, R. H. (1963). *Introduction to Knot Theory.* New York, NY: Ginn and Co.

Enomoto, H., & Kuwabara, T. (2004). *Personality Psychology.* Chiba City, JP: University of the Air. (A textbook; in Japanese)

Eysenck, H. J. (1997). *Dimensions of Personality, a Record of Research Carried Out in Collaboration With H. T. Himmelweit [and Others].* New Brunswick, NJ: Transaction Publishers. (Original work published 1947)

Eysenck, H. J. (2006). *The Biological Basis of Personality.* New Brunswick, NJ: Transaction Publishers. (Original work published 1967; 2006 reprint of 1967 edition published by Charles C. Thomas, Springfield, IL)

Eysenck, H. J., & Eysenck, S. B. G. (1969). *Personality Structure and Measurement.* London, GB: Routledge.

Eysenck, H. J., & Eysenck, S. B. G. (1976). *Psychoticism as a Dimension of Personality.* London, GB: Hodder and Stoughton.

Jang, K. L., & Livesley, W. J. (1996). Heritability of the Big Five personality dimensions and their facets: A twin study. *Journal of Personality*, *64*(3), 577–591.

Kawauchi, A. (1996). *A Survey of Knot Theory*. Cambridge, MA: Birkhäuser.

Kawauchi, A. (1998). Physical-chemical systems and knots. *Have Fun with Mathematics (Nihon-hyoron-sha)*, *5*, 72–80. (In Japanese)

Kawauchi, A. (2007). A knot model in psychology. In A. Kawauchi (Ed.), *Knot Theory for Scientific Objects* (Vol. 1, pp. 129–141). Osaka, JP: Osaka Municipal University Press.

Lewin, K. (1938). *Principles of Topological Psychology*. Durham, NC: Duke University Press.

Menasco, W., & Rudolph, L. (1995). How hard is it to untie a knot? *The American Scientist*, *83*, 38–49.

Milnor, J. W. (1954). Link groups. *Annals of Mathematics*, *59*, 177–195.

Reidemeister, K. (1926). Elementare Begründung der Knotentheorie [Substantiation of elementary knot theory]. *Abhandlungen aus dem Mathematischen Seminar der Hamburgischen Universität*, *5*, 24–32.

Rucker, R. (1984). *The Fourth Dimension, a Guided Tour of the Higher Universes*. Boston, MA: Houghton-Mifflin Company.

Rudolph, L. (2006a). The fullness of time. *Culture & Psychology*, *12*, 157–186.

Rudolph, L. (2006b). Mathematics, models and metaphors. *Culture & Psychology*, *12*, 245–265.

Rudolph, L. (2006c). Spaces of ambivalence: Qualitative mathematics in the modeling of complex, fluid phenomena. *Estudios Psicologías*, *27*, 67–83.

Rudolph, L. (2008). A unified topological approach to *Umwelt*s and life spaces, Part II: Constructing life spaces from an *Umwelt*. In J. Clegg (Ed.), *The Observation of Human Systems: Lessons from the History of Anti-Reductionistic Empirical Psychology* (pp. 117–140). New Brunswick, NJ: Transaction Publishers.

Rudolph, L. (2009). A unified topological approach to *Umwelt*s and life spaces, Part I: *Umwelt*s and finite topological spaces. In R. I. S. Chang (Ed.), *Relating to Environments: A New Look at* Umwelt (pp. 185–206). Charlotte, NC: Information Age Publishing.

Schubert, H. (1956). Knoten mit zwei Brücken [Two-bridge knots]. *Mathematische Zeitung*, *65*, 133–170.

Stewart, B., & Tait, P. G. (1894). *The Unseen Universe or Physical*

Speculations on a Future State. London, GB: Macmillan.

Thomson, B., Bruckner, J., & Bruckner, A. (2011). *Mathematical Discovery*. Santa Barbara, CA, and Vancouver, BC: ClassicalRealAnalysis.com.

Tomotake, M., Harada, T., Ishimoto, Y., Tanioka, T., & Ohmori, T. (2003). Temperament, character, and eating attitudes in Japanese college women. *Psychological Reports, 92,* 1162–1168.

Valsiner, J. (1998). *The Guided Mind: A Sociogenetic Approach to Personality*. Cambridge, MA: Harvard University Press.

Zeeman, E. C. (1962). The topology of the brain and visual perception. In M. K. Fort (Ed.), *Topology of 3-Manifolds and Related Topics* (pp. 240–256). Englewood Cliffs, NJ: Prentice Hall.

9

A Novel Generic Conception of Structure: Solving Piaget's Riddle

Yair Neuman

"Structure" is a key term in fields ranging from psychology to biology. In his seminal monograph *Structuralism* (1970), Jean Piaget discussed the challenge of formalizing structure. Piaget considers a structure to be "a system of transformations" that comprises three key ideas (*ibid.*, p. 5): *wholeness*, *transformation*, and *self-regulation*. "Wholeness" is a defining mark of structure, which is, to use the famous Gestalt phrase, more than the sum of its parts. The idea of "transformation" emphasizes the dynamic nature of structure and the fact that it is constituted (*ibid.*, p. 13) as a system of transformations that map certain components of the system from t to $t' > t$. According to this suggestion, a structure is never static but, like Heraclitus's river, is a constant flux that under certain transformations appears invariant to a specific observer. (As "structure" is sometimes associated with a time-invariant form, the term "pattern" will be used interchangeably.) The third basic property, self-regulation, entails *self-maintenance* and *closure* (*ibid.*, p. 14). It means that the internal dynamics of the system constitutes the structure as a phenomenologically differentiated form. The structure is therefore not a mold passively formed by external forces but a form that actively maintains its existence from 'within'.

Piaget attempted to formulate the general meaning of 'structure' in algebraic terms, using a concept from mathematics known as

Keywords: category theory, groupoid, interdisciplinary research, mathematical group, morphism, Piaget, pattern, structuralism, structure.

a *group* (Piaget, 1970).[1] He was aware of the difficulty of applying this notion to irreversible systems that evolve over time and maintain their time-dependent structure (*i.e.*, pattern) through feedback loops. In this chapter, I propose a novel generic conception of structure—based on the mathematical concepts of *category theory*—that addresses the above difficulty. Although this conception is applicable to a variety of cognitive and biological phenomena, the chapter focuses only on the abstract notion of structure, leaving implications and applications for future studies. As the aim is to address readers with a wide range of mathematical backgrounds, the formalism is self-contained.

The group as a prototype of structure

One of Piaget's insights concerns how self-regulation is achieved in a structure which he conceived in algebraic terms, using the concept of a mathematical group. Self-regulation is important, as it explains self-maintenance and closure; the idea of a group is important, as it "may be viewed as a **kind of prototype of structures in general**" (Piaget, 1970, p. 19; emphasis added). At this point, a brief introduction to group theory is in order.

The defining properties of groups

In mathematics, a *group* is an algebraic structure consisting of (i) an *underlying set* G of elements and (ii) a *binary operation* $\star$ on G, *i.e.*, an operation on ordered pairs of elements of G. It is the following four properties of G and $\star$ together that make the structure a group.

(G1) *Closure with respect to $\star$.* The system is *closed* with respect to the given operation $\star$, in the sense that $\star$ yields only elements of G. In formal terms, for all a and b in G, $a \star b$ is in G: $a \in G$ and $b \in G$ imply $a \star b \in G$.

(G2) *Identity element with respect to $\star$.* The system contains an *identity* (or *neutral*) element j for the binary operation $\star$. That is, for each a in G, $a \star j = a$ and $j \star a = a$.

(G3) *Inversion with respect to $\star$.* There is an operation of *inversion* on G: each element a in G has an *inverse element* $a^{\leftarrow}$ in G

---

[1][These "groups" are unrelated to "groups" in the social sciences. (Ed.)]

which, when combined with a using the binary operation $\star$, yields the identity element. Formally, $a \star a^{\leftarrow} = a^{\leftarrow} \star a = j$.

(G4)   *Associativity with respect to* $\star$. For every a, b, and c in G, $(a \star b) \star c = a \star (b \star c)$ (*i.e.*, $a \star b \star c$ is unambiguous).

To help fix ideas, here are some examples and counterexamples, all with the familiar set $\mathbb{Z} = \{\dots, -2, -1, 0, 1, 2, \dots\}$ of *integers* (*i.e.*, the natural numbers, their negatives, and 0) as underlying set.

(G1')   *Closure*. If a and b are any two integers whatsoever, their sum $a + b$ is an integer (*e.g.*, $1 + 3 = 4$, $2 + (-4) = -2$). Thus $\mathbb{Z}$ is closed with respect to the operation $+$ of summation (addition). Similarly, their difference $a - b$ and product $a \times b$ are integers (*e.g.*, $5 - 3 = 2$, $(-4) \times (-2) = 8$), so $\mathbb{Z}$ is *also* closed with respect to the operations of subtraction ($-$) and multiplication ($\times$). Each of the operations $+$, $-$ and $\times$ keeps us within the boundaries of the system (algebraic structure) that it, together with $\mathbb{Z}$, defines. On the other hand, $\mathbb{Z}$ is *not* closed with respect to the operation $/$ of division: if a and b are integers, a/b may or may not be an integer (*e.g.*, $10/2 = 5$ is an integer, but $9/2$ is not; and for no integer a is $a/0$ defined, much less an integer). Thus $\mathbb{Z}$ *is not a group with respect to division* because property (G1) fails.

(G2')   *Identity*. Since $a + 0 = 0 + a = a$ for each integer a, we see that 0 is the identity for addition on $\mathbb{Z}$. Similarly, 1 is the identity for multiplication on $\mathbb{Z}$, since $1 \times a = a \times 1 = a$ for each integer a. However, there is no identity for subtraction on $\mathbb{Z}$. (Proof: suppose j *is* an identity for subtraction on $\mathbb{Z}$. Then $1 - j = j - 1$, so by arithmetic we calculate that $j = 1$. But also $2 - j = j - 2$, so $j = 2$. Therefore $1 = 2$, which is false; so the supposition that j is an identity for subtraction leads to a contradiction and is therefore false.) Thus $\mathbb{Z}$ *is not a group with respect to subtraction*: (G2) fails.

(G3')   *Inversion*. With respect to addition, every integer a has an inverse $a^{\leftarrow} \in \mathbb{Z}$, namely its *negative* $-a$, since $a + (-a) = 0 = -a + a$. However, with respect to multiplication, an integer a has no inverse $a^{\leftarrow} \in \mathbb{Z}$ unless $a = 1$ or $a = -1$: from arithmetic, for $a \neq 0$, $a \times (1/a) = 1$ and $(1/a) \times a = 1$, but unless $a = \pm 1$, the *reciprocal* $1/a$ of a is *not an integer* (reflecting the failure of $\mathbb{Z}$ to be closed under division). Thus $\mathbb{Z}$ *is not a group with respect to multiplication*: (G3) fails.

(G4')   *Associativity*. Addition is an associative operation on $\mathbb{Z}$ (*e.g.*,

$(1 + 2) + 3 = 1 + (2 + 3))$, as is multiplication (*e.g.*, $2 \times (3 \times 5) = (2 \times 3) \times 5$). On the other hand, subtraction is not an associative operation on the integers (*e.g.*, $1 - (2 - 3) \neq (1 - 2) - 3$). This provides a second reason that $\mathbb{Z}$ is not a group with respect to subtraction: (G3) fails.

Since the addition operation on $\mathbb{Z}$ has the closure, identity, inverse, and associative properties, *the integers are a group with respect to addition*. As noted in (G1')–(G4'), the integers are *not* a group under division (failure of closure), subtraction (failure of identity and associativity), or multiplication (failure of inverse).

Reversibility in groups

According to Piaget, the self-regulation of a structure, here interpreted in terms of groups, is built into the structure. To use modern terminology, the structure is *self-organizing* in the sense that it constitutes its closure and identity through its built-in internal logic. A key term for understanding this process is reversibility.

As Piaget points out (*ibid.*, p. 15), the binary operation in a group is *reversible* in the sense that it has an inversion operation. He argues that the reversibility results in *self-regulation of the system* and *constitution of the system's boundaries*, because an erroneous result is "simply not an element of the system (if $+n - n \neq 0$ then $n \neq n$)" (*ibid.*), which is a contradiction. In other words, the existence of inversion is closely associated with reversibility and identification of the structure.

Problems with the group as prototype of structure

Piaget was reflective and bold enough to realize the shortcomings of his analogy between a mathematical group and a "real world" structure, shortcomings that hinder even its heuristic value. He noticed that there is a crucial difference between mathematical structures and other structures, psychological or biological, whose transformations unfold over *time* (Piaget, 1970), and that there is a distinct class of structures whose transformations are governed by laws that "are not in the strict sense 'operations,' because they are not entirely reversible" (Piaget, 1970, p. 15). That is, the concept of group is limited as a "prototype of structure in general" because cognitive and biological systems are *irreversible*. We can better explain this point by using the concept of *symmetry*.

Symmetry and groups

The symmetry of an object describes its invariance under certain operations/transformations. The *symmetry group* of an object is a mathematical group: the underlying set G is the set of all maps (transformations) from the object to itself that preserve the object's structure (such a map is called an "automorphism" of the structured object), and the binary operation $\star$ is *composition of transformations* (*i.e.*, if the end result of transforming the object first by T_1 and then by T_2 is T_3, then $T_3 = T_2 \star T_1$). For example, certain objects are symmetrical under the operation of 'reflection': the exchange of each point of the object with its mirror image. Although our faces are not perfectly symmetrical, they are approximately symmetrical along the vertical axis, meaning that the left side of the face is a(n) approximate) mirror image of the right side.[2]

This notion of symmetry is intuitive when illustrated on simple objects that are reversible under certain transformations. However, the meaning of symmetry is less intuitive in the case of complex systems. It is sometimes convenient to describe a complex system by means of the notion of a network, but in networks "with no given geometrical embedding, these concepts [such as reflection] must be relaxed" (Holme, 2006, p. 1) and other definitions of symmetry may be proposed. For example, the symmetry group of the network is usually considered to be the automorphism group of the graph, *i.e.*, the maps of the graph onto itself that preserve edge-node conectivity or, in the case of a labeled directed graph (Stewart, 2004), the permutations of nodes that preserve edges, directions, and labels. In any case, the concept of group is limited in describing a complex cognitive system in terms of a network because in its simplest sense symmetry concerns a global property of a single object—in the above example the symmetry group of a network. Most natural or artificial systems (*e.g.*, text), however, are complex in that they are composed from heterogeneous objects organized in a hierarchy in which information is fed back and forth between different levels.

---

[2][The profusely illustrated book of Weyl (1952) is a classic discussion of symmetries and structures in mathematics and nature. (Ed.)]

Generalizing symmetry

In this context, reversibility is not a simple issue and we should study the "symmetry" of the system by adopting a different perspective. For example, instead of one ultimate reversible object we may look for "islands" of symmetric *relations* that are mapped across time and constitute the emerging and dynamic whole. However, as MacArthur, Sanchez-Garcia, and Anderson (2008, p. 3525) argue, "a systematic study of symmetry structure of real world complex networks—which typically contain ordered and disordered elements—has not yet been undertaken."

Despite the aforementioned difficulties, Piaget seems to point in the right direction—first, by rejecting both simple reductionist and holistic approaches to structure, adopting instead a kind of "relational epistemology" and arguing that "it is neither the elements nor a whole [. . .] *but the relations among elements that count*" (Piaget, 1970, pp. 9–10; emphasis added), and second, by describing a group as a prototypical structure, established through the inversion operation.

The shortcoming of the idea of a group with respect to formalizing reversibility and closure of patterns in complex dynamic systems may be addressed by using category theory as a mathematical modeling language to address the challenge of describing pattern.

Category theory: a self-contained introduction

The language of *category theory* describes similarities of phenomena that occur across many different mathematical fields (Adámek, Herrlich, & Strecker, 1990; Goldblatt, 1979; Lawvere & Schanuel, 2000). This language is ostensibly very simple since it deals with nothing but "objects" and "maps" (depicted by arrows) between those objects. In fact, however, its apparent simplicity can make category theory a general and powerful language for modeling beyond mathematics, specifically in fields where maps and transformations appear at different levels of abstraction. Nevertheless, with a few exceptions (*e.g.*, Ehresmann & Vanbremeersch, 2007; MacArthur et al., 2008; Neuman & Nave, 2008, 2009; Rosen, 2005), category theory has seldom been used to model cognitive or biological systems.

A category is specified by its *data* and its *rules*. The data comprise objects, maps, domain/codomain, identity map and composite map. The rules comprise identity and associative laws.

The data of a category

The first data component of a category C is its *objects* (denoted A, B, C, etc.). Objects can be anything you may think about, from people to groups. Interestingly, in category theory relations between objects take precedence over objects themselves (Adámek et al., 1990), making category theory perfect for modeling relational systems.

The second data component of a category C is *maps* between objects (denoted f, g, h, etc.). These maps (also called "morphisms") are merely a way of relating the objects to themselves and to each other.[3] Each map f of C relates one object of C, which serves as the *domain* of f, to one that serves as the *codomain* of f. The domain of f is the *source* object from which f originates, and the codomain of f is the *target* object where it ends. Typographically, map f from domain A to codomain B is represented as an arrow[4] from A to B,

$$f: A \to B \text{ or } A \xrightarrow{f} B.$$

It can happen that two particular objects A and B of C are unrelated (from the viewpoint of C), in the sense that there is neither any map $f: A \to B$ nor any map $g: B \to A$. It can also happen that A and B are multiply related, and have several maps between them (in one or both directions).

Each object A of C is related to itself by its *identity map* $A \overset{I_A}{\curvearrowright}$ or $I_A: A \to A$, which has domain A and codomain A. An identity map is evident in a tautology, when we say that $A = A$ (*e.g.*, "a dog is a dog"). The identity map $I_A: A \to A$ is the only map from A to A that is guaranteed to exist in C.

For each pair of maps $h: A \to B$ and $g: B \to C$ (where the domain of g is the codomain of h), there is a *composite map*

$$e = g \circ h: A \to C \tag{9.1}$$

---

[3][Another word for a "map", which seems to be used with equal frequency, is a "mapping". This terminological equivocation—between a nominalized verb form and an unambiguous noun—accurately reflects psychological equivocation (both within the mathematical community and by individual mathematicians) about the ontological status of "mappings"/"maps": are they processes or are they things? They are, of course, both. (Ed.)]

[4][By metonymy, "arrow" is sometimes used as another synonym for "map"/ "morphism"/"mapping". See Chap. 13, p. 412 *ff.*, for more on arrows. (Ed.)]

in the category C (the symbol "$\circ$" may be read as "following" or "after"). This generalizes the closure property (G1) of groups.

The rules of a category

The first rule of a category C, its *identity law*, is the requirement that for every object A, the manner in which I_A relates to all maps to or from A is analogous to the manner in which the identity of a group relates to all elements of that group: for all objects B and C of C, and all maps $f: A \rightarrow B$ and $g: C \rightarrow A$, we have

$$f \circ I_A = f \text{ and } I_A \circ g = g \qquad (9.2)$$

which generalizes the identity property (G2) of groups.

The second rule of a category C, its *associative law*, is the requirement that whenever $h: A \rightarrow B$, $g: B \rightarrow C$, and $f: C \rightarrow D$ are maps in C, then

$$f \circ (g \circ h) = (f \circ g) \circ h$$

i.e., $f \circ g \circ h: A \rightarrow D$ is an unambiguously defined map in C. This generalizes the associative property (G4) of groups.

An important type of map (within a category C) is isomorphism. A map $f: A \rightarrow B$ is called *an isomorphism* and the objects A and B are said to be *isomorphic* (in C), if there is a map $g: B \rightarrow A$ for which

$$g \circ f = I_A \text{ and } f \circ g = I_B \qquad (9.3)$$

(it follows that g also is an isomorphism). An isomorphism from A to itself is called an *automorphism of A in C*; e.g., I_A is an automorphism of A.[5]

Notice that we have not stated a generalization to categories of the inverse property (G3) of groups. That is because, in general, a category need not have any such property. The question of what it means for a category to have a generalized inverse property (as well as a generalized identity property stronger than equation (9.2)) is important; we address it below, p. 268 *ff.*

---

[5][The automorphisms of an object A in a category C are the underlying set of a mathematical group, the *automorphism group* of A in C, with $\circ$ as binary operation and I_A as identity element. (Ed.)]

Structures in categories

Despite its apparent simplicity, a category is a starting point that gives rise to incredibly abstract and complex structures. Following Piaget, the main objective here is to find an analogy in category theory for the inversion operation in group theory, and through this analogy to define structure in categorical terms. To achieve this aim, I introduce several basic definitions.

Initial objects and terminal objects

An object 0 in a category $\mathcal{L}$ is *initial* in $\mathcal{L}$ if, for every $\mathcal{L}$-object A, there is one and only one arrow from 0 to A in $\mathcal{L}$ (Goldblatt, 1979, p. 43). This arrow can be denoted 0_A.[6] If we think of a category as a directed graph, with a node for each object and a directed edge (arrow) for each map (as in Fig. 9.1), then an initial object is a node that sends one and only one arrow to each node, itself included. Since I_0 is an arrow from 0 to itself, and there is only one such arrow, we see that $I_0 = 0_0$. It is important to realize that multiple arrows may point *to* an initial object; the definition does not exclude the possibility that some arrow points to 0 from *another* object of $\mathcal{L}$.

As a simple example, consider the category $\mathcal{L}$ in Fig. 9.1. It has only three objects, 0, A, and B, related by eight maps. Identity maps I_0, I_A, and I_B are guaranteed to exist just because 0, A, and B are objects of $\mathcal{L}$. Initial maps 0_A, 0_B, and $0_0 = I_0$ exist by the assumption that 0 is an initial object in $\mathcal{L}$. The remaining maps, f, p, and q exist by decree. Given those assumptions, the data and rules of $\mathcal{L}$ force certain equations between maps to be true, *e.g.*, $f \circ 0_A = 0_B$ (because both have domain 0 and codomain B, and 0 is

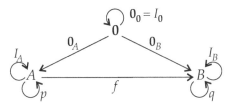

Figure 9.1: This directed graph, with three nodes and eight directed edges, depicts a category with three objects and eight maps.

---

[6][This notation can be read as "0 to A", with the preface "the unique map (or arrow) from" understood. (Ed.)]

initial), and $f \circ p = f = q \circ f$ (because f is assumed to be the only map with domain A and codomain B). Some freedom remains: it is consistent to assume/decree either $p \circ p = I_A$ or $p \circ p = p$; similarly, and independently, $q \circ q$ may be assumed/decreed to equal either I_B or q. In other words, $\mathcal{L}$ is not fully specified by its directed graph: more information about composition of maps in $\mathcal{L}$ (here, details of the "self-relations" of A and of B) is needed to pin down the exact structure of $\mathcal{L}$.

In category theory, many proofs are facilitated by use of a *diagram* , that is, a directed graph depicting only some of the objects and maps that belong to $\mathcal{L}$. To illustrate this proof technique (which mathematicians often call "diagram-chasing"), here is a proof of the useful fact that, while there may be multiple initial $\mathcal{L}$-objects, any two of them must be isomorphic in $\mathcal{L}$.

The diagram in Fig. 9.2 depicts two initial objects $\mathbf{0}'$ and $\mathbf{0}''$ of $\mathcal{L}$, along with just those maps of $\mathcal{L}$ that are required to exist by hypothesis: two identity maps (which exist simply because $\mathbf{0}'$ and $\mathbf{0}''$ are objects, and every object has an identity map), and two "initial maps" (which exist because we have assumed that $\mathbf{0}'$ and $\mathbf{0}''$ are initial). Since $\mathbf{0}'$ is initial, and $\mathbf{0}''$ is an $\mathcal{L}$-object, by definition there exists a map $\mathbf{0}'_{0''} : \mathbf{0}' \to \mathbf{0}''$. Similarly, since $\mathbf{0}''$ is initial and $\mathbf{0}'$ is an $\mathcal{L}$-object, there exists a map $\mathbf{0}''_{0'} : \mathbf{0}'' \to \mathbf{0}'$. Now, the composition $\mathbf{0}''_{0'} \circ \mathbf{0}'_{0''}$ is an arrow from $\mathbf{0}'$ to itself; but there is only *one* arrow from an initial object to itself, namely, its identity arrow, so we must have $\mathbf{0}''_{0'} \circ \mathbf{0}'_{0''} = I_{0'}$. Identical reasoning shows that $\mathbf{0}'_{0''} \circ \mathbf{0}''_{0'} = I_{0''}$. Thus $\mathbf{0}''_{0'}$ and $\mathbf{0}'_{0''}$ play the roles of f and g in the equation (9.3), so each is an isomorphism. This completes the proof.

Dually, an object $\mathbf{1}$ is *terminal* in category $\mathcal{L}$ if for every $\mathcal{L}$-object A there is one and only one arrow $_A\mathbf{1}$ from A to $\mathbf{1}$ in $\mathcal{L}$ (Goldblatt, 1979, p. 44)[7]; and chasing a diagram dual to that in Fig. 9.2 shows that any two terminal $\mathcal{L}$-objects must be isomorphic in $\mathcal{L}$.

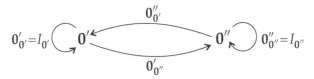

Figure 9.2: This diagram proves two initial objects are isomorphic.

---

[7][Dually to the notation for $\mathbf{0}_A$, the notation $_A\mathbf{1}$ can be read "A to $\mathbf{1}$", again with the preface "the unique map (or arrow) from" understood. (Ed.)]

The product in a category

The notion of product in a category $\mathcal{L}$ (Goldblatt, 1979, p. 47) is important because products are building blocks of hierarchical systems. Given two $\mathcal{L}$-objects A and B, their *product* is

(1) an $\mathcal{L}$-object denoted $A \times B$, equipped with

(2) two $\mathcal{L}$-maps $pr_A: A \times B \to A$ and $pr_B: A \times B \to B$ (called the projections from the product $A \times B$ to A and to B, respectively), as in Fig. 9.3(a), such that

(3) for any $\mathcal{L}$-object C, and any $\mathcal{L}$-maps $f: C \to A$, $g: C \to B$ from C to A and to B, respectively, there is exactly one map $\langle f, g \rangle: C \to A \times B$ such that the diagram in Fig. 9.3(b) is *commutative* in the sense that $pr_A \circ \langle f, g \rangle = f$ and $pr_B \circ \langle f, g \rangle = g$.

The map $\langle f, g \rangle$ is called the *product of f and g* with respect to the projections pr_A and pr_B.

The idea of a product of two objects A, B can easily be extended to a product $A \times B \times \cdots \times Z$ of any number of objects $A, B, \ldots, Z$.

Explaining the product

The product is described by Lawvere and Schanuel (2000, p. 269) as the "best thing of its type". Being "best" is encapsulated in the *universality* requirement (3): every other object equipped with maps to A and B must have *exactly one* map to the product $A \times B$ that makes the diagram Fig. 9.3(b) commutative. As Lawvere &

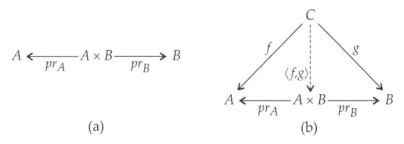

(a)   (b)

Figure 9.3: The product $A \times B$ of A and B is an object having projection maps pr_A to A and pr_B to B as in (a), such that, for any object C and maps f to A and g to B, there is a unique map $\langle f, g \rangle: C \to A \times B$ making diagram (b) commutative. (The arrow representing $\langle f, g \rangle$ is dashed to indicate that it is 'drawn later' than the solid arrows: they are 'data', it is a 'conclusion'.)

Schanuel explain, refuting the claim that our product is the "best thing of its type" would simply require that we show that there is another product object and that there is not exactly one map that makes the diagram commutative.

In an attempt to make sense of the product, Wikipedia (*Product (category theory)*, n.d.), as a non-academic compilation of knowledge, describes the "product of a family of objects" as the "most general" object admitting a morphism to each given object. This interpretation resonates with Lawvere and Schanuel's description of the product as "the best thing of its type". Following this idea the product can be considered a picture of the "most general object" in each family member (*i.e.*, object) to which it admits a morphism. In this sense, the product is a way to see each specific case from the perspective of the general object existing at a higher level of the hierarchy.

From a philosophical perspective, the product object is like the Platonic Idea or "form" represented in a concrete object (Ellerman, 1988). It is what Ellerman insightfully describes as a "concrete universal". For example, the idea of a Chair may be considered the best of its type with regard to the particulars armchair, barber chair, and wheelchair. These are concrete objects that somehow, in their limited way, represent the idea of Chairness. The universal property of the product and the idea that it is the best of its type may be expressed by the fact that any other object with maps to armchair, barber chair, and wheelchair must factor through Chair. The Chair is the "best of its type" in the sense that any higher-order conceptual category (*e.g.*, FURNITURE) to which the zero-level entities (*e.g.*, wheelchair) belong must uniquely factor through Chair.

The "most general object" that is the "best of its type" cannot simply be considered some kind of sum or average of the instances of that type; on the contrary, the idea of 'maximal generalization' is far from trivial. In cognitive sciences, one of the main problems in understanding concept formation is that different instances of a given concept (*e.g.*, Chair) do not share a fixed set of features that can be used to define the concept. This lack of essential features for that define a concept is usually referred to in the context of "family resemblance", a term coined by Wittgenstein (1953/2001) in his attack on *essentialism*, "the view that there must be something common to all instances of a concept that explains why they fall under it" (Glock, 1996, p. 120). For example, although chess, basketball, and throwing a ball in the air are all instances of Game, there is no set of features that uniquely defines the concept Game. Games are "united not by a single common defining feature, but by a complex network of overlapping and criss-crossing similarities, just as the different members of a family resemble each other in different respects (build, features, colors of eyes etc.)" (Glock, 1996, p. 121).

The idea of a product offers a unique perspective on the problem of family resemblance and the way concepts are organized. According to category theory, and as Wittgenstein argues, there is nothing "essential" about the "best of its type". It is the best of its type due to the universal property (3) of the product as an abstract structure, and the way it constrains information flow in the system. The objects have family resemblance in the structural-relational sense that they are all codomains of a single object (the product) and that any other object that functions as their codomain must factor uniquely through their product. This interpretation may help us to better understand the construction of a hierarchical system in which family resemblance is the 'glue' that holds lower-level objects together.

The co-product

The co-product is complementary to the product. To define this notion formally, we first introduce the idea of *duality* in a category (already mentioned informally in the definition of terminal objects, p. 264). The *dual* of a category $\mathcal{L}$, denoted $\mathcal{L}^{op}$, is constructed from $\mathcal{L}$ by (i) keeping the same objects and maps, while (ii) interchanging the domain and codomain of each map and (iii) replacing every equation $h = g \circ f$ (where "$\circ$" represents composition of $\mathcal{L}$-maps, as in equation (9.1), p. 261) by the equation $h = f \circ g$ (where now "$\circ$" represents composition of $\mathcal{L}^{op}$ maps; see Goldblatt, 1979, p. 45). In other words, to get $\mathcal{L}^{op}$ from $\mathcal{L}$, simply reverse all arrows.[8] Now we can define the *co-product* (or *sum*) of objects in $\mathcal{L}$ as the dual of the product, described identically except for the reversal of all arrows. That is, given two $\mathcal{L}$-objects A and B, their co-product is

(1) an $\mathcal{L}$-object $A + B$, together with
(2) two $\mathcal{L}$-maps $i_A : A \to A + B$ and $i_B : B \to A + B$ (called the *injections* into $A + B$ of A and of B, respectively) such that

---

[8][If, as above on p. 263, a category is thought of as a directed graph, then dualization consists of changing the direction of every edge. (Ed.)]

(3) for any $\mathcal{L}$-object C, and any $\mathcal{L}$-maps $f: A \rightarrow C$, $g: B \rightarrow C$ to C from A and from B, respectively, there is exactly one map $[f,g]: A + B \rightarrow C$ such that the diagram in Fig. 9.4 is commutative, *i.e.*, such that $[f,g] \circ i_A = f$ and $[f,g] \circ i_B = g$.

The map $[f,g]$ is called the *co-product* of f and g with respect to the injections i_A and i_B (Goldblatt, 1979, p. 54). As with the product, the idea of a co-product can easily be extended to multiple objects.

Wikipedia describes the co-product as the "least specific" object to which each object of the family admits a map (*Coproduct*, n.d.). Like the product, the co-product is a way to see the general object ("the best of its type") from the perspective of the specific objects. This distinction between "most general" and "least specific" may seem trivial because we usually and unconsciously conceive the particular in light of the general and the general in light of the particular. For example, the general concept DOG is constituted through our acquaintance with particulars (*e.g.*, German Shepherd or Poodle), while at the same time it recursively allows us to identify particulars as instances of the general (Neuman & Nave, 2008). However (*ibid.*), the dual is important for understanding recursive hierarchical systems with information flowing both bottom-up (*i.e.*, to the least specific) and top-down (*i.e.*, from the most general).

Identity and inversion in category theory

Our next step is to investigate the analogues, in a category in which an operation of multiplication (product) and/or addition (sum) is defined, of identity element and inverse with respect to that operation. The situation turns out to be rather different from that in groups (p. 256 *ff.*).

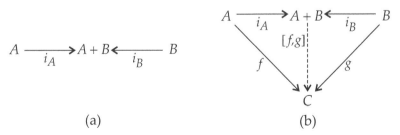

(a)   (b)

Figure 9.4: The co-product $A + B$ is the dual of the product $A \times B$; diagrams (a) and (b) are, respectively, dual to (a) and (b) in Fig. 9.3.

Identity objects in categories

Any terminal object **1** serves as an *identity object* for product (multiplication), in the sense that, for every object B, we have

$$B \times \mathbf{1} = B \tag{9.4}$$

with projection maps

$$pr_B = I_B : B \times \mathbf{1} \to B, \quad pr_1 = {}_B\mathbf{1} : B \times \mathbf{1} \to \mathbf{1} \tag{9.5}$$

(Lawvere & Schanuel, 2000); see Fig. 9.5(a). In other words, the product of any object with a terminal object is the object.

To prove this, we have to show that given any object C and maps $f : C \to B$, $g : C \to \mathbf{1}$ there is a unique map $\langle f, g \rangle : C \to B$ with

$$I_B \circ \langle f, g \rangle = f \text{ and } {}_B\mathbf{1} \circ \langle f, g \rangle = {}_C\mathbf{1}. \tag{9.6}$$

Because **1** is terminal, $g : C \to \mathbf{1}$ can only be ${}_C\mathbf{1}$, so (9.6) becomes

$$I_B \circ \langle f, {}_C\mathbf{1} \rangle = f \text{ and } {}_B\mathbf{1} \circ \langle f, {}_C\mathbf{1} \rangle = {}_C\mathbf{1} \tag{9.7}$$

The identity law for categories (equation (9.2) on p. 262), taken together with Fig. 9.5(b), shows that the first equation in (9.7) is satisfied if and only if $\langle f, {}_C\mathbf{1} \rangle = f$. The second equation in (9.7) is true for any choice of $\langle f, {}_C\mathbf{1} \rangle$, in particular, for f. So defining $\langle f, {}_C\mathbf{1} \rangle$ to be f makes (9.7), and therefore (9.6), true, which was to be proved.

The same logic applies to the co-product—dual of the product— and proves that an initial object **0** serves as an identity object for co-product (sum): $B + \mathbf{0} = B$. Note that these results are general and hold true in any category.

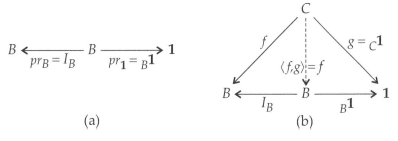

(a)   (b)

Figure 9.5: Equations (9.4) and (9.5), expressing the claim that **1** acts as an identity for ×, are represented diagrammatically in (a). The proof of the claim comes down to chasing the diagram in (b).

Is there an inversion operation?

After formulating the notion of an identity object in the language of categories, we move on to the next issue, which is to define the notion of an inversion operation in the language of categories. Recall that Piaget discusses structure in terms of the inversion operation in a mathematical group. Studying the inverse (reciprocal/ negative) in categorical terms may help us address the challenge of formalizing pattern in categorical terms, by looking for a notion of inversion that may be more adequate for describing pattern in complex systems.

Under the heading "Can objects have negatives?", Lawvere and Schanuel (2000, p. 287) insightfully suggest that if A is an object of a category, a *negative* of A should mean an object B such that $A + B \equiv 0$, where "+" represents the co-product of objects, "$\equiv$" represents isomorphism, and "0" is an initial object.

Let C be any object of the category. Because $A + B$ is isomorphic to an initial object 0, $A + B$ itself is an initial object, so by definition of initial object there is a unique map $(A + B)_C$. Now let f be any map from A to C, and g any map from B to C. Then $[f, g]$ is, by definition, a map $A + B \rightarrow C$, and therefore it is the only map $A + B \rightarrow C$, namely, $(A + B)_C$. Going back to the definition of initial object, there are injections $i_A : A \rightarrow A + B$ and $i_B : B \rightarrow A + B$ such that $f = [f, g] \circ i_A = (A + B)_C \circ i_A$ and $g = [f, g] \circ i_B = (A + B)_C \circ i_B$. These equations say that the only maps $f : A \rightarrow C$ and $g : B \rightarrow C$ are, respectively, $(A + B)_C \circ i_A$ and $(A + B)_C \circ i_B$. We have proved that A has the defining property of an initial object, namely, that the apparent wealth of choice in the phrase "let f be any map from A to C" is a Hobson's choice—no matter what C is, there is only one map to C from A. Exactly similarly, no matter what C is, there is

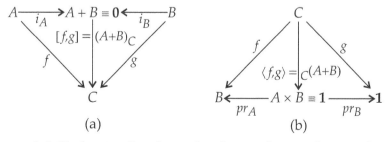

(a)   (b)

Figure 9.6: Understanding inversion for product and co-product.

only one map to C from B. We have proved that A and B are initial objects (see Fig. 9.6(a)).

The conclusion is that with regard to co-product, only initial objects have negatives; and, furthermore, their negatives/inverses are themselves initial objects. That is, $A + B \equiv 0$ if and only if $A \equiv B \equiv 0$. The dual conclusion also holds: for the product (dual of the co-product), only terminal objects have inverses (reciprocals), which are themselves terminal. That is, $A \times B \equiv 1$ if and only if $A \equiv B \equiv 1$ (see Fig. 9.6(b)).

These conclusions are striking. If only initial objects in a category have negatives, then reversibility is guaranteed only for them, and only under the condition of there being a co-product in that category. This statement and its dual (for product and terminal objects) mean that *reversibility through an inversion operation* can exist only in limited portions of a hierarchically structured category.

Interestingly, since with regard to product only terminals have reciprocals, a terminal object has a reciprocal only in the company of two other terminals (again, see Fig. 9.6(b)). Such a *triad* is composed of three isomorphic terminal objects that mutually constitute *equivalence relations*, as *each of them factors* through the others. They are "isomorphic subobjects" of each other (Goldblatt, 1979). This is an interesting situation, since with respect to the product each terminal is "the most general object" of the others. Similarly, in the dual case of the co-product, each initial is the "least specific object" of the others. What does this mean? In a hierarchical system, information is lost as we move up the ladder from the particular to the general. This triad, however, is a structure in which information is not lost, and therefore it is an "island" of reversibility and symmetry.

A novel concept of structure

A little more mathematical apparatus has to be introduced before proposing a generic solution to Piaget's riddle.

Groupoids

The mathematical structures depicted in Fig. 9.7, where the objects A, B,..., F all are terminal objects, and the heavy, striped, bidirectional arrows signify isomorphism, are so-called groupoids. In general, a *groupoid* is a small category[9] in which each map is an isomorphism; equivalently, it is "a [small] category in which each edge (morphism) is invertible" (Higgins, 2005, p. 5). In this context, the relation between group and groupoid is apparent. The symmetry group of an object is its automorphism group as defined on p. 262. A group is actually the specific case of a groupoid having just one object. In other words, "a group is a one object groupoid and a groupoid is a many object group" (Brown & Porter, 2006, p. 265). Note that some or all of the objects in a groupoid may be non-terminal objects (so that, *as groupoids*, the examples in Fig. 9.7 are atypically simple). Note also that Fig. 9.7 is *not* a "diagram" like those seen earlier: in it, the arrows do not denote isomorphisms, *i.e.*, maps, but rather the *relationship of isomorphism* between the objects they join—there must be at least one map-isomorphism in each direction, but there may be many. Also, Fig. 9.7 includes no visible indication of the possibly very extensive group of automorphisms at each object. We are operating at an even higher level of abstraction than before.

A general groupoid operation is a composition of maps between different objects and automorphisms of those objects, *e.g.*,

$$A \xrightarrow{\alpha} A \xrightarrow{f} B \xrightarrow{\beta} B \xrightarrow{g} C \xrightarrow{\gamma} C$$

where α, β and γ are automorphisms of A, B, and C respectively, and $f: A \to B$, $g: B \to C$ are isomorphisms. In a complex system, sequences of groupoid operations result in 'symmetric structure'; *i.e.*, there is a sequence of morphisms that ends back at the object where it originated. In a complex system it may be difficult to identify such

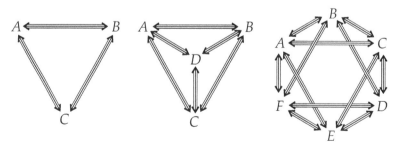

Figure 9.7: Groupoids with three, four, and six objects, respectively.

---

[9]A category is called "small" in case the collection of all its maps is a set (*i.e.*, not a so-called 'proper class' like the collection of all sets, which is not itself a set).

a path, but the notions of a groupoid and of a symmetric structure associated with it allow us to identify a pattern better than we could based on the concept of group *per se*. As Stewart (2004, p. 603) argues, "Groupoids have, on the whole, been somewhat removed from the mathematical mainstream [...]. However, they are the ideal tool for describing symmetries that apply only to *parts of the systems*" (emphasis in the original).

Groupoids as 'boundary objects'

A groupoid like those illustrated in Fig. 9.7 involves (1) product/co-product, and (2) at least three terminal/initial objects. I would like to suggest that such a structure be described as a *boundary object*, and the objects constituting it as *boundary sub-objects*. I would also like to suggest that, as islands of symmetry/reversibility, these boundary objects constitute and regulate the structure of the system from within. These groupoids are boundary objects because they are "islands of reversibility" that ensure the identity of the system from within. Following this suggestion, a structure/pattern cannot be considered in terms of a simple object that preserves its symmetry under transformations; rather, it is an abstract relational system that emerges from local symmetries between several objects. Moreover, the identity of the pattern is determined through the boundary objects and the morphisms that connect them. More specifically, the pattern is constituted by the mapping of this category across time.

To present a formal characterization of structure, we need to introduce one last concept from category theory, that of a *functor, i.e.,* a mapping from a category C_1 to a category C_2 that assigns objects to objects and maps to maps in such a way that (1) composition is preserved and (2) identity maps of C_1 map to identity maps of C_2.

A minimal definition of structure using category theory

As Piaget realized, cognitive and biological systems are dynamic systems that evolve over time. Therefore, structure simply does not exist in the static sense and would be better described in dynamic terms. Following Ehresmann and Vanbremeersch (2007) we can model such a dynamic system with the concept of a *state-category*. A state-category is actually a directed graph interpreted as a category. The change in the system from t to t' (for $t' > t$) can be modeled through a *transition functor* from the state-category at t to the state-category at t', *i.e.*, as a map of the *categorical structure* from t to t' (Ehresmann & Vanbremeersch, 2007).

Let $\mathcal{G}$ be the state-category at t. This state-category includes boundary objects. Let B_I denote boundary objects consisting of initials, B_T boundary objects consisting of terminals, and *Int* a set of *intermediate objects* that do not belong to B_I or B_T. In addition, assume $\mathcal{G}$ includes a set Φ of morphisms connecting B_I and B_T and having at least one object from *Int*.

The definition of the intermediate objects and the set of morphisms may be a matter of an *ad hoc* decision. For example, it was found (Zahn, 1971) that the "cut-point" of a graph is crucial for identifying wholes (*i.e.*, Gestalt structures) algorithmically; in that case, Φ may be defined as the cycle originating from B_T and passing through A, where the object A in *Int* functions as the cut-point of the graph.

We let *structure* as minimally defined in this chapter be a triple $(\mathcal{G}, \mathcal{G}', F)$ in which:

(1) $\mathcal{G}$ is a state-category at t, comprising
   (a) sets of boundary objects B_I and B_T,
   (b) a set of intermediate objects *Int*,
   (c) morphisms Φ between the boundary objects and having at least one object from *Int*;
(2) $\mathcal{G}'$ is a state-category at t', with analogous data; and
(3) F is a functor that maps $\mathcal{G}$ at t to $\mathcal{G}'$ at t'.

This generic conception is applicable to a wide variety of domains. Note, however, that moving from the abstract realm of mathematical formalism to noisy reality is far from simple. In the cognitive and biological realms, for example, terminal objects do not exist in the strict mathematical sense but can be identified only through softer (*e.g.*, statistical) measures. In this context, the abstract mathematical formalism introduced here should be considered a model with heuristic value only.

Acknowledgments

The author would like to thank Rob Goldblatt for a constructive reading of an early draft, Stephan Schanuel and Ophir Nave for clarifying some mathematical issues, Irun Cohen and Peter Harries-

Jones for ongoing dialogue and feedback, Jaan Valsiner for his support and Lee Rudolph for his intensive editorial work and thought provoking suggestions.

References

Adámek, J., Herrlich, H., & Strecker, G. E. (1990). *Abstract and Concrete Categories*. London, GB: John Wiley & Sons.

Brown, R., & Porter, T. (2006). Category theory: An abstract setting for analogy and comparison. In G. Sica (Ed.), *What is Category Theory?* (pp. 257–274). Monza, IT: Polimetrica Publisher. (Advanced Studies in Mathematics and Logic, Vol. 3)

Coproduct. (n.d.). Wikipedia, The Free Encyclopedia. Retrieved May 10, 2010, from http://en.wikipedia.org/wiki/Coproduct

Ehresmann, A. C., & Vanbremeersch, J.-P. (2007). *Memory Evolutive Systems*. New York, NY: Elsevier.

Ellerman, D. P. (1988). Category theory and concrete universals. *Erkenntnis, 28*, 409–429.

Glock, H.-J. (1996). *A Wittgenstein Dictionary*. Oxford, GB: Blackwell.

Goldblatt, R. (1979). *Topoi: The Categorial Analysis of Logic*. Amsterdam, NL: North Holland Publishing Company.

Higgins, J. P. (2005). Categories and groupoids. *Theory and Applications of Categories, 7*, 1–195.

Holme, P. (2006). Detecting degree symmetries in networks. *Physical Review E, 74*, 036107.

Lawvere, F. W., & Schanuel, S. H. (2000). *Conceptual Mathematics*. Cambridge, GB: Cambridge University Press.

MacArthur, B. D., Sanchez-Garcia, R. J., & Anderson, J. W. (2008). Symmetry in complex networks. *Discrete & Applied Mathematics, 156*, 3525–3531.

Neuman, Y., & Nave, O. (2008). A mathematical theory of sign-mediated concept formation. *Applied Mathematics & Computation, 201*, 72–81.

Neuman, Y., & Nave, O. (2009). Metaphor-based meaning excavation. *Information Sciences, 179*, 2719–2728.

Piaget, J. (1970). *Structuralism*. New York, NY: Basic Books.

Product (category theory). (n.d.). Wikipedia, The Free Encyclopedia. Retrieved May 10, 2010, from http://en.wikipedia.org/wiki/Product_(category_theory)

Rosen, R. (2005). *Life Itself*. New York, NY: Columbia University Press.

Stewart, I. (2004). Networking opportunity. *Nature*, *427*, 601–604.

Weyl, H. (1952). *Symmetry*. Princeton, NJ: Princeton University Press.

Wittgenstein, L. (2001). *Philosophical Investigations*. Oxford, GB: Blackwell. (Original work published 1953)

Zahn, C. T. (1971). Graph-theoretical methods for detecting and describing Gestalt structures. *IEEE Transactions on Computers*, *20*, 68–86.

Dynamic Models

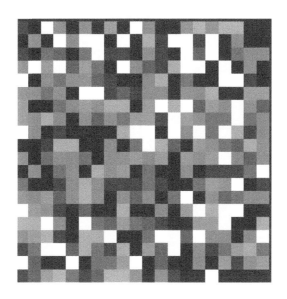

10

The Varieties of Dynamic(al) Experience

Lee Rudolph

> The knife-edge along which he must crawl
> [...] divided two kingdoms of force which had
> nothing in common but attraction. They were as
> different as a magnet is from gravitation,
> supposing one knew what a magnet was, or
> gravitation, or love. [...] This problem in
> dynamics gravely perplexed an American
> historian. (Adams, 1918, pp. 383–384)

In 2010, if any historians or other social scientists still embraced "Vitalism", it was not under that name; the once lively debate between "Evolutionists" and "Degradationists" described by Henry Adams (1910) seemed as dead as the debaters; and the alarm inspired in the human sciences by the conclusions for humanity (and the human sciences) that Faye (1885) and others had drawn from the Second Law of Thermodynamics (enunciated by physicists *c.* 1850) had apparently been dissipated or displaced by 20th century physicists' discoveries of nuclear fission and fusion, and the concomitant extension of the upper bound on how long life as we have come to know it can be energetically sustained by Earth's (fusion-powered) insolation and (fission-powered) internal heat production.

Yet despite such changes in particulars since the day in 1900 (recounted in Adams, 1918) when Adams, "an American historian", stood so "gravely perplexed" in the hall of dynamos at the Paris

Keywords: agency, dynamic system, dynamical system, evolution, laws of motion, laws of thermodynamics, Mechanism, space, statistical mechanics, time, *Weltformel*.

Exhibition, there remains room for grave perplexity—certainly in the mind of one American mathematician—at the use and meaning of 'dynamics', 'dynamic[al] systems', etc., in the human sciences. My aim in this chapter is to ponder and partly dispel this perplexity.

'Dynamical systems' or 'dynamic systems'?

I need first to clarify a point of usage and establish a convention.

Mathematicians speaking or writing English use the phrase 'dynamical system' much more often than the phrase 'dynamic system',[1] which in mathematical English (henceforth *math-Eng*) is its exact synonym. Precisely, I claim that in math-Eng the phrases are identically polysemous: there are several closely related, but distinct, families of mathematical entities whose members all can correctly be referred to by either phrase; and no mathematical entity is correctly referred to by one phrase but not the other. I take as evidence for this claim my understanding, as a speaker and writer of math-Eng, that both phrases are idioms—lexical atoms. That is, neither is correctly parsable in math-Eng as a putative math-Eng noun, 'system', modified by a putative math-Eng adjective, 'dynamic[al]'; for instance, math-Eng assigns no general, standard, mathematical meaning to the word 'system' standing alone, nor to the phrases 'non-dynamic system'/'non-dynamical system', as pragmatics would demand if the putative parsing were correct.[2]

Like mathematicians, social scientists writing English use both phrases; but they in their case favor 'dynamic system' over 'dynamical system', though by a smaller margin than that by which

---

[1]A search of the American Mathematical Society's *Mathematical Reviews* (MR) database of (mostly brief) reviews and authors' abstracts of more than 2,600,000 mathematical and paramathematical publications suggests that during the 20th and 21st centuries English-writing mathematicians have favored 'dynamical system' over 'dynamic system' by a factor between 6 and 7. (In other languages, mathematicians make do with no such choice; *e.g.*, French has no alternative to *système dynamique*, and German none to *dynamisches System*.) The JSTOR and MUSE databases (smaller, but containing full texts of articles) show even greater disparity.

[2]MR attests these phrases four times, all in 'systems engineering' contexts where 'system' is presumably a defined term. JSTOR attests a single use, written for (and by) ecologists: "However, these experiments were carried out with a non-dynamic system (i.e. 'static' press experiments with no feedback between predator and prey)" (Nelson, McCauley, & Wrona, 2001, p. 1223)   clearly a nonce coinage deployed rhetorically to emphasize a contrast in experiment types.

mathematicians favor the latter over the former.[3] Unlike mathematicians, at least some social scientists use both phrases and do not treat them as synonyms.

> There is an important distinction between a dynamic and a dynamical system. A dynamic system is one in which there is change — possibly continuous change. But there is a direct, predictable relationship between the input into the system and its outputs. [...] Reality and complex systems do not fit this mold. [...] *A dynamical system is a system which changes but for which changes in outcomes appear to bear no relationship to the changes in system input.* (Michaels, 1995, p. 21)

Even social scientists who use just one of the phrases may give it a definition much broader than any of those used in mathematics; *e.g.*, according to Lichtwarck-Aschoff, Kunnen, and van Geert (2009, p. 1365),

> A dynamic system is defined as a set of interconnected elements that affect each other over the course of time (Smith, Thelen, Titzer, & McLin, 1999; van Geert, 1994; van Gelder, 1998). No general rule exists about how to define a system. This choice depends on the universe of discourse and the phenomena in which the researcher is interested (van Geert, 2003).

Worse yet for the prospects of useful communication between groups are cases of 'dynamical system' and another *non-synonymous* term from math-Eng being treated as synonyms in, *e.g.*, psychological English:

> The theory of dynamical systems (sometimes called chaos theory) [...]. (Galatzer-Levy, 1995, p. 1085)

To simplify my exposition, I will reserve 'dynamical system' for mathematical meanings, and 'dynamic system' for other meanings.

---

[3]Psychological English (as represented by the American Psychological Association's databases PsycINFO, PsycARTICLES, and PsycCRITIQUES, searched on October 7, 2011) favors 'dynamic system' to 'dynamical system' by 3 to 2. Sociological English (as represented by EBSCO Publishing's SocINDEX with Full Text, "the world's most comprehensive and highest-quality sociology research database", EBSCO, 2011, same date) favors 'dynamic system' to 'dynamical system' by 7 to 2.

The role of time in dynamical systems

Not uncommonly, folklore, mythology, philosophy, and literature have depicted time as an *agent*, as, in its way, has mathematics. (See Table 10.1.) The way of mathematics being generally strewn with long trails of explicit definitions that all lead back sooner or later to either axioms, 'undefined terms', or (mathematically well-trained!) intuitions, mathematical depictions of time as an agent may not at first resemble those from those other distant universes of discourse. To begin with, in the more colloquial registers of mathematical discourse (*e.g.*, undergraduate textbooks, lectures at all levels, or conversations among mathematicians), occasional personification of mathematical entities is quite common ("the curve wants to be a circle", Garrity, 2002, p. 151); sometimes such usages become normalized into mathematical terms of art appropriate to graduate textbooks and scholarly publications ("the element $\iota \otimes \iota'$ [...] kills the element $\iota \cup \iota'''$", Peterson & Stein, 1959, p. 289). However, when it comes to "time" in mathematics, as far as I know personification occurs only in such etiolated forms as 'time acts/acting' and 'time action' (as illustrated in the last section of Table 10.1); it never bears the negative attributes —lethality, destructive voracity, mischievousness, or at best callous indifference —common to personifications of time outside mathematics (as in most of the preceding items in the table).

Further, whereas outside mathematics the patient (or object) of time-as-agent is *matter*, mathematical time-as-agent (with the possible exception of some—certainly not all—applications of mathematics in physics) has *space* as patient; or rather, *mathematical* spaces of many and various sorts, most of which do not, and are not intended to, model *physical* space at all (see Chap. 6 and Rudolph, 2006b, for more on the meanings of "space" in mathematics).

Finally, in mathematics an "action" of time (on a mathematical space) is invariably regular and non-capricious—in short, *lawful*.

The nature and scope of dynamical laws

Recollect from our Introduction (p. 13 *ff.*) that it is essentially universal among contemporary mathematicians[4] to take the Peano Postulates for the structured set $\mathbb{N}$ as a license to assume as *already*

---

[4]Notes 11 and 12 on p. 14 give references to some exceptions.

Table 10.1: Time as an agent.

Universe	Depiction
Folklore	Time "is such a glutton that he does not spare even his own children, and when all fails, he devours himself and then springs up anew" (Basile, 1636/1894, p. 171, and generally Thompson motif Z122, Thompson, 1960, p. 798).
Mythology	Greek, 476 B.C.E.: χϱονος ο παντων πατηϱ (Pindar, 1882, p. 7), *i.e.*, "time the father of all things".
	Hindu, *c.* 400 B.C.E.: Kala (time) plays a game of dice with Ka-li (death) to determine men's destinies (Bhartṛhari, 100?/1886, p. 43; Lüders, 1907, p. 66).
	Roman, 8 C.E.: *tempus edax rerum* (Ovid), *i.e.*, "tyme the eater up of things" (Golding, 1567/1806, p. 300).
Philosophy	Αἰὼν παῖς ἐστι παίξων πεσσεύων· παιδὸς ἡ βασιληίη (He-raclitus, *c.* 500 B.C.E.), *i.e.*, "Time is a child playing at draughts, a child's kingdom" (Patrick, 1888, pp. 688 and 662).
	"Partout où quelque chose vit, il y a, ouvert quelque part, un registre où le temps s'inscrit. Ce n'est là, dira-t-on, qu'une métaphore. — Il est de l'essence du mécanisme, en effet, de tenir pour métaphorique toute expression qui attribue au temps une action efficace et un réalité propre." [*Wherever anything lives, there is, open somewhere, a registry where Time signs in.* Nothing to that but a metaphor, they'll say. — It is of the essence of Mechanism, in fact, to take as metaphorical any expression that attributes to Time effectual activity and a reality of its own.] (Bergson, 1908, pp. 16–17.)
Literature	"Since I last had the misfortune of leaving you, time has bin strangely active" (Boyle, 1655, p. 345).
	"[M]ischief-making Time would never dare / Play his ill-humored tricks upon us two" (Brown, 1932/1996, p. 119).
Axiomatics	"We will use the modern axiomatic definition of a dynamical system. [...] *A* (real) *flow on a space X is a* (time) *action on the points of X satisfying the group property:* $x \in X$, $t,s \in \mathbb{R}$ imply that $x_{(t+s)} = (x_t)_s$." (Stroyan & Luxemburg, 1976, p. 155).
	"the infinitesimal generator of a new time action" (Echeverría-Enríquez, Muñoz-Lecanda, & Román-Roy, 1995, p. 5560).
	"The action of a continuous map $f:M \to M$ on a topological space M is called a discrete time dynamical system." "A discrete time dynamical system can be thought of as a 'flow' in which time is measured in discrete time increments" (Burns & Gidea, 2005, p. 94; the quoted sentences appear in the opposite order).

"generated" (and thus available *now*) the entire infinite sequence of counting numbers, faultlessly and without omission or redundancy, by extra-temporal iteration of a single, lawful operation "next!".[5] By "lawful" in this case I mean simply that, on *every* occasion of counting, what is next after | is ||, what is next after || is |||, ..., what is next after ||||| ||||| ||||| || is ||||| ||||| ||||| |||, and so forth and so on, independent (except for a choice of notation) of the occasion.

Rather similarly, the "modern axiomatic definition of a dynamical system" implicitly assumes that the infinitely extended future and past of the system are already "generated" (and thus available now), faultlessly and without omission or redundancy, by extra-temporal application of a single, lawful operation, the operation in this case being a distinctive (indeed, defining) feature of whichever dynamical system is under consideration.[6]

Thus, for a given "discrete time dynamical system" (Burns & Gidea, 2005), iteration of a single operation—the given "continuous map $f: M \to M$"—lays down a law necessary and sufficient to determine the entire *discrete flow* at every click $t = 0, 1, \ldots$ of a discrete-time clock: for any point $x_0 \in M$, its (forward) *trajectory*

$$x_0, x_1 := f(x_0), x_2 := f(x_1) = f(f(x_0)), \ldots,$$

$$x_n := f(x_{n-1}) = \overbrace{f(\cdots(f(x_0))\cdots)}^{n \text{ times}}, x_{n+1}, \ldots$$

is completely and exactly determined by f. In modeling, the mathematical space M often represents the *state space* of some system, *i.e.*, each mathematical point $x \in M$ represents a single *state* of that system, and every state of the system is represented by exactly one point; in such a case, the trajectory of x represents the 'development' of the system (in discrete time) through successive states x_n, starting from the *initial condition* $x = x_0$.

Similarly, "the group property [...] $x_{(t+s)} = (x_t)_s$" assumed by Stroyan and Luxemburg (1976) readily implies an enormously

---

[5]Once arithmetical operations on the counting numbers have been defined (making essential use of the fifth postulate, Mathematical Induction), "next!" can be replaced by "add 1!"; but *nextness*, the essence of mere counting, is a feature of the (primitive) *order structure* of $\mathbb{N}$, unlike *addition*, a feature of the (constructed) *arithmetic structure* of $\mathbb{N}$.

[6]A minor dissimilarity is that there is an intrinsic "beginning" to counting. A major dissimilarity is that the Peano Postulates are "categorical", *i.e.*, any two mathematical objects that satisfy them are isomorphic (in those aspects addressed by the postulates), whereas there are many non-isomorphic dynamical systems.

strong regularity property of a *real flow* on X: for any time interval $a < t < b$, no matter how short its duration $b - a$, the corresponding collection of data $\Phi^{(a,b)} := \{\varphi^{\{t\}} : a < t < b\}$ (where $\varphi^{\{t\}}$ denotes the 'time-t map' $X \to X : x \mapsto x_t$ of the real flow) *determines the flow completely*, that is, determines the much larger collection of data $\Phi^{\mathbb{R}} := \{\varphi^{\{t\}} : t \in \mathbb{R}\}$. Here, the (complete) *trajectory* of a point $x = x_0 \in X$ is $\{x_t : t \in \mathbb{R}\}$, and the translation of this regularity property into the vocabulary of development is the proposition that an arbitrarily short segment $\{x_t : a < t < b\}$ of the development of any given state of a system governed by a real flow suffices to determine that state's entire development, $\{x_t : t \in \mathbb{R}\}$, both future and past. If the space X is assumed to be a *differentiable manifold* (which is nearly always so in applications), and the time action on X is "sufficiently smooth", then the regularity property can be made stronger yet: there is, as in the case of discrete-time actions, a *single* datum called (as by Echeverría-Enríquez et al., 1995, in Table 10.1) an *infinitesimal generator* of the flow,[7] which lays down the law determining the flow—and the trajectories (developmental histories) of each x—for all times, past and future.[8]

---

[7] The infinitesimal generator is a certain "vector field" on X, obtained from any $\Phi^{(a,b)}$ by a limiting process in which $b - a$ shrinks to 0.

[8] A refinement of the "group property" is both useful and necessary if many systems that are important in practice are to be included as "dynamical" systems. Like the "group property", this 'semi-group property' occurs in two variants, discrete and continuous.

In the discrete case, as Burns and Gidea point out (*ibid.*, p. 95), unless f is assumed to have a (continuous) *inverse*, the discrete-time clock can only be assumed to run forward. In technical language (not used by Burns and Gidea) what they have defined is a *discrete-time semi-flow*. An example of a discrete-time non-flow semi-flow (from 1878!) is given in note 25 (p. 298). Other examples include many cellular automata, like the one designed by Poddiakov (2000/2006) that is described and illustrated in Chap. 12, pp. 355–358. For that automaton, a state is an assignment of one of 9 colors to each of the 3,600 cells, so M is a discrete topological space containing $3{,}600^9 = 1.01559956668416 \times 10^{32}$ points; the function $f : M \to M$ can be derived straightforwardly from the "transition rules" given on p. 355.

In the continuous case, for a *continuous-time semi-flow* the *1-parameter semi-group* $\mathbb{R}_{\geq 0}$ of non-negative real numbers plays the role that the 1-*generator semi-group* $\mathbb{Z}_{\geq 0} = \{0, 1, 2, \dots\} = \mathbb{N} \cup \{0\}$ of non-negative integers plays in the discrete case. Continuous-time semi-flows are the subject of a distinct subfield of dynamical systems theory, and are particularly suited to quantum mechanics (see, *e.g.*, Beltrametti & Cassinelli, 1981) and the study of "dissipative systems" (see, *e.g.*, Milani & Koksch, 2005). Semi-flows that are not flows do not determine the past.

Explications of the system of the world

Stroyan and Luxemburg's phrase "modern axiomatic definition of a dynamical system" suggests that mathematicians had studied "dynamical systems" under that name before the "modern axiomatic definition" had been formulated. As a matter of (unsurprising) fact, mathematicians had begun to study "dynamical systems" before they named them. In this section I present some pre-history of "dynamical systems" that is particularly relevant to this book.

Mathematical principles of natural philosophy

G. D. Birkhoff, addressing the 1919 annual meeting of the American Association for the Advancement of Science as "vice-president and chairman of Section A—Mathematics and Astronomy", asserted

> The concept of a dynamical system did not exist prior to Newton's time. [...] Newton was able to deal with the Earth, Sun, and Moon as essentially three mutually attracting particles, and [...] to formulate their law of motion by means of differential equations. [...] *Such a set of ordinary differential equations form the characteristic mathematical embodiment of a dynamical system* [...]. (Birkhoff, 1920, p. 51; emphasis added)

This is hard not to see as a clear case of putting "Newton's Dynamics in Modern Mathematical Dress" (a chapter title from Brackenridge, 1996). Birkhoff, an innovative expert in the theory of dynamical systems,[9] was not an historian of science. Nor am I; but he seems to me to be projecting into "Newton's time" a modern understanding of "[t]he concept of a dynamical system" as a structure characteristically embodied by "a set of ordinary differential equations".

Newton certainly did both "formulate" and "treat" the "law of motion" of "the Earth, Sun, and Moon"; he did so, as Birkhoff (*ibid.*) points out a few sentences later, "from a geometrical point of view."

---

[9]One of Birkhoff's great discoveries was that "In a very deep sense the periodic motions [of a dynamical system in his sense] bear the same kind of relation to the totality of motions that repeating doubly infinite sequences of integers 1 to 9 such as ...2323... do to the totality of such sequences" (Birkhoff, 1920, p. 54; Birkhoff's ellipsis points "..." indicate bidirectional indefinite extension). As might have been expected, Birkhoff's "symbolic dynamics" has long since come to lead a life of its own, with differential equations far in the background or entirely invisible.

And it is the glory of Geometry that from those few principles, fetched from without, it is able to produce so many things. Therefore Geometry is founded in mechanical practice, and is nothing but that part of universal Mechanics which accurately proposes and demonstrates the art of measuring. [...] In this sense Rational Mechanics will be the science of motions resulting from any forces whatsoever [...], accurately proposed and demonstrated. [...][W]e give an example of this in the explication of the System of the World [...] the motions of the Planets, the Comets, the Moon, and the Sea. (Newton, 1686/1729, Author's Preface, Vol. 1, pp. viii–x).

While writing *Principia Mathematica*, Newton certainly had in mind his 'fluxional' version of what now (following Leibniz's terminology) is generally called 'differential calculus', alongside geometry. But neither there nor (as far as I can tell from the complete, heavily and helpfully annotated five volumes of Newton's mathematical papers Newton & Whiteside, 1967–1981) did he ever publicly describe his laws of motion in terms of sets of "differential equations",[10] (whatever he may have had privately in mind), nor ever use any phrase (in Latin or English) similar to "dynamical system".

A specimen of dynamics

In fact, Newton seems never (for whatever reason) to have used the word "dynamics" or its variants in his publications, nor can I find any contemporary evidence (or later claim) that any other British mathematician did so in Newton's lifetime or during several decades after his death in 1727. John Freind, a British chemist who was ardently Newtonian both in his methodology and in his support for Newton against Leibniz in the long feud over the invention of calculus, quotes the word in Latin (as "*Dynamicum*") in 1711.

---

[10] According to Ince (1926, p. 3), "The term *æquatio differentialis* or differential equation was first used by Leibniz in 1676". This Latin phrase appears (duly attributed to Leibniz) in an English book by John Craig (1685, p. 28) who had "a privileged relationship with Newton" (Guicciardini, 1989, p. 12); so Newton surely knew that meaning of the phrase by then, though his own few uses of it refer to what are now called "difference equations". The first use in print of the English phrase "differential equation" may be in 1704, in Hayes's *A Treatise of Fluxions*.

> *Non me latet quod Cl. L. quem quasi Numen aliquod suspiciunt* Editores, *in* Specimine *illo, quod* vocabulâ Eleganter sonante *nuncupat,* Dynamicum, *planissimè scripserit,* Vim Activam seu nisum intimam corporum Naturam constituere.[11] [I am well aware that the learned *Leibnitz*, whom the Editors regard as a kind of Deity, in that Specimen which he calls by a pretty sounding Name, *Dynamic*, has very plainly told us, that active Power or Endeavour constitutes the intimate Nature of Bodies.] (Freind, 1711/1749, p. 336, 1711, pp. 431–432; original emphases)

Similarly, Samuel Clarke (who translated Newton's *Opticks* from Latin to English at 22, but spent most of his life as a cleric) quotes *"dynamiques"* from a letter written him by "the late learned Mr. Leibniz", rendering it in English as "Dynamick":

> Now, by that single Principle, *viz.* that *there ought to be a sufficient Reason why things should be so, and not otherwise*, one may demonstrate [...] in some Measure, those Principles of *Natural Philosophy*, that are independent upon *Mathematicks* : I mean, the *Dynamick* Principles, or the *Principles of Force*. (Clarke & Leibniz, 1717, pp. 21–23; the italics, as well as the translation, are Clarke's)

Yet Clarke doesn't use the word in his reply, or (it seems) elsewhere.

Perhaps the word "dynamics" was poisoned for English scholars by its associations—with Leibniz, his monadic cosmology (which was highly incompatible with Newton's), and especially his claim that the *"Principles of Force"* (*i.e.*, the three familiar "laws of motion" that Newton, 1726/1871, pp. 13–15, took as *"Axiomata"*, and concerning the origins of which he famously wrote *"hypothesis non fingo"*) are derivable from the Principle of Sufficient Reason alone.

Geometrical lectures

Kenneth Boulding (1956, p. 79) found it "not surprising [...] that in an age of clocks Newton fashions the universe in the likeness of

---

[11]Freind refers to *Specimen Dynamicum* (Leibniz, 1696/1846); the final phrase *"Vim Activam ... constituere"* does not appear there (or elsewhere in Leibniz's works), but it paraphrases *"ut intimam corporum naturam constituat, quando agere est character substantiarum"* ("force 'is the deepest nature of bodies,' since it is a 'characteristic of substances to act'"; Boudri, 2002, pp. 84–85), which does (p. 145).

an orrery", *i.e.*, of "a mechanical model, usually clockwork, devised to represent the motions of the earth and moon (and sometimes also the planets) around the sun" (orrery, 2011).[12] This is far more elegant and evocative than its paraphrase (or plagiary) by Marshall McLuhan (1964, p. 25)—"Newton, in an age of clocks, managed to present the physical universe in the image of a clock"—but I think both authors are misleading (or misled). The background to my doubts is the *Geometrical Lectures* of Newton's teacher, Isaac Barrow.

Newton states his well-known distinction between "Absolute and Relative, True and Apparent, Mathematical and Common" time succinctly in a scholium near the beginning of *Principia Mathematica*.

> Absolute, True, and Mathematical Time, of itself, and from its own nature flows equably without regard to any thing external, and by another name is called Duration: Relative, Apparent, and Common Time is some sensible and external (whether accurate or inequable) measure of Duration by the means of motion, which is commonly used instead of True time; such as an Hour, a Day, a Month, a Year. (Newton, 1686/1729, Vol. I, p. 9)

This is a highly condensed version of approximately one third of Lecture I of Barrow (1674/1735, pp. 1–26), which Newton certainly read, and to which he may have contributed some words or notions.[13] A less condensed outline follows (all italics are Barrow's).

(A) Barrow begins by distinguishing "Time absolutely" from the "longer and shorter Time" that "is common in every Body's Mouth" (p. 5).

(B) Expanding on (A), Barrow poses the rhetorical question "does not Time imply motion?" and answers "no, as to its absolute

---

[12]The first orrery was built, in England, *c.* 1700 (14 years after the first edition of *Principia Mathematica*), and duplicated soon thereafter at the behest of Charles Boyle, fourth Earl of Orrery, who thus became its eponym. Charles's great-grandfather was Roger Boyle, the first Earl (quoted in Table 10.1, in a line that appears to have been plagiarized from Suckling, 1638/1994, Act III, Scene 1) and the father of Robert Boyle (eponym of "Boyle's Law of gases"; *cf.* Potter, 2001), whose work on chemistry was an important intellectual influence on Newton.

[13]In his Preface, Barrow wrote that "Mr. *Isaac Newton*, my Collegue, a Man of great Learning and Sagacity, [...] revised my Copy and noted such things as wanted Correction, and even gave me some of his own, which you will see here and there interspersed with mine, not without their due Commendations" (*ibid.*, pp. iv–v), but unfortunately the "due Commendations" are in fact nowhere given.

and intrinsic nature" (p. 6). In contrast to Time as "absolute", "Time as measurable signifies motion; for if all Things were to continue at Rest, it would be impossible to find out by any Method whatsoever how much Time has elapsed", and so "we shou'd not perceive how Time flows" (p. 7).

(C) Next, Barrow addresses how, "for common Use, the most remarkable Motion possible ought to be taken", to wit, "the Motion of the Stars, and particularly of the *Sun* and *Moon*" (p. 9). Celestial motion is not, however, any more 'absolute' than terrestrial motion.

> But you may ask, how we *can know that the Sun moves with an equal Motion, that the Time of one Day or Year, for Supposition Sake, is exactly equal to that or another?* I answer, we cannot know this any otherwise (except what we gather from the divine Testimony) but by comparing the Motion of the Sun with other equal Motions. (p. 10; author's italics)

(D) Barrow's paradigmatic example of an "artificial Time-Keeper accurately made" to be a source of "other equal Motions" is

> an Hour-Glass, destined to measure an Hour; and because the Water or Sand contain'd in it remain entirely the fame as to Quantity, Figure, and Force of descending, and the Vessel that contains them, as likewise the little Hole they run thro' don't undergo any Kind of Mutation, at least in a short Space of Time, and the State of the Air much the same; there is no Manner of Reason for us not to allow the Times of every running out of the Water or Sand to be equal (p. 10–11).

(E) Barrow next states that, if "the solar Motions [. . .] are found to correspond entirely with the repeated Motions of such an Instrument" as an hour-glass, then we "may very justly conclude"

> that the celestial Bodies are not essentially, and properly speaking, the primary and original Measures of Time; but rather those Motions which are near us, that strike upon our Senses, and fall under our Experience, since by their Means, we discover the Regularity of the celestial Motions. Nor is even the Sun itself a proper Judge or Witness in this Affair, any farther than as its Veracity is shown, by

the Attestation of an Horary Machine.[14] (p. 11)

(F) Barrow sums up his remarks on time and its measurement with the key observation that "we first assume Time from some Motion, and afterwards judge thence of other Motions, which in Reality is no more than comparing some Motions with others, by the Assistance of Time", as "*Aristotle* doubtless knew, and has plainly taught. [...] *We not only measure Motion by Time, but also Time by Motion, because they determine each other*" (p. 13).

This completes the background for the following section.

Newton and Leibniz in an age of accuracy

Aside from the scholium quoted on p. 289, Newton writes very little *about* "Relative, Apparent, and Common Time" or its measurement —he simply *uses* it, "whether accurate or inequable", relying (mostly without explicit mention) on his profound capacity for calculation to make that use as accurate as he can. The chief "artificial Time-Keeper" mentioned in his acknowledged writings is the pendulum clock (invented by Huygens in 1656), and in most of its appearances it functions *not* primarily as an instrument to measure time, but as an instrument "To find and compare together the weights of bodies in the different regions of our Earth" (Newton, 1686/1729, Vol. 2, p. 245).[15] He mentions geared "Clocks and such like instruments, made up from a combination of wheels"[16] once, while discussing his laws' application to 'simple machines' (balance, pulley, screw, and wedge; Newton, 1686/1729, Vol. 1, pp. 38–39). His writings seem to be free of references to hour-glasses, sundials, and pocket-watches.

Some historians have suggested that Newton may have been not just a pre-publication reader, but an unacknowledged author of at least some passages in the letters to Leibniz that were published as by Samuel Clarke alone (Clarke & Leibniz, 1717). Even if that is so,[17] it adds at most half a dozen mentions of clocks (*ibid.*, pp. 15,

---

[14]Did the near-homophony of "orrery" and "horary" help name the former?

[15]The Laws of Motion imply that a pendulum subjected to varying gravitational forces (e.g., at varying latitudes) oscillates with varying periods, and so gives varying measures of time as compared with an hour-glass or sundial alongside it.

[16]Assemblies of gears, wheels, etc., commonly called 'clockwork', had been used in clocks—including water-clocks like those mentioned by Barrow—for several centuries before Huygens patented the pendulum clock.

[17]Hall (1980, p. 220), quoting Koyré and Cohen (1962), writes "[T]he declaration

45, 145, 151, 359, and 363) to Newton's writings, all very like the first (p. 15):

> The Notion of the World's being a great *Machine*, going on *without the Interposition of God*, as a Clock continues to go without the Assistance of a Clockmaker ; is the Notion of *Materialism* and *Fate*, and tends, (under pretense of making God a *Supra-Mundane Intelligence*,) to exclude *Providence* and *God's Government* in reality out of the World.

Whoever wrote that passage (in response to a claim of Leibniz's, in his first letter to Clarke[18]), it surely—in light of the long and unabated friendship between Newton and Clarke—must express Newton's attitude towards the notion of a clock-work universe (commonly conflated with Deism) that Boulding (1956) and McLuhan (1964) (mis)attribute to him.

In fact, Newton and Leibniz—despite their differences, and regardless of their respective positions (whatever they may have been) on the mathematical nature of we now call "dynamical systems"— had this in common: they both believed in the efficacy, for understanding matter changing over time, of *accurate calculation*.

> [...] Mechanics is so distinguished from Geometry, that what is perfectly accurate is called Geometrical, what is less so is called Mechanical. But the errors are not in the art, but in the artificers. He that works with less accuracy, is an imperfect Mechanic, and if any could work with

---

of the two most recent students of the matter that 'there is no doubt that Newton took part in the fight between Leibniz and Clarke,' literally read, goes beyond what was doubtless their intention to express. Certainly Newton was in Clarke's corner, but he did not on more than an isolated occasion put lead in Clarke's gloves"; in his end-notes (*ibid.*, p. 328) Hall adds that "I was myself formerly of the opinion that Newton had drafted arguments for Clarke's benefit, i.e., taken 'part in the fight.'" See also Vailati and Yenter (2009).

[18] "Sir *Isaac Newton*, and his Followers, have also a very odd Opinion concerning the Work of God. According to their Doctrine, God Almighty ⋆ wants to wind up his Watch from Time to Time : Otherwise it would cease to move. [...] [T]he Machine of God's making, is so imperfect, according to these Gentlemen ; that he is obliged to *clean* it now and then [...] and even to *mend* it, as a Clockmaker mends his Work", Clarke and Leibniz (1717, pp. 3 and 5, Clarke's translation and emphases; the note marked by ⋆ is Clarke's citation of *"The Place Mr.* Leibnitz *here seems to allude to,"* in Newton's *Opticks*).

perfect accuracy, he would be the most perfect Mechanic of all. (Newton, 1686/1729, p. xiv).

[E]verything proceeds mathematically—that is, infallibly —in the whole wide world, so that if someone could have sufficient insight into the inner parts of things, and in addition had remembrance and intelligence enough to consider all the circumstances and to take them into account, he would be a prophet and would see the future in the present as in a mirror. (O. T. Benfey's translation, in Cassirer, 1937/1956, pp. 11–12, of Leibniz, 1690?/1840, pp. 49–50, as quoted by Cassirer, 1937, pp. 23–24)[19]

The perfected science of mechanics

The theory of differential equations (based on differential and integral calculus), especially as applied to mathematical physics and "celestial mechanics", grew enormously deeper and richer in the 1700s through the work of dozens of mathematicians. At the end of that century a major contributor to this growth, Pierre-Simon Laplace (who had been doing important mathematics since 1770, when he was 21), published the first volume of his immense *Traité de mécanique céleste* (Laplace, 1798–1825). It had been preceded by *Exposition du système du monde* (Laplace, 1795), published "in effect as preface addressed to non-professional readers" which "gives the classic verbal description of a universe operating by determinant natural law [. . .]" (Gillispie, 1972, p. 2).

In his *Exposition*, Laplace credited two in particular of his senior contemporaries with providing the final ingredients necessary to the realization that "the characteristic mathematical embodiment of a dynamical system" is "a set of ordinary differential equations"— as Birkhoff would put it 125 years later.

This method of reducing the laws of motion to those of equilibrium, for which we are principally indebted to

---

[19]It is interesting to compare this version of Leibniz's vision of infallible calculation (which is explicitly mathematical, and must include the use of differential calculus, and likely differential equations) to the version (Leibniz, 1684?/1840) described on pp. 46–47: the application of a projected *calculus rationator* (an algorithm, according to Rogers, 1963, or perhaps a 'library' of algorithms) to a projected *characteristica universalis* (essentially, a formal—but not necessarily 'purely mathematical' or fully mathematized—language), which is more purely 'logical'.

d'Alembert, is very luminous and universally applicable. [...] It still remained to combine the principle which has been just explained, with that of virtual velocities, in order to give to the science of mechanics all the perfection of which it appears to be susceptible. This is what Lagrange has atchieved [*sic*], and by this means has reduced the investigation of the motion of any system of bodies, to the integration of differential equations. The object of mechanics is by this means accomplished, and it is the province of pure analysis to complete the solution of problems. (Laplace, 1795/1830, p. 288)[20]

The pre-history of "dynamical systems" was over, and the first stage of their history was underway.

Certainty and probability

Laplace's *Exposition du système du monde*, called "the classic verbal description of a universe operating by determinant natural law" by Gillispie on the preceding page, was followed by the preface (in all editions after the first) to his *Théorie analytique des probabilités* (Laplace, 1816), "often republished separately under the title *Essai philosophique sur les probabilités*, [...] which does verbally for its subject what the *Système du monde* had done for celestial mechanics." (Gillispie, 1972, pp. 2–3). In fact, that essay also did something for dynamical systems, as I describe in this section.

Laplace's vast intelligence

Near the beginning of his essay, Laplace reworks Leibniz's vision of infallible calculation (quoted on on the previous page), as follows.

Une intelligence qui pour un instant donné, connaîtrait toutes les forces dont la nature est animée, et la situation respective des êtres qui la composent, si d'ailleurs elle était assez vaste pour soumettre ces données à l'analyse, embrasserait dans la même formule, les mouvemens des plus grands corps de l'univers et ceux du plus léger atome : rien ne serait incertain pour elle, et l'avenir

---

[20]In the context of differential equations, "integration" means 'solution'.

comme le passé, serait présent à ses yeux. [An intelligence that for a given instant could know all the forces animating nature, and the respective places of the beings that compose it, were it also vast enough to submit these data to analysis, should embrace in one formula the movements of the largest bodies in the universe and of the lightest atom: nothing would be uncertain for it, and the future like the past would be present to its eyes.] (Laplace, 1816, pp. 3–4; my translation)

This is in at least two ways not a mere restatement of Leibniz's claim that someone having "sufficient insight into the inner parts of things", and "remembrance and intelligence enough", "would be a prophet and would see the future in the present as in a mirror".

First, Laplace makes "sufficient insight" explicit: it is knowledge of (a) the laws of "all the forces animating nature" (these laws are assumed to be certain 'second order' differential equations), and (b) the states at a "given instant" of "all the beings composing" nature (these states are the 'initial conditions' of the differential equations); the "inner parts of things" are relevant only insofar as they enter into the differential equations—further "insight" is irrelevant and redundant.[21]

Second, Laplace makes "intelligence enough" explicit, too: it is intelligence "vast enough" *to solve those differential equations*. But how vast, exactly, is that? And—speaking quantitatively—how well can an intelligence of a specified degree of vastness expect to fare with the business of collating an instant's universe-full (or a universe's instant's-worth) of data, analyzing it, and fitting it all into "one formula" adequate to make just some limited portion of "the future like the past [...] present to its eyes"?

---

[21] For Laplace there was only one general force, Newtonian 'universal gravitation', and "the forces animating nature" were all, and only, those instances of that force that relate each pair of "beings" by subjecting them to a mutual attraction proportional to the *mass* of each "being" (its only relevant "inner part": operationally, this eliminates Leibnizian monadology from the equation) and inversely proportional to the squared distance between them. The ensemble of mass, position, and net force data for all "being"s is essentially *identical* to the "infinitesimal generator" of the flow of the system of differential equations (note 7, p. 285).

Later, other general forces (beginning with electricity, with electric *charge* as a newly relevant "inner part") were found to fit equally well into Laplace's scheme, and could be assimilated to it as long as the vast intelligence remained in business.

Formulating the world

Early reactions to the passage just quoted from Laplace (1816) accepted his hypothetical sufficiently vast intelligence for the sake of argument, then argued for a variety of conclusions. By 1821, the Scottish philosopher Dugald Stewart was worrying that

> the unrivalled splendour of [Laplace's] mathematical genius may be justly suspected [...] to throw a false lustre on the dark shades of his philosophical creed (Stewart, 1824, p. 138)

which incorporates (no doubt among other undesirable qualities)

> the very essence of Spinozism [...] studiously kept by the author out of the reader's view; and hence the facility with which some of his propositions have been admitted by many of his *mathematical* disciples, who, it is highly probable, were not aware of the consequences which they necessarily involve. (*ibid.*, p. 139 *n.*; original emphasis).

On the other hand, Charles Babbage (now best known for his "difference engine", a precursor of the programmable computer) applied Laplace's idea[22] in support of an argument in "Natural Religion":

> Whilst the atmosphere we breathe is the ever-living witness of the sentiments we have uttered, the waters, and the more solid materials of the globe, bear equally enduring testimony of the acts we have committed.

> If the Almighty stamped on the brow of the earliest murderer,—the indelible and visible mark of his guilt, he has also established laws by which every succeeding criminal is not less irrevocably chained to the testimony of his crime; for every atom of his mortal frame, through whatever changes its severed particles may migrate, will still retain, adhering to it through every combination,

---

[22]In his text proper, Babbage quotes another sentence from Laplace (1816, p. 6), stating that the path of "a simple molecule of air" is as completely determined (for all time) as "the planetary orbits"; but he thinks enough of the longer passage quoted on p. 294 to include it whole (with its immediately succeeding paragraph) in an Appendix (Babbage, 1837, pp. 173–174).

some movement derived from that very muscular effort, by which the crime itself was perpetrated. (Babbage, 1837, pp. 116–117)

It appears that the first author to raise the quantitative question posed on p. 295 (though only incidentally to another point) was Émil du Bois-Reymond (1872), a Swiss physiologist [23]

[T]he impossibility of stating and integrating the differential equations of the universal formula, and of discussing the result, is not fundamental, but rests on the impossibility of getting at the necessary determining facts, and, even where this is possible, of mastering their boundless extension, multiplicity, and complexity. (Du Bois-Reymond, 1874, p. 20)

However, du Bois-Reymond did not attempt actual quantification (his "boundless" is, if anything, vaguer than Laplace's "vast"). In the end, his main contribution to the discussion may have been to introduce the term *Weltformel* (literally, "world-formula"; translated as "universal formula" in du Bois-Reymond, 1874).

Indeed (despite a cameo appearance by "the Almighty" in Babbage, 1837), for nearly a hundred years the focus of that discussion was on the formula—that is, the "immense system of simultaneous differential equations" ("*unermessliches System simultaner Differential-*

---

[23]The point to which all du Bois-Reymond's arguments is directed is his purported proof that "the investigator of Nature", who "has long been wont to utter his 'Ignoramus' with manly resignation",

as regards the enigma what matter and force are, and how they are to be conceived, [...] must resign himself once for all to the far more difficult confession—"IGNORABIMUS!"

(du Bois-Reymond, 1874, p. 32), that is, "We shall be ignorant!". In the human sciences, this stance was more or less congenial, or at least debatable (see, *e.g.*, Marique, 1886, p. 118; Dubois, 1905, p. 31; James, 1906, p. 44 and 1912, p. 86; Wundt, 1920, pp. 129–131; etc.), but it was infuriating to some mathematicians, particularly David Hilbert (1900, p. 262), who in 1900 told the assembled International Congress of Mathematicians in Paris that "*in der Mathematik giebt es kein Ignorabimus!*": "in mathematics there is no *Ignorabimus!*". On the other hand, Paul du Bois-Reymond, a mathematician, sided with his brother in maintaining that (even) in mathematics there are limits to human knowledge (McCarty, 2005). History has shown that Paul was right, for the wrong reasons.

gleichungen";[24] du Bois-Reymond, 1872, p. 5)—much more than on the "intelligence". Other writers soon adopted *Weltformel* (or a translation), using it in du Bois-Reymond's sense whether or not they agreed with (or cited) him. Laplace's "formula" had assumed the form of Birkhoff's "characteristic" (and impersonal) "mathematical embodiment of a dynamical system".[25]

William James, in particular, was as taken with this idea of du Bois-Reymond as he had been with the man when attending physiology lectures in Berlin.[26]

> The "universal formula" of Laplace which Du Bois-Reymond has made such striking use of in his lecture [...] (James, 1879, p. 337)

> The ideal which this ["modern mechanico-physical"] philosophy strives after is a mathematical world-formula [...] (James, 1890, vol. 2, p. 667)

> There can *be* no difference which doesn't *make* a difference—no difference in abstract truth which does not

---

[24]The mistranslation (du Bois-Reymond, 1874, p. 18) of *unermessliches* as "infinite" rather than "immense" is unfortunate; in mathematics, the distinction is fundamental and extremely significant.

[25]I have found one very interesting variant. Vaihinger (1878, p. 447) wrote

> If we break the flow of sensations into a series of states, then the task (of epistemology, psychology, and what is not yet known) is to find from this series the law which the individual states $a, b, c, \ldots$ of the world-flow follow, much as if, given the law $y = x^3 - x + 1$, an arithmetic progression flows from it so that each member is calculated from the others. This law would be the world formula. I am aware of only one attempt to establish such a formula, made independently and almost identically by Spencer, Zöllner and Avenarius [...].

(My translation is very loose, but I believe it captures Vaihinger's meaning. I have corrected an obvious typographical error, "$y = x^3 - y + 1$".) Here, the *Weltformel* and its associated *Weltfluss* (world-flow) are not manifested as a *continuous* dynamical system, but as a *discrete* one. In fact, if Vaihinger intended his polynomial example to be a typical (though much simplified) *Weltformel*, then the mathematized *Weltfluss* might be a discrete *semi*-flow that is not a flow; see note 8, p. 285). Apparently Vaihinger's idea was never taken up (nor were the "attempt"s he ascribed to Spencer, Zölner, and Richard Avenarius mathematized).

[26]In an 1867 letter to Henry P. Bowditch (grandson of Nathaniel Bowditch, the American translator of Laplace, 1798–1825), James wrote that "Du Bois-Raymond [*sic*], an irascible man of about forty-five, gives a very good and clear, yea, brilliant, series of five lectures a week" (James, 1920, pp. 120–121).

> express itself in a difference of concrete fact, and of con-
> duct consequent upon the fact, imposed on somebody,
> somehow, somewhere and somewhen. [...] [T]he whole
> function of philosophy ought to be to find out what
> definite difference it will make to you and me, at def-
> inite instants of our life, if this world-formula or that
> world-formula be the one which is true. (James, 1904,
> p. 674–675)

Given James's "rooted dislike for mathematics" (see p. 15), it is not surprising that, like du Bois-Reymond, he attempted no quantification: but neither did such distinctly non-mathematics-averse writers as Friedrich Lange (1877, pp. 149–156), author of a book (Lange, 1865) on the foundations of mathematical psychology; Wilhelm Wundt (1886, pp. 200, 206), whose *Logik* (Wundt, 1883, vol. 2, *Methodenlehre*) contains 146 pages devoted to "The Logic of Mathematics", up to and including calculus; and others, for years.

Dynamics dæmonized

In 1913, Friedrich Kuntze, a philosopher at the University of Berlin whose many interests included the philosophy of mathematical mod-eling (Sveistrup, 1929, p. 295), changed the balance of the discussion (though not towards quantification) with an apparently off-handed phrase in a wide-ranging article[27] published in *Kunstwart und Kul-turwart*, a twice-monthly popular journal of arts, culture, philosophy, *usw.* Kuntze introduced the *"Weltformel"* as "a well-known fiction, going back to Laplace", then summarized the situation of Laplace's "intelligence":

> From any given time and place [...] Laplace's dæmon
> [*der Dämon des Laplace*] would see both of Nietzsche's
> paths—one into the endless past, the other into the end-
> less future—in the full light of Lawfulness; truly the past
> would lie before him, and the future have no secrets.
> (Kuntze, 1913, p. 489; my translation here and below)

---

[27]For instance, in a single paragraph, Kuntze invoked the mathematicians Gauss, Riemann, Weierstrass, Abel, von Staudt, Hamilton, Grassmann, and Boole, the physical scientists Faraday, Maxwell, and Hertz, and the social scientists Dilthey, Windelband, Rickert, Bergson, and James—tagging each with his nationality.

Just short of its 100th birthday, Kuntze had personified Laplace's "intelligence" as a "dæmon": in other words, *as a being with nominal agency but without will*: that is, *as the Time of dynamical systems theory*.

The year before, in a scholarly article where he had occasion to discuss the First and Second Laws of Thermodynamics, Kuntze had written of

> the Maxwellian fiction of a dæmon gifted with infinitely sharp senses, who can watch all natural phenomena as what they actually are in the mechanistic theory, motions. [...] Well, that Maxwellian demon is nothing but the fiction of mechanistic explanation at its highest strength, nothing but a personification of the ideal of Kantian "possible experience". (Kuntze, 1912, pp. 383–384)

Now, Kuntze certainly knew that for Maxwell the essence of *his* "demon" was that it *does* act on the universe, and *does* exercise will in the process (of sorting out faster molecules from slower ones, eventually confuting the Second Law after some statistically significant number of molecules have been sorted); it does not simply "watch".[28] Still, for Kuntze *der Dämon des Laplace* remained a mere observer:[29] omniscient, *computationally* omnipotent, but *effectually* omni-impotent (*cf.* the last sentence quoted from Bergson, 1908, in Table 10.1, p. 283).

After a decade and a half, other writers began to use Kuntze's term: von Mises (1928, 1930/1931), Herrmann (1935), Peter (1938), etc., in its original German; Lichtenstein (1932) and a reviewer

---

[28] In the section "Limitation on the Second Law of Thermodynamics" in his *Theory of Heat,* Maxwell (1871, p. 308) wrote only of "a being whose faculties are so sharpened that he can follow every molecule in its course [...] a being, whose attributes are still as essentially finite as our own". Thomson (1874, p. 326 *n.*) stated that "a demon, according to the use of this word by Maxwell, is an intelligent being endowed with free-will" (etc.). In editorial footnotes to a letter of Maxwell (2002, pp. 185–186), P. M. Harman details how thereafter Thomson and Maxwell (who were close friends) each credited the other with the coinage.

[29] "Dum calculat Deus ...", the title of Kuntze's *Kunstwart* article, is the first part of an aphorism of Leibniz—"*Dum calculat Deus, fit mundus*": "when God calculates, a world comes into being". In that article (1913, p. 488), Kuntze—varying a theme from Goethe—leaves open the possibility of viewing "the cosmos as a vast mathematical vision of the deity" while still drawing human comfort in "the cold night" from *die Gott-Vater*'s "great, eternal, brazen [...] tone of harmony [...] played on the harp of life"; but this (possible) *Weltgeist* or Creator Spirit is entirely distinct from *der Dämon des Laplace*.

(Leclerc (?), 1937) of Cassirer (1937),[30] etc., in French; Margenau (1931, 1934), etc., in English. Perhaps by contagion from its Maxwellian relatives, Laplace's demon began to acquire will—and perhaps none too soon, given the state that the physical universe had been falling into from the mid-19th century onward.[31]

Thermodynamics: Time's revenge

When we last saw Henry Adams, he was standing "gravely perplexed" in the great hall of dynamos at the Paris Exhibition of 1900. *A Letter to American Teachers of History* (Adams, 1910) is a recapitulation, at greater length, of his "problem in dynamics".

(A) Adams began his *Letter* with a brief overview of the beginnings of thermodynamics "[t]owards the middle of the nineteenth century", when it was "made famous by the names of William Thomson" (mentioned in note 28 on the facing page) "in England, and of Clausius and Helmholz in Germany" (*ibid.*, p. 2) with their Second Law of Thermodynamics, called by Adams "the Law of Dissipation", and "quoted exactly" from Thomson (1852/1882, p. 514):

> Within a finite period of time past, the earth must have been, and within a finite period of time to come, the earth must again be, unfit for the habitation of man as at present constituted, unless operations have been, or are

---

[30] Cassirer referred to *"Maxwellschen Dämons"* only (*ibid.*, p. 94); it was the reviewer who brought in the *"démon de Laplace"*.

[31] I am obliged to note that William Sorley (1905, p. 392), Professor of Ethical Philosophy at the University of Cambridge, had committed the "Laplacian demon" to English print eight years before Kuntze's article. I cannot prove that none of the later uses of such a term (particularly those in English) are descended from his article, printed in the collected proceedings of the Congress of Arts and Science that was convened as a satellite of the 1904 Universal Exposition in St. Louis, Missouri. For that matter, I cannot *prove* that any of the uses are descended from Kuntze's article, printed in a biweekly widely read by the German intelligentsia. (Neither article appears ever to have been cited, by anyone, anywhere in the scholarly literature. Kuntze's is at least mentioned, as "für Kuntzes Denken ungemein charakteristische", *i.e.*, "very characteristic of Kuntze's thought", in his *Kant-Studien* death notice by Sveistrup, 1929.) If I had to bet, my money would be on Kuntze; to avoid betting, I invoke the *Zeitgeist*, that other useful dæmon. —Incidentally, Louis Menand (2002, p. 195 *ff.*) gets the story of the two demons just the wrong way around, implying (*e.g.*, with the phrase "Laplace introduced his demon", *ibid.*, p. 196) that Laplace's "intelligence" had been thought of as a demon/ dæmon from its first appearance onwards.

to be performed, which are impossible under the laws to which the known operations going on at present in the material world, are subject.

Adams's remark on this was that

[w]hen this young man of twenty-eight [Thomson, in 1852] thus tossed the universe into the ash-heap, few scientific authorities took him seriously; but after the first gasp of surprise physicists began to give him qualified support which soon became absolute. (Adams, 1910, p. 4)

(B) After several pages sketching the intellectual controversies attendant on "the Law of Dissipation", Adams (speaking of himself, as usual, in the third person) described the position that he—as an historian—found himself in, a few decades into the controversies.

[H]e was clear that the energy with which history had to deal could not be reduced directly to a mechanical or physicochemical process. He was therefore obliged either to deny that social energy was an energy at all; or to assert that it was an energy independent of physical laws. Yet how could he deny that social energy was a true form of energy when he had no reason for existence, as professor, except to describe and discuss its acts? He could neither doubt nor dispute its existence without putting an end to his own; and therefore he was of necessity a Vitalist [. . .]. (*ibid.*, p. 12)

(C) Adams's perplexity increased, inexorably as the entropy of the universe.

Vital energy was, perhaps, an intensity;—so, at least, he vaguely hoped;—he knew nothing at all!

No one knew anything; and yet the analogy between Heat and Vital Energy, suggested by Thomson in his Law of Dissipation,—and received by the public with sleepy indifference,—was insisted upon by the physicists in accents that became sharper with every generation, until it began to pass the bounds of scientific restraint. [. . .] Thus, it seemed, that whatever the universities thought or taught, the physicists regarded society as an organism in the only respect which seriously concerned historians :—
It would die! (*ibid.*, pp. 16–17)

(D) Of course, while physicists continued to study the physical world (and sharpen their accents), geologists and biologists—notably Lyell and Darwin—were also at work. Soon Adams found,

> at the same moment, three contradictory laws of energy [...] in force, all equally useful to science:—1. The Law of Conservation [...]. 2. The Law of Dissipation [...]. 3. The Law of Evolution, that Vital Energy could be added, and raised indefinitely in potential, without the smallest apparent compensation (*ibid.*, p. 24)

to which (*ibid.*, p. 25) he appended a fourth "Law of Degradation", a strong form of the Law of Dissipation that "applies to all vital processes even more rigidly than to mechanical."

(E) In the next 100 pages of "The Problem" (Chap. I of the *Letter*), Adams documented a half century of conflict between Degradationists and Evolutionists, up to the time of his writing.

> The historian of human society has hitherto, as a habit, preferred to write or to lecture on a tacit assumption that humanity showed upward progress, even when it emphatically showed the contrary, as was not uncommon; but this passive attitude cannot be held against the physicist who invades his territory and takes the teaching of history out of his hands. (*ibid.*, p. 87)

(F) Chap. II, "The Solutions", runs to another 87 pages. Boiled down (perhaps further than Adams would have found acceptable), its primary suggestion was

> to treat [...] humanity as a volume of human molecules of unequal intensities[...]. History would then become a record of successive phases [...]. (*ibid.*, pp. 126–127).

Adams followed his suggestion up, inconclusively, in the posthumously published essay "The Rule of Phase applied to History" (1919, written in 1909). The title refers to Josiah Willard Gibbs, who had enunciated his eponymous "rule" in 1876–1878 (his "abstract" Gibbs, 1878, is easier to find), early in his development of *statistical mechanics* (independently of its essentially independent development by Maxwell and by Ludwig Boltzman, but named by him alone, in 1902), to which I turn briefly next.

Statistics and mechanics: from matters of state to states of matter

Adams's hope was that statistical mechanics—which had been developed to allow physical scientists to model the macroscopic behavior of gases (and other fluid forms of matter, even electricity) via the study not of the detailed microscopic behavior of each and every particle's motion (assumed to be under the strict control of the "world formula") but rather of the *statistical*, macroscopic behavior of the ensemble of particles—might somehow lead to an analogous tool for social scientists, in particular, historians. Had that hope been fulfilled, it would have provided an interesting example in the transfer of (intellectual) modeling technology.

Gillispie (1963, 1972) has amply demonstrated that the origins of physicists' statistical mechanics lay in the applications of probability theory (in particular, the "law of errors" of Laplace, 1774) to statistics, a discipline with (pre-mathematical) origins in the state, its politics, and its rationalized governance.[32] Certainly Adams was aware of this—as Keith Burich (1989) has pointed out, after quoting from Adams (1894, p. 112) the decree that "Any science of history must be absolute, like other sciences, and must fix with mathematical certainty the path which society has got to follow",

> As an exemplar of this type of scientific history, Adams cited Thomas Henry Buckle, who applied to history the rigid, mechanical determinism that scientists were applying to an increasing number of phenomena in the nineteenth century (Burich, 1989, p. 112)

while Adams knew that Maxwell had acknowledged Buckle:

> The statistical method of investigating social questions has Laplace for its most scientific and Buckle for its most

---

[32]Statistics was, "[i]n early use, that branch of political science dealing with the collection, classification, and discussion of facts (especially of a numerical kind) bearing on the condition of a state or community" (statistics, 2011, **1. a.**). According to the etymological notes to statistic (2011), Modern Latin *statisticum collegium* was coined (before 1747) by a German professor whose meaning for it was roughly that just given (not excluding, but not emphasizing, facts "of a numerical kind"); German *Statistik* was commonplace by 1748; French *statistique* was used to translate *Statistik* by 1771; and the modern English meaning of "statistics" as "Numerical facts or data collected and classified", statistics (2011, **2. a.**), was in use by 1837. Porter (1986, p. 23 *ff.*) gives many historical details of how this development of the terminology was "accompanied by a subtle mutation of concepts".

> popular expounder. [. . .] [T]hose uniformities which we
> observe in our experiments with quantities of matter con-
> taining millions of millions of molecules are uniformities
> of the same kind as those explained by Laplace and won-
> dered at by Buckle, arising from the slumping together
> of multitudes of cases, each of which is by no means uni-
> form with the others. (Maxwell, 1873/1882, pp. 438–439)

What Adams was not, it seems, sufficiently aware of (or sufficiently
willing to acknowledge to himself) was that from the viewpoint of
mathematical modeling the most notable feature of Buckle's theory
is that it is *not* "rigid"ly "mechanical" *on the level of individuals*: it
makes no attempt to "fix with mathematical certainty the path" of
any particular person. Thus Adams's desired analogy to statistical
mechanics fails crucially (insofar as it intends to provide a "science
of history" and historical dynamics with a mathematized foundation
in some 'science of individual histories' and individual dynamics).[33]

> What does it matter? I've been studying science for
> ten years past, with keen interest, noting down my
> phases of mind each year; and every new scientific
> method I try, shortens my view of the future. The last—
> thermodynamics—fetches me out on sea-level within ten
> years. I'm sorry Lord Kelvin [that is, William Thomson]
> is dead. I would travel a few thousand-million miles
> to discuss with him the thermodynamics of socialistic
> society. His law is awful in its rigidity and intensity of
> result. (Adams, 1992, p. 517; from a letter of May 2, 1909)

With this, I abandon Adams to his perplexity.

---

[33]The foundation of thermodynamics—of which the "laws" had originally been
derived empirically, and justified likewise: steam engines *worked*—upon statistical
mechanics, by applying probabilistic and statistical methods to "mechanics" (the
Newtonian dynamics of particles), as summed up by Maxwell above, is an example
of successful reductionism that is as rare as it is undisputed. This is so even in the
physical sciences, where its success contrasts with, *e.g.*, the reduction-in-principle of
chemistry to physics (via *exact* solutions of the Schroedinger equations), which has
not led to any (mathematically strict) reduction-in-practice for chemical systems
much more complicated than a hydrogen atom (although of course computational
approximations are used, often successfully). As to the social sciences, a typical case
is that of the continuing attempt, by various authors, to 'reduce' macroeconomics
to microeconomics.

Dynamic systems and the human sciences

The three quotations from the human science literature on p. 281 are sufficiently representative to establish to my satisfaction the lack of any single, dominant, mathematizable definition of the phrase 'dynamic system' as used in human sciences. I see no point in trying to impose one here, or even in suggesting a candidate for imposition. Instead, I conclude this chapter with some scattered observations, in the hope that one or another of them may prove useful.

What acts on what? And who knows how?

I have shown that the underlying metaphor of the term 'dynamical system' (as mathematicians use it) is that of *time acting lawfully* on a mathematical space. It seems to me that the underlying metaphor of the term 'dynamic system' (as used in the human sciences) is most often that of *a system acting on itself, in* time but not *as* time.

Now, despite being 'lawful', a 'dynamical system' can be (and many are) more or less obscure in its details, when inspected by limited human intelligence. Yet mathematicians have developed (mathematical) tools and techniques that are useful (1) to describe that obscurity both quantitatively and qualitatively, and even in some cases (2) to roughly quantify the relationship between the particular ways in which a particular 'dynamical system' is obscure, and the expenditure of human intelligence (and other resources) required to understand that 'dynamical system' to a specified degree. As to (1), we see G. D. Birkhoff observing that

> Nearly all fields of mathematics progress from a purely formal preliminary phase to a second phase in which rigorous and qualitative methods dominate. From this more advanced point of view [. . .] we may formulate the aim of dynamics as follows: to characterize completely the totality of motions of dynamical systems by their qualitative properties. (Birkhoff, 1920, p. 52)

As to (2), we see I. J. Good, a great probabilist and mathematical statistician, explaining in a journal of philosophy that

> if a flea is deterministic, it is like an unbreakable cipher machine. To predict its future, under all normal circumstances, for a time T ahead, assuming some deterministic

theory analogous to Newtonian mechanics, Laplace's demon would need the initial conditions expressed to a
number of decimal places proportional to T. (Good, 1971,
p. 384)

What we do *not* see is mathematicians declaring (some, or most)
'dynamical systems' off limits and altogether out of the range of
human (mathematical) understanding; and this is so even though
(for instance) the particular hopes that G. D. Birkhoff expressed, that
his own great successes in characterizing "the totality of motions of
dynamical systems" with two degrees of freedom "by their qualitative properties" would be extensible to general systems, were to be
proved in the 1960s to be generally impossible, while (again, for instance) the particular cautions urged by Good, on the practical computational intractability of long-term accurate predictions, may yet
be shown—given recent developments in both *quantum computing*
(Pitowsky, 1996, 2002) and flea theory (Buksh, Chen, & Wang, 2010)
—to be over-cautious.

   In contrast, as the quotations from Galatzer-Levy (1995) and
Michaels (1995) on p. 281 demonstrate, some (I suspect many) of the
persons who work with 'dynamic systems' in the human sciences[34])
have, in effect, put them "off limits and altogether out of the range"
of any understanding of a *mathematical* sort, to their (and the human
sciences generally) potentially great loss.[35]

Agents, observers, and modelers

In "agent-based modeling" (Chap. 11), though the "agents" are ficti-
tious beings that exist only as their complex of (computer-simulated)

---

[34]Not all; *cf.* the quotation from Lichtwarck-Aschoff et al. (2009) on that page.

[35]I decline to count mere *allusion* to mathematics as contributing to *mathematical*
understanding (indeed, I am dubious that it contributes to any form of what I
would call 'understanding', rather than some other kind of meaning-making). By
"mere allusion" I mean, for instance, invocation of some body of mathematical work
(*e.g.*, "chaos theory") unaccompanied by any evidence of *engagement* with that body
as a Gestalt embracing definitions, proofs, and theorems—and extra-mathematical
elements—in a meaningful unity (which may, as in the case of "chaos theory",
incorporate lively contradictions and contestations: see, *e.g.*, the exchange of letters
between David Ruelle (2009a, 2009b) and James Yorke (2009), whose amicable but
deeply felt disagreements on definitions correspond to very different conceptions
of what should be the use and interpretation of "chaos theory" in mathematical
modeling).

behaviors, the "operator" of the model (more precisely, of a particular computer program instantiating the model)—who may be, but need not be, the modeler—typically does more than simply push a 'start button'. The operator is, in fact, a *non*-fictitious (human) being, who exists (in part) as his or her complex of those (human) behaviors involved in 'operating' the model (and, perhaps, in modeling it originally). In short, the operator, too, is an agent; and the manner of that agency is not obviously irrelevant to the project of modeling. *Mutatis mutandis*, the same applies to *any* model, its user(s), and its author(s). I do not think that this observation is entirely trite, but I have neither space nor time to develop it here.

Different kinds of time

Rudolph (2006a) surveyed a wide variety of mathematical structures that have been proposed as models of time—far wider than those used in the "modern axiomatic definition of a dynamical system" (Stroyan & Luxemburg, 1976, p. 155) and its variants described early in this chapter. One of the non-standard models of time mentioned by Rudolph (2006a) was explicitly proposed in the context of a new kind of dynamical system (Kennison, 2002), but to my knowledge no one has yet used it in an extra-mathematical model; in it, as in the "modern axiomatic definition", time acts (on various mathematical spaces) via a group (the so-called *p-adic integers*) that shares some properties with both the real numbers (ordinary 'continuous time') and the integers (ordinary 'discrete time') but is fundamentally very different from either. Another, the *branching time* developed by Burgess as an aspect of one version of temporal logic (Burgess, 1979, 1980), doesn't fit well with the 'group action' paradigm, but it (or perhaps the variant, in which "branches" are allowed not only to diverge as in Burgess's version, but also to converge—as indeed they did in his avowed inspiration, Borges, 1962—proposed in a logic-free way by Rudolph, 2006a) feels to me as if it *ought* to support a notion of 'dynamical system', yet to be devised, particularly suitable for models in the human sciences.

References

Adams, H.   (1894).   *The tendency of history*.   Retrieved March 8, 2011, from http://www.historians.org/info/aha

_history/hbadams.htm (Presidential Address to the American Historical Association, delivered by letter from Guadaj-lajara, dated December 12, 1894)

Adams, H. (1910). *A Letter to American Teachers of History.* Baltimore, MD: J. H. Furst Co.

Adams, H. (1918). The dynamo and the Virgin. In *The Education of Henry Adams* (pp. 379–390). Boston, MA: Houghton Mifflin Company.

Adams, H. (1919). The rule of phase applied to history. In *The Degradation of the Democratic Dogma* (pp. 267–311). New York, NY: Macmillan. (Dated January 1, 1909)

Adams, H. (1992). *Henry Adams: Selected Letters* (E. Samuels, Ed.). Cambridge, MA: Harvard University Press.

Babbage, C. (1837). *The Ninth Bridgewater Treatise. A Fragment.* London, GB: John Murray.

Barrow, I. (1735). *Geometrical Lectures.* London, GB: Stephen Austen. (Original work published 1674; "Translated from the *Latin* Edition, revised, corrected and amended by the late Sir *Isaac Newton*")

Basile, G. (1894). *The Pentamerone or The Story of Stories* (H. Zimmern, Ed.). New York, NY: Macmillan. (Original work published 1636; translated from the Neapolitan by John Edward Taylor)

Beltrametti, E. G., & Cassinelli, G. (1981). *The Logic of Quantum Mechanics* (Vol. 15; G.-C. Rota, Ed.). Reading, MA: Addison Wesley.

Bergson, H. (1908). *L'Évolution Créatrice [Creative Evolution].* Paris, FR: Félix Alcan.

Bhartṛihari. (100?/1886). *The Śatakas of Bhartṛihari* (B. H. Wortham, Trans.). London, GB: Trübner & Co. (From the Sanskrit original, believed to have been written *c.* 100 C.E.)

Birkhoff, G. D. (1920). Recent advances in dynamics. *Science, LI*(1307), 51–55. ("Address of the vice-president and chairman of Section A—Mathematics and Astronomy—American Association for the Advancement of Science, St. Louis, December, 1919.")

Borges, J. L. (1962). The garden of forking paths (D. A. Yates, Trans.). In D. A. Yates & J. E. Irby (Eds.), *Labyrinths* (pp. 123–138). New York, NY: New Directions.

Boudri, J. C. (2002). *What Was Mechanical About Mechanics: the Concept of Force Between Metaphysics and Mechanics From Newton to Lagrange*. New York, NY: Springer.

Boulding, K. E. (1956). *The Image: Knowledge in Life and Society*. Ann Arbor, MI: University of Michigan Press.

Boyle, R. (1655). *Parthenissa, a Romance. In Four Parts. Part Four*. London, GB: Henry Herringman.

Brackenridge, J. B. (1996). *The Key to Newton's Dynamics: the Kepler Problem and the* Principia. Berkeley, CA: University of California Press. Retrieved June 15, 2011, from `http://ark.cdlib.org/ark:/13030/ft4489n8zn/` ("[C]ontaining an English translation of sections 1, 2, and 3 of book one from the first (1687) edition of Newton's *Mathematical principles of natural philosophy*; [. . .] with English translations from the Latin by Mary Ann Rossi.")

Brown, S. A. (1996). Southern Road : Poems. In M. S. Harper (Ed.), *The Collected Poems of Sterling A. Brown*. Evanston, IL: Northwestern University Press. (Original work published 1932)

Buksh, S. R., Chen, X., & Wang, W. (2010). Study of flea jumping mechanism for biomimetic robot design. *Journal of Biomechanical Science and Engineering*, 5(1), 41–52.

Burgess, J. P. (1979). Logic and time. *Journal of Symbolic Logic*, 44, 566–582.

Burgess, J. P. (1980). Decidability for branching time. *Studia Logica*, 39, 203–218.

Burich, K. R. (1989). "our power is always running ahead of our mind": Henry Adams's phases of history. *The New England Quarterly*, 62(2), 163–186.

Burns, K. H., & Gidea, M. (2005). *Differential Geometry and Topology: With a View to Dynamical Systems*. Boca Raton, FL: Chapman & Hall/CRC Press.

Cassirer, E. (1937). *Determinismus und Indeterminismus in der modernen Physik [Determinism and Indeterminism in Modern Physics]*. Göteberg, SV: Wettergren & Kerbers Förlag.

Cassirer, E. (1956). *Determinism and Indeterminism in Modern Physics; Historical and Systematic Studies of the Problem of Causality* (O. T. Benfey, Trans.). New Haven, CT: Yale University Press. (Original work published 1937)

Clarke, S., & Leibniz, G. W. (1717). *A Collection of Papers, Which Passed Between the Late Learned Mr. Leibnitz, and Dr. Clarke, in the Years 1715 and 1716. Relating to the Principles of Natural*

Philosophy and Religion. (S. Clarke, Ed.). London, GB: James Knapton.

Couturat, L. (1900). L'algèbre universelle de M. Whitehead [Review of the book *A Treatise on Universal Algebra, with Applications. Vol. 1*]. *Revue de Métaphysique et de Morale, 8*(3), 323–362.

Craig, J. (1685). *Methodus figurarum lineis rectis et curvis comprehensarum quadraturas determinandi [The Method of Straight and Curved Lines, Including the Determination of Quadrature].* London, GB: M. Pitt.

Dubois, P. (1905). *Die Psychoneurosen und ihre psychische Behandlung [Psychoneuroses and their Psychotherapy].* Bern, CH: A. Franke.

du Bois-Reymond, E. (1872). *Über die Grenzen des Naturerkennens [On the limits of knowledge of nature].* Leipzig, DE: Veit.

du Bois-Reymond, E. (1874). The limits of our knowledge of nature. *The Popular Science Monthly, V*(1), 17–32. ("An Address delivered at the Forty-fifth Congress of German Naturalists and Physicians at Leipsic", "Translated from the German, by J. Fitzgerald, A. M.")

EBSCO. (2011). *EBSCOhost Online Research Databases: SocINDEX*™ *with Full Text.* Retrieved October 7, 2011, from http://www.ebscohost.com/academic/socindex-with-full-text

Echeverría-Enríquez, A., Muñoz-Lecanda, M. C., & Román-Roy, N. (1995). Non-standard connections in classical mechanics. *Journal of Physics A: Mathematical and General, 28*(19), 5553–5567.

Faye, H. (1885). *Sur l'origine du monde: théories cosmogoniques des anciens et des modernes [On the Origin of the World: Cosmogonic Theories of Ancient and Modern Man].* Paris, FR: Gauthier-Villars.

Freind, J. (1711). Prælectionum Chymicarum Vindiciæ, in quibus Objectiones, in Actis Lipsiensibus Anno 1710. Mense Septembri, contra Vim materiæ Attractricem Allatæ, diluuntur [Defence of his *Chemical Lectures*, in which are refuted the objections to the attractive force of matter brought in *Actis Lipsiensibus*, September, 1710]. *Philosophical Transactions, 27*(331), 330–342.

Freind, J. (1749). Of the doctrine of attraction, &c (Dr. J. Freind's defence of his chemical lectures, &c). In H. Jones (Ed.), *The Philosophical Transactions (From the Year 1700, to the Year 1720) Abridg'd, and Disposed under General Heads* (Vol. V). London, GB: W. Innys [and 10 others]. (Original work published 1711; Jones's is "The Third Edition Corrected, In which the Latin

Papers are now first translated into English"; no translator is named)

Galatzer-Levy, R. M. (1995). Psychoanalysis and dynamical systems theory: Prediction and self similarity. *Journal of the American Psychoanalytic Association*, *43*, 1085–1113.

Garrity, T. A. (2002). *All the Mathematics You Missed: But Need to Know For Graduate School.* Cambridge, GB: Cambridge University Press.

Gelder, T. van. (1998). The dynamical hypothesis in cognitive science. *Behavioral and Brain Sciences*, *21*, 615–665.

Gibbs, J. W. (1876–1878). On the equilibrium of heterogeneous substances. *Transactions of the Connecticut Academy*, *III*, 108–248 and 343–524.

Gibbs, J. W. (1878). On the equilibrium of heterogeneous substances. *The American Journal of Science and Arts*, *XVI*(XCVI), 441–458. ("Abstract by the author" of Gibbs, 1876–1878)

Gibbs, J. W. (1902). *Elementary Principles in Statistical Mechanics: Developed with Especial Reference to the Rational Foundations of Thermodynamics.* New York, NY: C. Scribner's Sons.

Gillispie, C. C. (1963). Intellectual factors in the background of analysis by probabilities. In A. C. Crombie (Ed.), *Scientific Change* (pp. 431–453). New York, NY: Basic Books.

Gillispie, C. C. (1972). Probability and politics: Laplace, Condorcet, and Turgot. *Proceedings of the American Philosophical Society*, *116*(1), 1–20.

Golding, A. (1806). *Shakespeare's Ovid* (W. H. D. Rouse, Ed.). London, GB: De La More Press. (Original work published 1567)

Good, I. J. (1971). Untitled [Review of the book *The Freedom of the Will*]. *The British Journal for the Philosophy of Science*, *22*(4), 382–387.

Guicciardini, N. (1989). *The Development of Newtonian Calculus in Britain, 1700–1800.* Cambridge, GB: Cambridge University Press.

Hall, A. R. (1980). *Philosophers at War: The Quarrel between Newton and Leibniz.* Cambridge, GB: Cambridge University Press.

Hayes, C. (1704). *A Treatise of Fluxions.* London, GB: Midwinter & Leigh.

Herrmann, G. (1935). Die naturphilosophischen Grundlagen der Quantenmechanik [The foundations of quantum mechanics in the philosophy of nature]. *Die Naturwissenschaften*, *42*, 718–735.

("Translated from the German, with an Introduction, by Dirk Lumma" in Lumma, 1974, *The Harvard Review of Philosophy* VII, 35–44)

Hilbert, D. (1900). Mathematische Probleme. Vortrag, gehalten auf dem Internationalen Mathematiker-Kongreß zu Paris 1900 [Mathematical Problems. Lecture delivered before the International Congress of Mathematicians at Paris in 1900], *Nachrichten von der Königliche Gesellschaft der Wissenschaften zu Göttingen*, 253–297.

Ince, E. L. (1926). *Ordinary Differential Equations*. London, GB: Longmans, Green and Company.

James, W. (1879). The sentiment of rationality. *Mind*, 4(15), 317–346.

James, W. (1890). *Principles of Psychology* (Vol. 1). New York, NY: Henry Holt & Company. (In two volumes)

James, W. (1904). The pragmatic method. *The Journal of Philosophy, Psychology and Scientific Methods*, 1(25), 673–687.

James, W. (1906). *Human Immortality: Two Supposed Objections to the Doctrine*. London, GB: Archibald Constable.

James, W. (1912). Present philosophical tendencies. In R. B. Parry (Ed.), (pp. 83–109). London, GB: Longmans, Green and Company.

James, W. (1920). *The Letters of William James* (H. James, Ed.). Boston, MA: The Atlantic Monthly Press.

Kennison, J. (2002). The cyclic spectrum of a Boolean flow. *Theory and Applications of Categories*, 10, 392–409.

Koyré, A., & Cohen, I. B. (1962). Newton and the Leibniz–Clarke correspondence. *Archives Internationales d'Histoire des Sciences*, 15, 64–126.

Kuntze, F. (1912). Natur- und Geschichtsphilosophie [Natural philosophy and philosophy of history]. *Vierteljahresschrift für wissenschaftliche Philosophie und Soziologie*, XXXVI(3), 381–412.

Kuntze, F. (1913). Dum calculat Deus .... *Kunstwart & Kulturwart*, 27(6), 483–491.

Ladd, G. T. (1897). *Philosophy of Knowledge: An Inquiry Into the Nature, Limits, and Validity of Human Cognitive Faculty*. New York, NY: Charles Scribner's Sons.

Lange, F. A. (1865). *Die Grundlegung der mathematischen Psychologie [Foundations of Mathematical Psychology]*. Duisburg, DE: Falk & Volmer.

Lange, F. A. (1877). *Geschichte des Materialismus seit Kant [History of*

Materialism Since Kant] (Vol. II). Iserlohn, DE: J. Baedeker.

Laplace, P.-S. (1774). Mémoire sur la probabilités des causes par les évènements [Memoir on the Probabilities of the Causes of Events]. *Mémoires de l'Académie Royale des Sciences de Paris, VI,* 621–656.

Laplace, P.-S. (1795). *Exposition du systême du monde [Exposition of the System of the World]* (Vols. I, 1st ed.). Paris, FR: Cercle-Social. (Published in "l'an IV de la République Français")

Laplace, P.-S. (1798–1825). *Traité de mécanique céleste [Treatise on Celestial Mechanics].* Paris, FR: Crapelet. (Five volumes published between 1798 and 1825)

Laplace, P.-S. (1816). *Essai philosophique sur les probabilités [Philosophical Essay on Probabilities]* (3rd ed.). Paris, FR: Courcier. ("Troisième Édition, revue et aumentée par l'auteur")

Laplace, P.-S. (1830). *The System of the World* (Vol. 1; H. H. Harte, Ed. & Trans.). London, GB: Longman, Rees, Orme, Brown, and Green. (Original work published 1795)

Leclerc, J. (?). (1937). Review of Determinismus und Indeterminismus in der modernen Physik, historische und systematische Studien zum Kausalproblem. *Revue de Métaphysique et de Morale,* 44(4), 3–4. (All reviews are unsigned; Jacques Leclerc was "éditeur-gérant")

Leibniz, G. W. (1684?/1840). De Scientia Universali seu Calculo Philosophico [On universal science, or, The philosophical calculus]. In J. E. Erdmann (Ed.), *Opera Philosophica* (pp. 82–85). Berlin, DE: G. Eichler. (From an unpublished manuscript in the Royal Library of Hanover. A precise dating to 1684 is ascribed to Trendelenburg, 1867, p. 18, note 1, by Windelband, 1893, p. 382, but this is an evidently careless reading of Trendelenburg. The same date is later published without ascription by Ladd, 1897, p. 69 and Couturat, 1900, p. 362, *inter alia*)

Leibniz, G. W. (1690?/1840). Von dem Verhängnisse [On determinism]. In G. E. Guhrauer (Ed.), *Leibnitz's deutsche Schriften* (Vol. 2, pp. 48–55). Berlin, DE: Veit. (Written between 1690 and 1700, according to Guhrauer)

Leibniz, G. W. (1696/1846). Probeschrift über die Dynamik, die Entdeckung der bewundernswerten Naturgesetze der Kräfte der Körper und ihrer Wechselwirkungen, und die Zurückführung derselben auf ihre Ursachen betreffend [A specimen of dynamics: On the discovery of the admirable laws of nature about the forces of bodies and their interactions, and the reduction of the same to their ultimate causes]. In G. Schilling (Ed. & Trans.), *Leibniz als Denker* (pp. 18–36). Leipzig, DE: Hermann Fritzsche. (Translation of Specimen Dynamicum, pro admirandis Natura legibus circa Corporum vires & mutuas actiones deregendis, & adsuas causa revocandis, *Acta Eruditorum Lipsiensibus*, 1696, 145–157)

Lichtenstein, L. (1932). La philosophie des mathématiques selon M. Émile Meyerson [Philosophy of mathematics according to M. Émile Meyerson] (A. Metz, Trans.). *Revue Philosophique de la France et de l'Étranger, 113*, 169–206. (The author's surname has been misspelled "Lightenstein" in the publication)

Lichtwarck-Aschoff, A., Kunnen, S. E., & van Geert, P. (2009). Here we go again: A dynamic systems perspective on emotional rigidity across parent—adolescent conflicts. *Developmental Psychology, 45*(5), 1364–1375.

Lüders, H. (1907). *Das Würfelspiel im alten Indien [Dice Games in Ancient India]*. Berlin, DE: Weidmannsche Buchhandlung.

Margenau, H. (1931). Causality and modern physics. *The Monist, 41*(1), 1–36.

Margenau, H. (1934). Meaning and scientific status of causality. *Philosophy of Science, 1*(2), 133–148.

Marique, J.-M.-L. (1886). *Recherches Expérimentales sur le Mecanisme de Fonctionnement des Centres Psycho-moteurs du Cerveau [Experimental Research on the Mechanism of the Function of the Psycho-motor Centers of the Brain]*. Brussels, BE: Gustave Mayloez.

Maxwell, J. C. (1871). *Theory of Heat*. London, GB: Longmans, Green and Co.

Maxwell, J. C. (1873/1882). Does the progress of physical science tend to give any advantage to the opinion of necessity (or determinism) over that of the contingency of events and the freedom of the will? In L. Campbell & W. Garnett (Eds.), *The Life of James Clerk Maxwell: With Selections From His Correspondence and Occasional Writings* (pp. 431–444). London, GB: Macmillan. (Address to the Erānus Club, Cambridge, February 11, 1873)

Maxwell, J. C. (2002). *The Scientific Letters and Papers of James Clerk Maxwell* (Vol. III. 1874–1879; P. M. Harman, Ed.). Cambridge, GB: Cambridge University Press.

McCarty, D. C. (2005). Problems and riddles: Hilbert and the du Bois-Reymonds. *Synthese*, *147*(1), 63–79.

McLuhan, M. (1964). *Understanding Media: The Extensions of Man*. New York, NY: McGraw-Hill.

Menand, L. (2002). *The Metaphysical Club*. New York, NY: Macmillan.

Michaels, M. (1995). Seven fundamentals of complexity for social science research. In A. Albert (Ed.), *Chaos and Society* (pp. 15–33). Amsterdam, NL: IOS Press.

Milani, A., & Koksch, N. J. (2005). *An Introduction to Semiflows*. Boca Raton, FL: Chapman & Hall/CRC Press.

Mises, R. von. (1928). *Wahrscheinlichkeit, Statistik und Wahrheit [Probability, Statistics and Truth]*. Vienna, AT: Julius Springer.

Mises, R. von. (1930/1931). Über kausale und statistische Gesetzmäßigkeit in der Physik [On causal and statistical regularity in physics]. *Erkenntnis*, *1*, 189–210.

Nelson, W. A., McCauley, E., & Wrona, F. J. (2001). Multiple dynamics in a single predator-prey system: Experimental effects of food quality. *Proceedings of the Royal Society: Biological Sciences*, *268*(1473), 1223–1230.

Newton, I. (1729). *The Mathematical Principles of Natural Philosophy* (Vol. I & II; A. Motte, Trans.). London, GB: Benjamin Motte. (Original work published 1686)

Newton, I. (1871). *Philosophiæ Naturalis Principia Mathematica [Mathematical Principles of Natural Philosophy]* (W. Thomson & H. Blackburn, Eds.). Glasgow, GB: James Maclehose. (Original work published 1726; "reprint [of] Newton's last edition without note or comment")

Newton, I., & Whiteside, D. T. (1967–1981). *The Mathematical Papers of Isaac Newton* (Vols. I–VII; D. T. Whiteside, Ed.). New York, NY: Cambridge University Press.

orrery. (2011). *OED Online*. New York, NY: Oxford University Press. Retrieved September 30, 2011, from http://www.oed.com/view/Entry/132764

Patrick, G. T. W. (1888). A further study of Heraclitus. *The American Journal of Psychology*, *1*(4), 557–690.

Peter, H. (1938). Keynes' neue Allgemeine Theorie [Keynes's new general theory]. *FinanzArchiv/Public Finance Analysis, New Series*, *5*(1), 51–84.

Peterson, F. P., & Stein, N. (1959). Secondary cohomology operations:

Two formulas. *American Journal of Mathematics, 81*(2), 281–305.

Pindar. (1882). *Selected Odes of Pindar* (T. S. Seymour, Ed.). Boston, MA: Ginn, Heath & Co.

Pitowsky, I. (1996). Laplace's demon consults an oracle: The computational complexity of predictions. *Studies in the History and Philosophy of Modern Physics, 27*, 161–180.

Pitowsky, I. (2002). Quantum speed-up of computations. *Philosophy of Science, 69*, 168–177.

Poddiakov, A. N. (2000/2006). *Issledovatel'skoe povedenie: strategii poznaniia, pomosht', protivodeĭstvie, konflikt [Exploratory Behavior: Cognitive Strategies, Help, Counteraction, and Conflict]*. Moscow, RU: PER SE. Retrieved July 15, 2011, from http://hse.ru/data/866/913/1235/ip.rar

Porter, T. M. (1986). *The Rise of Statistical Thinking 1820–1900*. Princeton, NJ: Princeton University Press.

Potter, E. (2001). *Gender and Boyle's Law of Gases*. Bloomington and Indianapolis, IN: Indiana University Press.

Rogers, H. J. (1963). An example in mathematical logic. *The American Mathematical Monthly, 70*(9), 929–945.

Rudolph, L. (2006a). The fullness of time. *Culture & Psychology, 12*, 157–186.

Rudolph, L. (2006b). Spaces of ambivalence: Qualitative mathematics in the modeling of complex, fluid phenomena. *Estudios Psicologías, 27*, 67–83.

Ruelle, D. (2009a). Reply to Yorke [Letter to the Editor]. *Notices of the American Mathematical Society, 56*(10), 1233. Retrieved December 11, 2011, from http://www.ams.org/notices/200910/rtx091001232p.pdf

Ruelle, D. (2009b). Some comments on "Period Three Implies Chaos" [Letter to the Editor]. *Notices of the American Mathematical Society, 56*(6), 688. Retrieved December 11, 2011, from http://www.ams.org/notices/200906/rtx090600688p.pdf

Smith, L. B., Thelen, E., Titzer, R., & McLin, D. (1999). Knowing in the context of acting: The task dynamics of the A-not-B error. *Psychological Review, 106*, 235–260.

Sorley, W. R. (1905). The relations of ethics. In H. J. Rogers (Ed.), *International Congress of Arts and Science, Vol. I. Philosophy and Mathematics* (pp. 391–414). Boston, MA and New York, NY: Houghton Mifflin Company.

statistic. (2011). *OED Online*. New York, NY: Oxford University

Press. Retrieved December 25, 2011, from `http://www.oed.com/view/Entry/189317`

statistics. (2011). *OED Online*. New York, NY: Oxford University Press. Retrieved December 25, 2011, from `http://www.oed.com/view/Entry/189322`

Stewart, D. (1824). *Dissertation First [etc.]. Part II*. Edinburgh, GB: Archibald Constable. (Completed at "Kinneil House, August 7, 1821." Published and bound with *Supplement to the Fourth, Fifth, and Sixth Editions of the Encyclopædia Britannica. With Preliminary Dissertations on the History of the Sciences*, Vol. V [HUN TO MOL]; 257 pp., separately numbered)

Stroyan, K. D., & Luxemburg, W. A. J. (1976). *Introduction to the Theory of Infinitesimals*. New York, NY: Academic Press.

Suckling, J. (1994). *Aglaura*. Cambridge, GB: Chadwyck-Healy. (Original work published 1638)

Sveistrup, H. (1929). Friedrich Kuntze †. *Kant-Studien*, *34*(1–4), 291–299.

Thompson, S. (1960). *Motif-index of Folk-literature: A Classification of Narrative Elements in Folktales, Ballads, Myths, Fables, Mediaeval Romances, Exempla, Fabliaux, Jest-books and Local Legends* (Vol. 6.2). Bloomington, IN: Indiana University Press.

Thomson, W. (1874). The kinetic theory of the dissipation of energy. *Nature*, *9*, 441–444.

Thomson, W. (1882). On a universal tendency in nature to the dissipation of mechanical energy. In *Mathematical and Physical Papers, Vol. I* (pp. 511–514). Cambridge, GB: Cambridge University Press. (Original work published 1852)

Trendelenburg, F. A. (1867). *Historische Beiträge zur Philosophie [Historical Survey of Philosophy]* (Vol. III). Berlin, DE: G. Bethge.

Vaihinger, H. (1878). Das Entwickelungsgesetz der Vorstellungen über das Reale. Zweiter Artikel [The law of development of ideas about reality. Second article]. *Vierteljahrsschrift für wissenschaftliche Philosophie*, *2*(4), 415–448.

Vailati, E., & Yenter, T. (2009). Samuel Clarke. In E. N. Zalta (Ed.), *The Stanford Encyclopedia of Philosophy* (Summer 2009 ed.). Stanford, CA: Metaphysics Research Lab, CSLI, Stanford University. Retrieved October 5, 2011, from `http://plato.stanford.edu/archives/sum2009/entries/clarke`

van Geert, P. (1994). *Dynamic Systems of Development: Change Between Complexity and Chaos*. New York, NY: Harvester / Wheatsheaf.

van Geert, P. (2003). Dynamical systems approaches and modeling of developmental processes. In J. Valsiner & K. J. Connolly (Eds.), *Handbook of Developmental Psychology* (pp. 3–17). London, GB: Sage.

Windelband, W. (1893). *A History of Philosophy* (J. H. Tufts, Trans.). New York, NY: Macmillan. (Translation of *Geschichte der Philosophie*, 1892, Freiburg im Breisgau, DE: J. C. B. Mohr)

Wundt, W. (1883). *Logik [Logic].* Stuttgart, DE: Ferdinand Enke Verlag.

Wundt, W. (1886). *Ethik [Ethics].* Stuttgart, DE: Ferdinand Enke Verlag.

Wundt, W. (1920). *Erlebtes und Erkanntes [Experienced and Known].* Stuttgart, DE: Albert Kröner.

Yorke, J. (2009). Reply to David Ruelle [Letter to the Editor]. *Notices of the American Mathematical Society, 56*(10), 1232–1233. Retrieved December 11, 2011, from `http://www.ams.org/notices/200910/rtx091001232p.pdf`

11

Complex Dynamical Systems and the Social Sciences

Ralph Abraham, Dan Friedman, and Paul Viotti

The symbiosis of mathematics and the sciences has been described as an hermeneutical circle. That is, experimental data determines a model, the model suggests new experiments, new data refines the model, and so on. This endless loop is the motor for the advance of science.

Mathematics and the sciences became joined in this hermeneutical circle four hundred years ago, when mathematical modeling came of age: mathematical physics since 1600, mathematical biology since 1930 or so, and the mathematical social sciences since the 1950s. *Agent-based modeling (ABM)*, a tool for the modeling and simulation of complex dynamical systems, has pumped this latter into an excited state. ABM may be regarded as an evolution of object-oriented programming (OOP); whereas OOP introduced objects, reusable modules, protected data, interfaces, and so on, in ABM we have objects which may act as independent agents. Chris Langton, of Artificial Life fame, pioneered the way with his Swarm language at the Santa Fe Institute in the 1980s. An early application (Kohler, West, Carr, & Langton, 1996) considered the native cultures of the American Southwest as a complex dynamical system.

The hermeneutical hope of ABM models in the social sciences is to discover rules for the behavior of social systems. In physics,

Keywords: agent-based modeling, computer simulation, dynamical systems, evolutionary game theory, fitness gradient, global analysis, landscape dynamics, NetLogo, spatial voting models, strategy space.

we have the archetype of Galileo rolling billiard balls down inclined tracks, collecting data points timed with his isochronous pendulum, and fitting a quadratic curve. In the social sciences we have no Newton's law, $F = ma$, but if we can fit behavioral data with a model, we may discover a rule. ABM makes this easy, and yet for any given computer model, successful as such, there may be as yet no corresponding mathematics. In this chapter we briefly review the history of ABM, and then describe some of our work on embedding social models into mathematical frameworks.

Survey of ABM environments and literature

A number of programming environments for agent-based modeling have been developed recently, mostly at graduate schools of the social sciences, and there are now many. We may mention six.
- Swarm (Center for the Study of Complex Systems, University of Michigan; see *SwarmWiki*, 2005)
- Ascape (Center on Social & Economic Dynamics, The Brookings Institution; see *Ascape Guide*, 2010)
- MASON (Center for Social Complexity, George Mason University; see *MASON Multiagent Simulation Toolkit*, 2010)
- RePast (University of Chicago Social Science Research Computing Center; see *RePast Agent Simulation Toolkit*, 2008)
- StarLogo TNG (Media Laboratory, Massachusetts Institute of Technology; see *StarLogo TNG*, 2008)
- NetLogo (Center for Connected Learning & Computer-Based Modeling, Northwestern University; see Wilensky, 1999/2010)

Two of these—NetLogo and StarLogo TNG—evolved from LOGO, the language developed in 1967 by Seymour Papert (co-founder of MIT's AI Lab and its successor, the Media Lab) and Wallace Feurzeig, as a tool for teaching children programming concepts (recursion, modularity, ...). After the ABM concept emerged, LOGO evolved into StarLogo, then into NetLogo, both designed to accommodate large numbers of independent, interacting agents.[1]

The original StarLogo was developed at the MIT Media Lab in 1989–1990 and ran on a massively parallel super-

---

[1][LOGO originally ran on mainframes (*e.g.*, a PDP-10) and featured only a single agent. The number of agents available in StarLogo TNG and NetLogo is limited only by effectively negligible hardware considerations. (Ed.)]

computer called the Connection Machine. A few years later (1994), a simulated parallel version was developed for the Macintosh computer. That version eventually became MacStarLogo. StarLogoTNG (1997), developed at the Center for Connected Learning and Computer-Based Modeling (CCL), is essentially an extended version of MacStarLogo with many additional features and capabilities. Since then two multi-platform Java-based multi-agent Logos have been developed: NetLogo (from the CCL) and a Java-based version of StarLogo (from MIT). (Wilensky, 1999/2010, p. 127)

In this chapter we will focus on the one that we have used extensively in our research, NetLogo.

Consulting the libraries of sample models at the NetLogo website, we find the categories Art, Biology, Chemistry & Physics, Computer Science, Earth Science, Games, Mathematics, Networks, Social Science, and System Dynamics. Under Social Science are exemplary models such as AIDS, Altruism, Cash Flow, Ethnocentrism, Party, Prisoner's Dilemma, Rumor Mill, Scatter, Segregation, Simple Birth Rates, Traffic, Voting, Wealth Distribution, and so on. Hundreds more models are in the library of NetLogo User Community Models maintained on the NetLogo website.[2]

Landscape dynamics

Advances in mathematics in the 1960s made available a host of new modeling strategies for all the sciences. The framework of *global analysis*—applied to dynamical systems (for an early, elementary survey by a founder of the field, see Smale, 1969), calculus of variations (the

---

[2][NetLogo models have been used in recent publications on artificial emotions (Maria & Zitar, 2007), primate social behavior (Bryson, Ando, & Lehmann, 2007), urban gentrification (Torrens & Naraa, 2007), crowd behavior and social comparison theory (Fridman & Kaminka, 2007), qualitative approaches to the study of newcomer socialization in groups (Yang & Gilbert, 2008), and the semiotics of mathematics education (Abrahamson, 2009). Recent social science publications using other ABM environments include models of bond pricing (Masson, 2003, in Ascape), collective intentionality (Cioffi-Revilla, Paus, Luke, Olds, & Thomas, 2004, using MASON), dynamics of public opinion (Suo & Chen, 2008, using SWARM), emergence (North et al., 2009, using RePast), and financial crises (Arciero, Biancotti, D'Aurizio, & Impenna, 2008, using StarLogo TNG). (Ed.)]

thesis of Karen Uhlenbeck, 1968, is one of the first examples), partial differential equations of evolution type (Browder, 1970; Carroll, 1969; Marsden, 1974, etc.), game theory (see Smale, 1973, and its five sequels), and so on—brought us catastrophe theory (Thom, 1974), chaos theory (*cf.* Abraham & Shaw, 1988), complexity and simplexity (Cohen & Stewart, 1994), neural network theory (a recent survey is given by Veltz & Faugeras, 2010), evolutionary game theory (Dawid, 2007), and more. The computer and computer graphic revolutions brought new possibilities of computational modeling, simulation, and scientific visualization. Of all the sciences, those with the greatest potential to benefit from these new methods are the social, behavioral, and economic sciences.

One effort to embed a social model into a mathematical framework is the gradient hill-climbing scheme called *landscape dynamics* developed by Daniel Friedman at the University of California at Santa Cruz in the 1990s (see, *e.g.*, Friedman & Yellin, 1997, 2000). This mathematical modeling and computer simulation technology, based on evolutionary games with continuous spaces of strategies, is an application of global analysis methods to game theory. It extends the class of models called *evolutionary games* and opens it to new applications in the social sciences. Landscape dynamics is intended to advance the arts of mathematical modeling, computer simulation, and scientific visualization of complex dynamical systems encountered in the social, behavioral, and economic sciences.

Evolutionary game models analyze the interactions of strategies over time. Equilibrium emerges, or fails to emerge, as players adjust their strategies in response to the payoffs they earn. Early models mainly considered situations in which players chose among only a few discrete strategies. Landscape dynamics allows players to choose within a continuous strategy space, A. The basic axioms of landscape dynamics are that

(1) myopic agents adopt a strategy in a continuous strategy space,
(2) such agents alter their own strategies only incrementally, and
(3) via these alterations agents ultimately also affect the strategies of other agents.

In this setup, the current state is the distribution of all players' choices over A. In any particular application, the current state defines a *payoff function* on A, whose graph is called the *adaptive landscape*. Players respond to the landscape in continuous time by adjusting their strategies towards higher payoff. Hence the current state (the distribution of chosen strategies) changes, and this in turn alters the landscape. The interplay between the evolving state and the landscape gives rise to nontrivial dynamics. In particular, when players follow the gradient (steepest ascent in the adaptive landscape), the evolving state can be characterized as the solution to a nonlinear partial differential equation, or equivalently, a dynamical system on an infinite-dimensional space (for more details, see Abraham & Friedman, 2007a).

We describe three models developed since 2005 in joint research on landscape dynamics by Daniel Friedman and Ralph Abraham, and implemented in NetLogo.

Conspicuous consumption

Our first example of landscape dynamics is a model for Veblen's (1912) notion of conspicuous consumption. Veblen shows our approach in the context of an intuitive, well-known problem. Fig. 11.1 depicts a streamlined version of the Veblen interface (for descriptions of the various visual elements available in a general NetLogo on-screen workspace, see Fig. 11.2). As described in the literature, there are two basic flavors of conspicuous consumption: *envy* and *pride*. In this chapter, we restrict attention to the envy model.

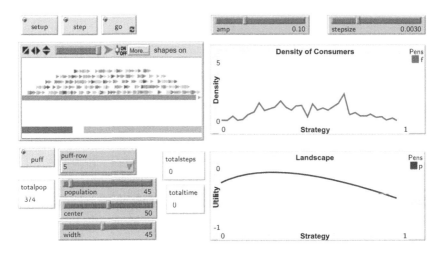

Figure 11.1: The graphical user interface of Veblen 5.2.

The initial setting. In this model the strategy space or *action set A* is the unit interval $[0,1]$. The *agents* (called *turtles* in Net-Logo[3]) are consumers. Each consumer is shown as a triangle on the strategy space. They have different colors for the visual effect, and to facilitate interpretation of the display: when several consumers are on the same *patch*[4] (discretized interval of the strategy space), only the top one can be seen in its entirety, but the x position is a floating point number, so lower turtles may be seen in part.[5] The

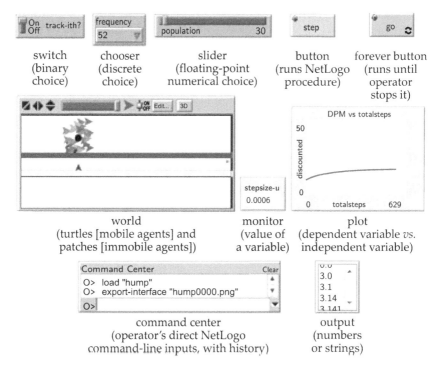

Figure 11.2: Visual elements of a NetLogo workspace.

---

[3][The single agent in LOGO (note 1, p. 322) was called a "turtle" for historical robotics-related reasons (*cf.* Abelson, Goodman, & Rudolph, 1974). (Ed.)]

[4][An interesting feature of NetLogo is that patches are also (non-mobile) agents, so interactions between mobile agents and the "space" through which they move can modify the space as well as the mobile agents. Compare this to John Archibald Wheeler's aphorism "Space tells matter how to move[.] Matter tells space how to curve" (Misner, Thorne, & Wheeler, 1973, p. 5). (Ed.)]

[5][The editor regrets that the practical necessity of replacing colors with grayscale has made it somewhat harder to distinguish lower turtles. To experiment with the program in full color, see Abraham and Friedman (2004). (Ed.)]

graphical representation of the strategy space is the five horizontal rows in the upper half of the *world* graphic display. These are to be regarded as superimposed layers on a single unit interval in which each point x corresponds to a choice of strategy. All consumers have the same income, 1, but choose variously how much, x, to spend on *ordinary consumption*, the remaining $1 - x$ being spent on *conspicuous consumption*. Thus the strategy choice $x = 0$ represents 100% conspicuous consumption (*e.g.*, diamond rings), and $x = 1$ represents 100% ordinary consumption (*e.g.*, savings).

A chosen number of consumers begin at initial positions in the strategy space. This *initial density* is, *a priori*, important to the outcome of a run. The model is arranged so that the initial density is the sum of an arbitrary number of square waves; such sums can approximate any initial density. In our implementation the initial distribution may be constructed by the operator. Interesting choices include a single square wave or *herd*, two herds, a tent shape or *heap*, two heaps, and so on. In any case, the operator begins by adding square waves, or sub-herds, until a desired initial distribution is obtained. Each addition of a sub-herd, called a *puff*, is achieved by a NetLogo procedure executed by the puff button; puff uses parameters for the sub-herd that are set by the operator using the population, center, and width sliders.

The distribution $F(x)$. The instantaneous state of the system can be represented either by f, the density of consumers in the strategy space (denoted by ρ in Friedman, 2001), which is a probability measure, or equivalently by F, the cumulative distribution of f (denoted by D in Friedman, 2001), *i.e.*, the integral of f (from $-\infty$ to x), a monotone function increasing from 0 to 1.

The "Density of Consumers" plot window depicts f graphically: the independent variable is x, and the dependent variable is the average density of turtles on the strategy space patch containing x.

The payoff $\phi(x, F)$. The most important function in the model is the *payoff function* (or *fitness function*) ϕ. It is a real-valued function of the two variables x and F. It also depends on a non-negative constant c, set by the slider amp, that represents the importance of ordinary consumption relative to conspicuous consumption. The function ϕ is the landscape in this example of landscape dynamics. The definition of ϕ used here, called the *envy rule*, is the sum

$$\phi(x, F) = c \log(x) - \int_0^x F(y)\, dy$$

of two non-positive functions: $c \log(x)$ is negative for x in the half-open interval $(0, 1]$ (note that we avoid the troublesome value $x = 0$); $-F(y)$ is non-positive (being the negative of a non-negative function) and therefore so is its integral.

The "Landscape" plot window depicts ϕ graphically. Again, the independent variable is x, in strategy space. The two plot windows are aligned to emphasize that they share the same independent variable, and to facilitate visual assessment of the relationship between f and ϕ. Both plots are updated every 10th step.

The slope. The *slope* ϕ_x of the landscape,[6] also called the *gradient of ϕ* or simply the *fitness gradient*, is given by the formula

$$\phi_x(x, F) = c/x - F(x).$$

The term c/x represents the *direct utility* that a consumer receives for ordinary consumption; it is monotone decreasing to the value c at $x = 1$ (where all consumption is ordinary). The term $-F(x)$ is also monotone decreasing. A color bar in the world window shows the qualitative behavior of ϕ_x:

- +0.1 or higher [red], positive, step to the right,
- ○ −0.1 to 0.1 [yellow], small step to the left or right,
- −0.1 or lower [green], negative, step to the left.

We are especially interested in the zero-crossing of the monotone decreasing function $\phi_x(x, F)$ of x, which occurs at some point x in the yellow region. As the slope depends on F, and F is time-dependent, the yellow segment of the slope color bar will be expected to move about.

The step. Consumers step 'uphill' on the landscape. With each step of discrete time, each consumer adjusts her or his strategy x by an increment proportional to the slope. The proportion is set with the stepsize slider. Each turtle moves uphill by an increment stepsize * slope. This is the *Euler method* for integrating the partial differential equation representing the envy rule.

The behavior of the model. Simulations with Veblen conform to the expected result: all the consumers converge to a single attractive strategy, independent of the initial distribution (an example appears in Fig. 11.3). This is expected because of a convergence

---

[6][The authors' notational convention, by which subscripting the function ϕ with the variable x indicates the "partial derivative" of ϕ with respect to x, is common in the physical sciences. (Ed.)]

theorem obtained using global analysis (Friedman & Ostrov, 2008). It is possible to prove this theorem because landscape dynamics is a bridge between agent-based modeling and pure mathematics. The agreement between the behavior of the simulations and the conclusion of the theorem provides a degree of validation of Veblen.

Financial markets

Our next example is actually a series of related NetLogo models for financial markets developed at the University of California at Santa Cruz since 2005; they represent joint work with a number of students, to whom we owe a large debt of gratitude. The goal has been to discover the landscape dynamics underlying the phenomena of bubbles and crashes. We describe two models in this series: Market 8.0, a baseline model with equilibrium behavior, and

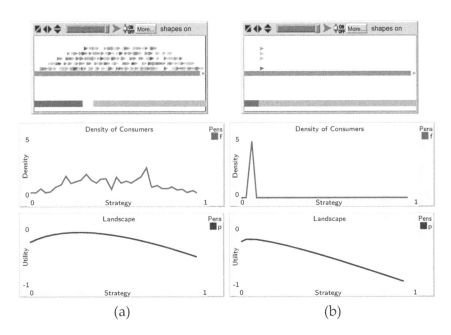

Figure 11.3: Snapshots of world and plot windows from one run of Veblen 5.2 with `totalpop` = 374. (a) Initial state (`totaltime` = 0) with population distributed in a rough hump. (b) Static attractive state (`totaltime` = 42.0, `totalsteps` = 14,000) with population in a point distribution ("delta function") concentrated at $x \approx 0.65$.

Market 8.1, in which bubbles and crashes emerge. For background from finance, and further details, see Friedman and Abraham (2009), where Market 8.0 and Market 8.1 are called Model0 (or "the basic model") and Model1, respectively. For the NetLogo models themselves, through Market 8.3.05, see Abraham and Friedman (2007b). Fig. 11.4 illustrates the interface of Market 8.1

The initial setting. For all models in the Market series, the state space is 2-dimensional. Like Veblen, Market has a 1-dimensional

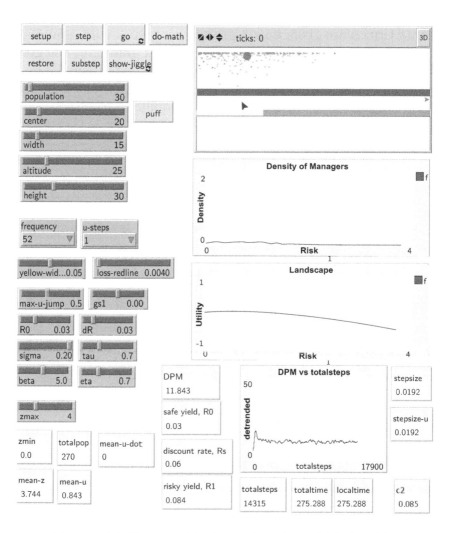

Figure 11.4: The graphical user interface of Market 8.1.

strategy space represented by an interval on the horizontal axis; each value of x in this interval represents the degree to which a fund manager is willing to invest in risky assets, so that moving to the right increases risk. In addition, Market has a *value space*, represented by an interval on the vertical axis; each value of z in this interval can be the size of a fund manager's portfolio. (The operator sets the upper limit on portfolio size with the zmax slider; its default value is 4. A value of 1 is considered normal.) The state space is the Cartesian product of the strategy and value spaces.

The distribution. Agents represent money market fund managers. Each agent is shown as a small triangle in the world window, located according to her strategy and the value of her fund. Using the population, center, width, altitude, and height sliders and the puff button, the operator sets the number of managers and their initial positions in state space. As before, this initial distribution is, *a priori*, important to the outcome of a run. The setup button selects a random initial distribution. Like Veblen, Market depicts the density of agents (managers) graphically in a plot window.

The payoff $\phi(x, F)$. The payoff function is

$$\phi(x, F) = x(R_1 - R_0) - \frac{1}{2}c_2 x^2$$

where R_0 is the rate of return of the risk-free asset and R_1 that of the risky asset. The financial parameter c_2 "can be interpreted as the market price of risk. In the basic model c_2 is an exogenous constant, but it can vary in extensions of the model" (Friedman & Abraham, 2009, p. 925). Like Veblen, Market depicts ϕ graphically in a plot window aligned with the "Density of Managers" window.

The slope. In Market 8.0, where c_2 is a constant set by a slider (actually, a piecewise-constant function of time, since the operator can vary c_2 during a run), the slope is simply

$$\phi_x(x, F) = R_1 - R_0 - c_2 x \qquad (11.1)$$

but in higher numbered versions of Market, c_2 depends on x (etc.) in a complicated way. It depends, among other things, on a random "surprise" variable, denoted jiggle and local to each manager, which models the observation (Abraham, n.d., p. 2) that

> even if two managers have chosen the same risk strategy, say u_0, so that they have the same mix of riskless and

risky investments, they most likely have chosen different risky investments. Thus, their payments at the end of [each] step may differ by a random variable.

It also depends highly nonlinearly on financial parameters alpha (a measure of "buying pressure"), beta (reflecting investors' sensitivity to perceived loss), eta (investors' "memory decay rate"), sigma (volatility), and tau ("streak decay") that are set by sliders, as well as on the values of losses. All of this means that in these versions of the equation, (11.1) is replaced by a complicated expression, which we do not record here (see Friedman & Abraham, 2009).

The step. *Stepsize* is a unit of time for periodic reports of financial data. It is set with the frequency *chooser.*[7] For instance, the choice of "52" signifies a frequency of 52 steps per year (*i.e.*, weekly), and the variable stepsize in the program is set to the value 1/52 years. Another chooser, u-steps, sets the number of substeps in a step. Increasing u-steps decreases the substepsize stepsize-u to the ratio stepsize / u-steps, decreasing the numerical error in the Euler integration.

At each step (or every u-step substeps), the manager's horizontal coordinate (location in strategy space) moves up the slope of the payoff function, and her vertical coordinate (location in value space) is adjusted to reflect the change in the size of her portfolio due to

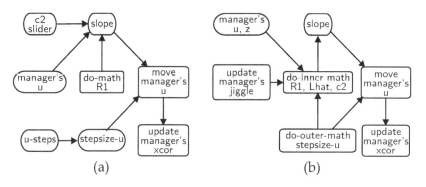

Figure 11.5: Outlines of the horizontal substep procedures in (a) Market 8.0 and (b) Market 8.1 (adapted from Abraham, n.d.).

---

[7][Functionally, a chooser is a drop-down menu; conceptually, it is a way for the operator to set the value of a variable on a nominal or ordinal scale. (Ed.)]

payoffs. The flowcharts in Fig. 11.5 outline the first of these procedures in Market 8.0 and Market 8.1; even in outline, it is clear that the horizontal substep in Market 8.1 is a "massive process" (Abraham, n.d., p. 5) compared to the corresponding substep in Market 8.0.

In contrast, vertical movement of agents is comparatively simple across the Market series. Fig. 11.6 gives the NetLogo code for this in Market 8.0 (slightly edited for added clarity); the corresponding Market 8.1 procedure simply adds a line incorporating the effect of each manager's jiggle on her annual-gross.

Behavior of the model. The differences between Market 8.0 and Market 8.1 are striking.

> [Market 8.0] is very stable. Analytic results suggest, and simulations confirm for a wide range of exogenous parameters, that the asset price converges quickly and reliably to a level proportional to the fundamental value. The proportion is a decreasing function of the risk cost parameter c_2. The model never bubbles or crashes.

```
to update-managers-z ;;; update turtle variable z
                     ;;; and move its ycor
   ask managers [
      let annual-gross ( u * ( R1 - R0 ) + R0 )
                     ;;; this is gross return
      let pay cut annual-gross
                     ;;; annual yield reduced
                     ;;; to stepsize
      set z ( z * ( 1 + pay ) )
                     ;;; increment size of portfolio
                     ;;; for timestep
      if (z < zlim) [ set z zlim ]
                     ;;; z not allowed to fall below zlim
      if (z > zmax) [ set z zmax ]
                     ;;; z not allowed to rise above zmax
      let ytemp (z - zmin) * (width-y -1 ) / ( 2 * width-z )
                     ;;; it's a float
      set ycor ytemp ;;; convert z to ycor, keep as float
   ]
end
```

Figure 11.6: A fragment of commented NetLogo code from Market 8.0 (adapted from Abraham, n.d.).

[Market 8.1], an extension of [Market 8.0], features an endogenous risk premium driven by constant-gain learning. It also has a unique steady state, but its dynamics are quite different. Although asset price usually is near its steady state value, there are recurrent episodes in which it rises substantially (typically 20–50% above normal levels) and then crashes (often to a third or less of normal levels within a few months). Such episodes occur over a wide range of "realistic" parameter values.

The episodes become rarer when parameter configurations give investors longer memories or give fund managers smaller (or more fleeting) streaks of luck. In opposite configurations the episodes become more common and, with extreme parameter values, normalcy becomes rare. (Friedman & Abraham, 2009, p. 932)

For a detailed exposition of the dynamics of the Market family, including mathematical and statistical analyses, the reader is referred to the just-quoted paper.

Two-party voting

Harold Hotelling (1929) developed a seminal microeconomic theory of spatial competition in which firms tend to be attracted to the center of a one-dimensional bounded space. The classical Hotelling model illustrates, using a colorful example, that two competing hotdog firms on a one-dimensional stretch of beach (bounded at each end) tend to move toward the middle of the beach, given a normal distribution of customers (*i.e.*, a greater concentration in the middle with fewer and fewer people towards the edges). Drawing from Hotelling's work, Black (1948) and Downs (1957) constructed spatial voting models. Black's model treats candidates as policy alternatives or points in Euclidean space; voters are the principal actors. In the Downsian tradition, under which the bulk of research in the spatial voting literature has been conducted, candidates are analogous to the firms in Hotelling's model.

The result of the Downsian model is that if a voter supports the nearest candidate, then candidates seeking to win an election ultimately will locate themselves at the position of the "median voter". Thus in Downsian models of a two-party system, political competition is often viewed as a fight for the middle. A political agent, *e.g.*, a party leader running for office, hopes to attract more votes by touting a center-of-the-road platform, as a large majority of potential votes lies near the "median voter". Since the early work of Black and Downs, the literature on spatial voting models has become quite rich with many variants and refinements—see, *e.g.*, Adams and Merrill (2006); Cox (1989); Davis and Hinich (1966); Eaton and Lipsey (1975); Hinich (1977); Riker and Ordeshook (1973); Roemer (1997); and Wittman (1973, 1977, 1983).

Given the axiomatic assumptions (1)–(3) of landscape dynamics (p. 324), we may set the stage for a particular application of landscape dynamics to a two-party voting model in which parties locate in a 2-dimensional *issue space* and adjust their platforms incrementally so as to increase their vote shares. Suppose that two salient issues dominate the political discourse of the country: Issue X and Issue Y. We then array these issues on two distinct axes. Think of this portion of a Cartesian plane as a space in which political agents compete for blocks of voters.

Each voter (who is assumed to have fixed positions on the two issues, and thus to occupy a fixed point in issue space) has an "attraction" to a given party (also conceived of as occupying a point in issue space, which can change with time) based on the Euclidean distance between the voter and the party. That is, a voter's attraction to Party A is assumed to diminish, in a specified manner, as the distance in issue space between the voter and Party A increases. For some voters, the attraction to all the parties may be so weak that they will not vote at all. The model assumes, in this regard, that if a voter's perceived cost of voting (*e.g.*, time spent waiting in line at the polls[8]) exceeds the voter's attraction to all parties or candidates (as represented by issues), then that voter will not vote.

Following the gradient rule, on each iteration a party examines the points (*i.e.*, joint positions on Issues X and Y) immediately surrounding it, then moves to an adjacent point in case that move puts the party in a position to receive more votes. In doing so, the party must "imagine" its prospective payoffs in the "local" strategy space. The parties' respective payoffs are a function of the number of likely

---

[8][At various epochs and locales, other known costs of voting have included: the expense (in time or money) of child care; the expense (in time, money, or frustration) of obtaining Voter ID, passing "literacy tests", or paying poll taxes; the fear of intimidation (before voting) or reprisal (after voting); and so on. (Ed.)]

voters that will support them, and the parties may incrementally change their strategies in order to garner more support (*i.e.*, to increase their payoff by moving locally according to the gradient rule); a change in the strategy (or location) of one party will affect the strategies of its counterpart. Do these assumptions lead to different outcomes than in Anthony Downs's "median voter model"? Under what conditions do parties converge and what parameters give rise to non-convergent outcomes?

Downs is a two-dimensional voting model developed with Net-Logo (see Fig. 11.7). It illustrates that starting from some initial conditions one gets convergence of the two platforms (essentially

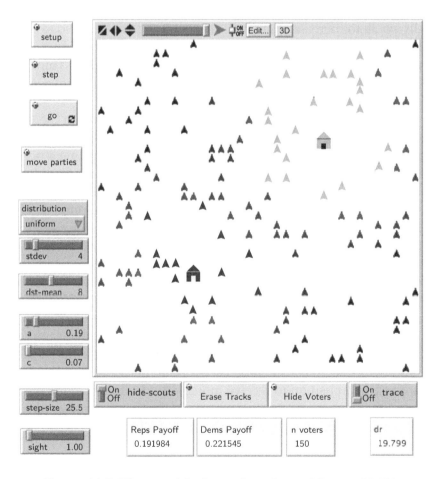

Figure 11.7: The graphical user interface of Downs 03-07.

to a bivariate median voter), whereas under other initial conditions the two parties remain far apart in steady state. A key parameter is the cost of voting described above. When voting is cost-free (the cost of voting is zero), parties converge to the point in issue space with the highest density of voters, an outcome consistent with the Downsian "median voter" model. As the cost of voting increases, however, the parties do not converge. The distance between parties when they reach a steady state is a function of the cost parameter.

Conclusion

As our three examples show, landscape dynamics combines the modeling ease of agent-based modeling with the powerful analytical tools of global analysis. Many more applications may be expected in the near future, and the scope of the complex dynamical systems coming under this approach may increase with the growing capability of computer hardware and software. A feature of NetLogo called HubNet provides a platform for experiments combining human subjects and robotic agents, a kind of science beyond the scope of mathematical modeling alone. Feedback from the frontiers of science is influencing the further development of ABM tools. This is a golden age for the social sciences.

Acknowledgments

This work was partially supported by the National Science Foundation under Grant SES-0436509.

References

Abelson, H., Goodman, N., & Rudolph, L. (1974). *LOGO Manual* (Tech. Rep. No. AIM-313). Cambridge, MA: Massachusetts Institute of Technology Artificial Intelligence Laboratory. Retrieved June 15, 2011, from ftp://publications.ai.mit.edu/ai-publications/pdf/AIM-313.pdf

Abraham, R. (n.d.). *Program Guide to Market 8.1.* Retrieved June 15, 2011, from http://www.vismath.org/research/landscapedyn/models/markets/market081.05Bguide.pdf

Abraham, R., & Friedman, D. (2004). *Veblen 5.3.* Retrieved June 15, 2011, from `http://www.vismath.org/research/landscapedyn/models/veblen/veblen053.nlogo` (NetLogo 2.0.0 code) and `http://www.vismath.org/research/landscapedyn/models/veblen/veblen05.html` (freestanding precompiled Java applet).

Abraham, R., & Friedman, D. (2007a). Landscape dynamics, complex dynamics, and agent based models. *Journal of the Calcutta Mathematical Society*, 3(1–2), 53–62.

Abraham, R., & Friedman, D. (2007b). *Models for Financial Markets.* Retrieved June 15, 2011, from `http://www.vismath.org/research/landscapedyn/models/markets`

Abraham, R., & Shaw, C. D. (1988). *Dynamics—the Geometry of Behavior. Part 4* (Vol. 4). Santa Cruz, CA: Aerial Press Inc. (Bifurcation behavior)

Abrahamson, D. (2009). Embodied design: Constructing means for constructing meaning. *Educational Studies in Mathematics*, 70(1), 27–47.

Adams, J., & Merrill, S. (2006). Why small, centrist third parties motivate policy divergence by major parties. *American Political Science Review*, 100(3), 403–417.

Arciero, L., Biancotti, C., D'Aurizio, L., & Impenna, C. (2008). Exploring Agent-Based Methods for the Analysis of Payment Systems: A Crisis Model for StarLogo TNG. *Journal of Artificial Societies and Social Simulation*, 12(1). Retrieved June 5, 2012, from `http://jasss.soc.surrey.ac.uk/12/1/2.html`

Ascape Guide. (2010). Retrieved from `http://ascape.sourceforge.net/` (Version 5.5.0, January 5, 2010)

Black, D. (1948). On the rationale of group decision making. *Journal of Political Economy*, 56, 23–34.

Browder, F. E. (1970). Existence theorems for nonlinear partial differential equations. In *Global Analysis (Proceedings of Symposia in Pure Mathematics, Vol. XVI, Berkeley, Calif., 1968)* (pp. 1–60). Providence, RI: American Mathematical Society.

Bryson, J. J., Ando, Y., & Lehmann, H. (2007). Agent-based modelling as scientific method: A case study analysing primate social behaviour. *Philosophical Transactions of the Royal Society B*, 362(1485), 1685–1699.

Carroll, R. W. (1969). *Abstract Methods in Partial Differential Equations.* New York, NY: Harper & Row Publishers. (Harper's Series in

Modern Mathematics)

Cioffi-Revilla, C., Paus, S. M., Luke, S., Olds, J. L., & Thomas, J. (2004). *Mnemonic Structure and Sociality: A Computational Agent-Based Simulation Model.* Online proceedings of Conference on Collective Intentionality IV, Siena, Certosa di Pontignano, Italy, October 13–15, 2004. Retrieved June 15, 2011, from http://www.istc.cnr.it/collintIV/authors/Cioffi Revilla.htm

Cohen, J., & Stewart, I. (1994). *The Collapse of Chaos: Discovering Simplicity in a Complex World.* New York, NY: Penguin Books.

Cox, G. W. (1989). Undominated candidate strategies under alternative voting rules. *Mathematical Computer Modeling, 12,* 451–459.

Davis, O. A., & Hinich, M. J. (1966). A mathematical model of policy formation in a democratic society. In J. L. Bernd (Ed.), *Mathematical Applications in Political Science, Vol. II* (pp. 175–208). Dallas, TX: S.M.U. Press.

Dawid, H. (2007). Evolutionary game dynamics and the analysis of agent-based imitation models: The long run, the medium run and the importance of global analysis. *Journal of Economic Dynamics and Control, 31*(6), 2108–2133.

Downs, A. (1957). *An Economic Theory of Democracy.* New York, NY: Harper Collins.

Eaton, B., & Lipsey, C. (1975). The principle of minimum differentiation reconsidered: New developments in the theory of spatial competition. *Review of Economic Studies, 42,* 27–49.

Fridman, N., & Kaminka, G. A. (2007). Towards a cognitive model of crowd behavior based on social comparison theory. In *Proceedings of the Twenty-Second National Conference on Artificial Intelligence (AAAI-07).* Menlo Park, CA: AAAI Press.

Friedman, D. (2001). Towards evolutionary game models of financial markets. *Quantitative Finance, 11,* 177–185.

Friedman, D., & Abraham, R. (2009). Bubbles and crashes: Gradient dynamics in financial markets. *Journal of Economic Dynamics and Control, 33*(4), 922–937.

Friedman, D., & Ostrov, D. N. (2008). Conspicuous consumption dynamics. *Games and Economic Behavior, 64*(1), 121–145.

Friedman, D., & Yellin, J. (1997). *Evolving Landscapes for Population Games.* (Draft manuscript, University of California, Santa Cruz)

Friedman, D., & Yellin, J. (2000). *Castles in Tuscany: The Dynamics of*

Rank Dependent Consumption. (Draft manuscript, University of California, Santa Cruz)

Hinich, M. J. (1977). Equilibrium in spatial voting: The median voter result is an artifact. *Journal of Economic Theory*, *16*, 208–219.

Hotelling, H. (1929). Stability in competition. *Economic Journal*, *39*(153), 41–57.

Kohler, T. A., West, C. R. V., Carr, E. P., & Langton, C. G. (1996). Agent-based modeling of prehistoric settlement systems in the northern American southwest. In *Proceedings of Third International Conference Integrating GIS and Environmental Modeling, Santa Fe, New Mexico.* Santa Barbara, CA: National Center for Geographic Information and Analysis.

Maria, K. A., & Zitar, R. A. (2007). Emotional agents: A modeling and an application. *Information and Software Technology*, *49*(7), 695–716.

Marsden, J. (1974). *Applications of Global Analysis in Mathematical Physics.* Boston, MA: Publish or Perish Inc. (Mathematical Lecture Series, No. 2)

MASON Multiagent Simulation Toolkit. (2010). Retrieved from http://cs.gmu.edu/~eclab/projects/mason/

Masson, P. R. (2003). *The Normal, the Fat-Tailed, and the Contagious: Modeling Changes in Emerging Market Bond Spreads* (Working Paper No. 32). Washington, DC: Center on Social and Economic Dynamics, Brookings Institution. Social Science Research Network. Retrieved June 15, 2011, from http://ssrn.com/abstract=1028085

Misner, C. W., Thorne, K. S., & Wheeler, J. A. (1973). *Gravitation, Part 3.* New York, NY: Macmillan.

North, M. J., Howe, T. R., Collier, N., Tatara, E., Ozik, J., & Macal, C. (2009). Search as a tool for emergence. In G. Trajkovski & S. G. Collins (Eds.), *Handbook of Research on Agent-Based Societies: Social and Cultural Interactions* (pp. 341–363). Hershey, PA: Idea Group Inc.

RePast Agent Simulation Toolkit. (2008). Retrieved June 15, 2011, from http://repast.sourceforge.net/ (Version 3, 2008)

Riker, H. W., & Ordeshook, P. (1973). *An Introduction to Positive Political Theory.* Englewood Cliffs, NJ: Prentice Hall.

Roemer, J. E. (1997). Political-economic equilibrium when parties represent constituents: The unidimensional case. *Social Choice and Welfare*, *14*, 479–502.

Smale, S. (1969). What is global analysis? *American Mathematical Monthly, 76*, 4–9.

Smale, S. (1973). Global analysis and economics. I. Pareto optimum and a generalization of Morse theory. In *Dynamical systems (Proceedings of a Symposium., University of Bahia, Salvador, 1971)* (pp. 531–544). New York, NY: Academic Press.

StarLogo TNG. (2008). Retrieved from http://education.mit.edu/drupal/starlogo-tng/ (Version 1.0, July 2008)

Suo, S., & Chen, Y. (2008). The dynamics of public opinion in complex networks. *Journal of Artificial Societies and Social Simulation, 11*(4), 41 paragraphs. Retrieved June 21, 2011, from http://jasss.soc.surrey.ac.uk/11/4/2.html

SwarmWiki. (2005). Retrieved from http://www.swarm.org/ (Version 2.2)

Thom, R. (1974). L'évolution temporelle des catastrophes [The temporal evolution of catastrophes]. In *Applications of Global Analysis. I* (pp. 61–69). Utrecht, NL: Mathematisch Instituut, Rijksuniversiteit Utrecht.

Torrens, P. M., & Naraa, A. (2007). Modeling gentrification dynamics: A hybrid approach. *Computers, Environment and Urban Systems, 31*(3), 337–361.

Uhlenbeck, K. (1968). *The Calculus of Variations and Global Analysis.* Unpublished doctoral dissertation, Brandeis University.

Veblen, T. (1912). *The Theory of the Leisure Class: An Economic Study of Institutions.* New York, NY: B. W. Heubsch.

Veltz, R., & Faugeras, O. (2010). Local/global analysis of the stationary solutions of some neural field equations. *SIAM Journal on Applied Dynamical Systems, 9*(3), 954–998.

Wilensky, U. (2010). *NetLogo.* Evanston, IL. Retrieved February 12, 2010, from http://ccl.northwestern.edu/netlogo/

Wittman, D. A. (1973). Parties as utility maximizers. *The American Political Science Review, 67*(2), 490–498.

Wittman, D. A. (1977). Candidates with policy preferences: A dynamic model. *Journal of Economic Theory, 14*(1), 180–189.

Wittman, D. A. (1983). Candidate motivation: A synthesis of alternative theories. *The American Political Science Review, 77*(1), 142–157.

Yang, L., & Gilbert, N. (2008). Getting away from numbers: Using qualitative observation for agent-based modeling. *Advances in Complex Systems, 11*(2), 1–11.

12

Intransitivity Cycles and Their Transformations: How Dynamically Adapting Systems Function

Alexander Poddiakov and Jaan Valsiner

Human rationality is often assumed to be based on the logical relation of transitivity. Yet, although transitivity fits relationships between physical objects or human decisions about targets that are independent of one another, it fails to fit the phenomena of systemic and developmental organization. Intransitivity has been shown to be present in various kinds of systems, ranging from the brain to society. In cyclical systems transitivity constitutes a special case of intransitivity. In this chapter, we examine different forms of emergence of intransitivity cycles, fixation of transitive parts in these cycles, and the organization of different levels of reflexivity within the systems. We conclude that reflexivity of cognitive processes—rather than transitivity in specific forms of thought—is the defining criterion of rationality.

The basics

In the most general terms, a binary relation R is *transitive* if "*A is related to B by relation R*" and "*B is related to C by relation R*" together necessitate that "*A is related to C by relation R*", more formally, if

$$A(R)B \wedge B(R)C \Longrightarrow A(R)C.$$

Keywords: emergence, intransitivity, probabilistic epigenesis, rationality, recursive (reflexive) thinking.

343

Some examples of transitive relations are the following:

> is equal to
> is parallel to
> is less than (or, is more than)

Not all relations are transitive. For example, the relation "to intersect" ("to be intersected by") is not transitive: if line A intersects line B, and line B intersects line C, A need not intersect C. Also, as is well known, the relation "to love" ("A loves B") is not transitive: if A loves B, and B loves C, A may or may not love C. Some meanings allow for multiple kinds of relations to emerge, among which transitivity of the relations is not a guaranteed given. Thus, the relation "to hate"—used often as an opposite of "to love"—is not guaranteed to be transitive. If A hates B, and B hates C, A need not hate C; it could even be just the opposite—'my enemy's enemy is my friend'. Translation of formal logical relations into psychological terms leads to some curious relations that need not have the transitivity property: 'is preferable to', 'is better than' (or, 'is worse than'), etc. Lots of paradoxes emerge in this domain.

Why is (in)transitivity important for psychology?

Transitivity has been the accepted cornerstone of rational thinking in many psychological theories. Mastery of transitive relations is considered as a main achievement of the developing human mind. In Jean Piaget's theory, understanding of transitivity is a necessary condition for acquisition of the concept of number, the ability to measure and make deductive conclusions, etc. At the same time, any developmental account of psychological functions implies intransitivity. Any systemic relation that entails cyclical (circular) processes in self-maintenance is intransitive. A most interesting area in transitivity/intransitivity relations consists of relations like dominance/subordination, superiority/inferiority, preferences, etc. If A dominates B and B dominates C, must it be so that A dominates C? If A is superior to B, and B is superior to C, must it be so that A is superior to C? What happens if superiority/inferiority (dominance /subordination, etc.) relations form a cycle, an intransitive loop?

The world is both transitive and intransitive (it contains both transitive and intransitive relations), and answers to these questions may vary in different domains and situations.

"Transitivity-" and "intransitivity-oriented" paradigms

Beginning with work of the Marquis de Condorcet (1785/2009), the history of science shows complex historical pathways in the study of this issue. Up to now arguments in that area are considered to be "powerfully appealing" and to display "a fundamental role in practical reasoning" (Temkin, 1996, p. 179). "Transitivity and intransitivity are fascinating concepts that relate both to mathematics and to the real world we live in" (Roberts, 2004, p. 63). A problem is "to discover which relationships are transitive, and which are not, and further to try to discover any general rules that might distinguish between the two" (*ibid.*).

The problem of transitivity/intransitivity is not an emotionally neutral question of formal logical relations: in its applications to psychological theorizing, it reflects our ideas and desires—about a major social value—that of superiority/inferiority relations in any society. Power relations are present in any social system, and their maintenance, change, and justification are major pastimes of human communication efforts. The question of transitivity of relations captures one of the most crucial socially desirable values of many human societies—that of equality. In general, it has been shown in many studies that in some types of situations transitivity of superiority relations is maintained (*e.g.*, in cases of comparison of unidimensional non-interacting objects: if stick A is longer than stick B, and B is longer than C, then necessarily A is longer than C), while in other, more complex types of situations, transitivity is violated. Distinguishing between these types can be an important and difficult problem. Yet perhaps a scientific and teaching problem of no less importance is that the issue of transitivity/intransitivity polarizes some groups of scholars (authors of handbooks, teachers, etc.), who seem to be "guided by different paradigms" (in the language of Kuhn (1962, p. 112)), or (to use terms closer to our issue) to have "creedal" differences: "creeds, whether concerned with physical reality, religion, ethics, or the principles of rational action, tend to be culturally conditioned" (Fishburn, 1991, p. 115). Analyzing the relation between some scholars' adherence to transitivity and others' rejection of it, Fishburn (*ibid.*, p. 117) notes that "analogous rejection of non-Euclidean geometry in physics would have kept the familiar and simpler Newtonian mechanics in place, but that was not to be."

In our context, we see that some scientists, teachers, etc., give 'proofs' that the transitivity of superiority (the relation "is better than", etc.) either is not, all things considered, *really* violated in their opponents' examples of intransitivity (Chan, 2003), or—much stronger—*cannot be violated even in principle* (Ivin, 1998). Such extremely 'transitivity-oriented' researchers use the term "Axiom of Transitivity" as a universally accepted principle and a normative rule for correct decision making without any discussion of possible intransitivity of superiority relations (Zinoviev, 1972) and as a key condition for rational actions (Kozielecki, 1979). At the same time, opponents of these 'transitivity-oriented' researchers present criticisms of the "Axiom of Transitivity" and "proofs" of its non-universality, and reach such conclusions as "rational choice theory does not obey the transitivity axiom" (Baumann, 2005, p. 238), "there is a genuine violation of transitivity in many situations" (Temkin, 1999, p. 780), etc. Yet these polemics of some 'intransitivity-oriented' researchers seem strange to others, who simply think transitivity of superiority is so self-evident in many domains that giving detailed explanations of the issue is like breaking down an open door.

Transitions between 'transitivity-' and 'intransitivity-oriented' approaches are possible: Anand (1993, pp. 344–345) demonstrates formally that "any 'intransitive' behaviour can be given a transitive description: conversely any 'transitive' behaviour can be given an intransitive description." Yet it looks as if the possibility of this universal formal transition does not satisfy and reconcile the different groups of researchers with each other.

Perspectives on transitivity/intransitivity

One can roughly distinguish among three interrelated groups of arguments concerning transitivity/intransitivity. The first group involves logical, formally exact proofs of (in)transitivity of superiority relations; the second, real (*i.e.*, empirical) transitivity/intransitivity of relations in different domains, and to domain-specific mechanisms causing the (in)transitivity; and the third, general-scientific and philosophical generalizations of the problem of transitivity/intransitivity and its significant consequences. We will consider all three of these interrelated groups of argumentation.

Our aim in this chapter is to analyze coexistence and interplay between transitivity and intransitivity in different areas and at different levels of organization of complex dynamical systems (including humans and their behavior), to show its role as a crucial factor of stability/novelty in these systems and to describe consequences of (mis)understanding of transitivity/intransitivity for problem solving and decision making, in stochastic and deterministic models.

(In)transitivity, formal logics, and logical fallacies

We start our analysis in the context of logical, formally exact proofs of (in)transitivity of superiority relations.

Transitive superiority relations and their (in)correct estimation by problem solvers (decision makers)

We first briefly mention two classes of situations of transitivity that are well-studied and do not require detailed presentation. The first class includes situations in which superiority relations are transitive (for example, stick *A* is longer than stick *B*, *B* is longer than *C*, and *A* is longer than *C*), and a decision maker (an animal, a human, a society, etc.) considers them transitive— there is no mistake of their understanding the situation. The second class includes those situations in which superiority relations are transitive, but a decision maker (a problem solver) wrongly considers them intransitive. Regularities of such wrong intransitive decisions are studied within the "heuristics" paradigm of Amos Tversky (Tversky, 1969; Tversky & Kahneman, 1986) and many others (for an overview, see Dawes, 1998). In Gerd Gigerenzer's approach, intransitive preferences of objectively transitive options, which are seemingly irrational in the normative framework, are considered functional and ecologically valid in terms of ecological rationality (Lages, Hoffrage, & Gigerenzer, 1999; Gigerenzer, 2002, especially Chap. II).

Intransitive superiority relations and their (in)correct estimation by problem solvers (decision makers)

Moving on to intransitive superiority relations, one should first note that most models of intransitivity of superiority relations concern *systems*, in which compared objects interact with one another. In contrast, decision making theory, which contains an axiom of transitivity of superiority relations, contains also an axiom about impossibility of interactions between outcomes (Kozielecki, 1975/1979). Clear as it is, the acceptance of an axiom of impossibility of interactions means that a theory cannot work with models of complex systems, in which interactions between outcomes are not only possible, but ordinary and to be expected. Actually, in complex systems with emergent, non-additive features neither an axiom of impossibility of interactions nor an axiom of transitivity is workable (Poddiakov, 2000, 2000/2006). In that context, intransitive loops in reasoning and behavior can be well-grounded and absolutely rational because the intransitivity in reasoning may reflect real intransitivity of superiority relations.

Thus we now consider models in which transitivity is violated because of interaction between compared options.

Stochastic models

Some very famous stochastic models of intransitivity are the intransitive dice models first invented by Bradley Efron of Stanford University. In such a model (a "dice combat club", in V. A. Petrovsky's happy phrase), the numbers on the faces of each die in a set of dice are specially chosen in such a way that Die 1 beats Die 2 (*i.e.*, shows a higher number) more often than not, Die 2 beats Die 3 more often than not, and so on—but Die N (the last of the set of $N \geq 3$ dice) wins against Die 1 more often than not (Deshpande, 2000; Roberts, 2004). Here is one example due to Ainley (1978, as cited by Roberts, 2004, p. 62). Four dice are marked as follows: Die A has markings 7, 7, 7, 7, 1, 1; Die B has 6, 6, 5, 5, 4, 4; Die C has 9, 9, 3, 3, 3, 3; and Die D has 8, 8, 8, 2, 2, 2. One can see that Die A beats Die B twice as often as not, Die B beats Die C twice as often as not, Die C beats Die D twice as often as not—but Die D beats Die A twice as often as not! Thus, the relation "X beats Y" is not transitive. "No one die is 'the best'. Rather, given that player 1 has selected a die, player 2 can always select a die that is more likely to win—which is of 'higher quality'" (*ibid.*).

This paradox leads to some other paradoxes, *e.g.*, the paradox of ambivalence of "money pump" reasoning. This reasoning is usually used to show the irrationality of intransitive choices:

> Suppose an individual prefers y to x, z to y, and x to z. It is reasonable to assume that he is willing to pay a

> sum of money to replace x by y. Similarly, he should
> willing to pay some amount of money to replace y by
> z and still a third amount to replace z by x. Thus, he
> ends up with the alternative he started with but with
> less money. (Tversky, 1969, p. 45)[1]

Yet Fishburn writes that money pump reasoning "is a clever device, but one that applies transitive thinking to an intransitive world" (Fishburn, 1991, p. 118). Indeed, in contrast with the pessimistic description of money pumping while making intransitive choices, suppose an advanced version of the intransitive dice, with rules that allow players to negotiate swaps of their dice by buying/selling ("I wish to buy your dice for 5 dollars"/"I wish to sell my dice for 7 dollars"). It can be profitable to buy and swap each successive 'more advantageous die' in an intransitive way. In that case, the money pump will paradoxically pump money not out of, but into the bank of the player who makes the intransitive choices.

Intransitivity can also arise from differential frequencies of appearance of pairs offered by the other participant in a sequential game (*e.g.*, Feed the Cat), where intransitive strategies can be both optimal and rational. Generalizations of such results are essential for economics and sociology (Piotrowski & Makowski, 2005).

A doctor for a doctor and other models

Between objects of well-differentiated structures, not only stochastic but also *completely deterministic* intransitive relations—both cooperative and competitive—are possible (Poddiakov, 2006). Suppose there are three physicians, A, B, and C, as follows.

> Physician A is a specialist in treating organs X, has healthy organs Y, and suffers from diseased organs Z.
> Physician B is a specialist in treating organs Y, has healthy organs Z, and suffers from diseased organs X.
> Physician C is a specialist in treating organs Z, has healthy organs X, and suffers from diseased organs Y.

It is evident that physician A should dominate—in expertise and power in the doctor/patient relationship—over B, B should dominate over C, and C should dominate over A. Intransitive loops of domination like the one in this a doctor for a doctor model appear

---

[1]See also Dawes (1998).

in analogous models like a psychotherapist for a psychotherapist, a teacher for a teacher, etc.

Such models show that intransitivity of dominance (superiority, etc.) is regular in interactions between *participants in differentiated structures* that have:

(a) *tools* to act on (influence) other participants (objects) in the system, by interactions within the system;
(b) *zones of sensitivity* (*receptivity*) to actions (influences) of other participants (objects);
(c) *zones of non-sensitivity* (*non-receptivity*) to actions (influences) of other participants (objects).

If these tools and zones of the interacting participants (objects) are composed in a Condorcet-like manner (considered in detail in the next section), they establish intransitivity of dominance (superiority) in the system (see Fig. 12.1). The models described above in this section (a doctor for a doctor, etc.) are all of *cooperative* type. Intransitive *competitive* relations of different natures (*e.g.*, physical, social, psychological) but the same structure are also possible. Concerning combative systems with differentiated subsystems of attack, defense, and vulnerable parts, Poddiakov (2000, 2006) has posited the model choice of weapons for the duel, and some formalisms revealing the mechanism of intransitive relations between weapons. It has been shown that the intransitive relations, in which weapon A beats weapon B, B beats C, but C beats A, can be absolutely regular in an exactly deterministic, non-stochastic way: if forced to choose between weapons A and B one should choose A, if forced to choose between B and C one should choose B, but if forced to choose between A and C one should choose C.

It is important that if one considers the total sums of all acts of giving and taking help (or winning and losing), interacting systems

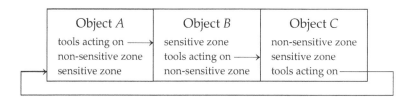

Object A	Object B	Object C
tools acting on ⟶	sensitive zone	non-sensitive zone
non-sensitive zone	tools acting on ⟶	sensitive zone
sensitive zone	non-sensitive zone	tools acting on ⟶

Figure 12.1: Intransitive relations between interacting objects of differentiated structures and functions.

can seem (absolutely) equal. Yet the result of an interaction between two systems can be unexpected and non-trivial, because it is
determined by specific features of the systems and their interaction.

Deterministic models of intransitivity can also be designed using
structural, not probabilistic, relations between elements of a set of
intransitive dice. One can see an essentially intransitive dice model
working its mischief in sports analogies. Let each die signify an arm-
wrestling team of six athletes, with each face of the die signifying an
athlete who has the power written on that face (measured in some
conventional units). In a tournament where all pairs of members
of different teams are matched, team *A* will beat team *B*, team *B*
will beat team *C*, and team *C* will beat team *A*, in terms of the
numbers of matches won. Which of the teams should one choose as
a potential winner? It depends on the pair considered.[2]

These examples show that in some kinds of situations rational
choices *must* be intransitive, and *keeping transitivity of choices is irrational*. Hence the determination of what is rational in a given
context is dependent on the context, and not on a pre-set equivalence rationality = transitivity (or rationality = intransitivity). Such
models can explain mass cooperative-and-combative relations between species, social groups, etc. The journal *Nature* has published
a series of articles with the words *Rock, Paper, Scissors* in their titles, concerning combative relations between species (Kerr, Riley,
Feldman, & Bohannan, 2002; Kirkup & Riley, 2004; Reichenbach,
Mobilia, & Frey, 2007; Semmann, Krambeck, & Milinski, 2003).
Articles with other titles but about the same intransitivity have been
published in other journals; for example, it has been shown that
in agar culture, *Phallus impudicus* replaces *Megacollybia platyphylla*,
M. platyphylla replaces *Psathyrella hydrophilum*, but *P. hydrophilum*
replaces *P. impudicus* (Boddy, 2000); some fungal strains are "engaged in an interaction that mimics the game 'rock–paper–scissors'"
(Nowak & Sigmund, 2007, p. 138); and so on. These authors prove
that intransitive relations support biodiversity in ecological systems.

---

[2]The numbers on the faces of the dice can be also interpreted in terms of
combinations of sensitive zones, non-sensitive zones, and tools, as follow. A face
of Die *X* that is marked with the number *n* is *sensitive* to a face of Die *Y* that is
marked with a number larger than *n*, it is *non-sensitive* to a face of Die *Y* that is
marked with *n*, and it is a *tool* that acts by winning (terminating, etc.) on a face
of Die *Y* that is marked with a number smaller than *n*. An appropriate table or
diagram can be built for the members of the arm-wrestling teams.

Also one can think that intransitivity of relations between lower-organized biological organisms can explain some features of behavior of more highly organized organisms. For example, this could explain experimental results of Shafir (1994), that honey bees make intransitive choices among flowers. The bees may 'know' that flowers of one class, when they are near flowers of another class, can become unpleasant/dangerous, or alternatively can become especially 'tasty', and then test this opportunity by combining transitive and intransitive choices (Poddiakov, 2006).

In general, the existence of such models as intransitive dice, tools for help, choice of weapons, etc., confirms the idea that transitivity may be violated in "contexts where the choice is intrinsically comparative, namely, where the utility from any chosen alternative depends intrinsically on the rejected alternative(s) as well (typically, certain competitive contexts)" (Bar-Hillel & Margalit, 1988, p. 118); that "there is a genuine violation of transitivity" (Temkin, 1999, p. 780) in the presence of a comparative, context-dependent, view, on which how bad or good a situation may be "depends on the alternative compared to it" (*ibid.*, p. 777).

Interplay: transitivity resulting in intransitivity

Let us return to Condorcet's work, mentioned above without details. Coming out of the fixed social hierarchies of the medieval world, in which the power relations of the lord, his soldiers, and his peasants were clearly fixed (and transitive), the Enlightenment mindset resulted in the dynamics of preferences—political choices. In the run-up to the French Revolution, two aristocrat-scholars —Jean-Charles, Chevalier de Borda (a military engineer and constructor of the standard metre, who survived the Revolution), and Marie-Jean-Antoine-Nicolas de Caritat, Marquis de Condorcet (an early advocate of human rights, who did not)—initiated the application of mathematical methods to the analysis of voting and 'social choice' (Borda, 1781; Condorcet, 1785/2009).

> [Condorcet's] chief contribution has been what might be called the Condorcet criterion, that a candidate who receives a majority as against each other candidate should be elected.[...]It was in this context that Condorcet discovered that pairwise majority comparisons might lead

> to intransitivity and *hence to an indeterminacy in the*
> *social choice.* (Arrow, 1963, pp. 94–95; emphasis added)

> It was perfectly possible that with three candidates, *A*
> be preferred to *B* by a majority, *B* to *C* by a majority, and
> *C* to *A* by a majority. An example is if one-third of the
> voters prefer *A* to *B* and *B* to *C*, one-third prefer *B* to
> ∪ and ∪ to *A*, and one-third prefer *C* to *A* and *A* to *B*.
> This possibility has become known in the literature as
> the "paradox of voting," or the Condorcet effect. [...]
> ([P]airwise) majority voting defines a relation which is
> connected (there must be a majority for one or the other
> of two alternatives, if the number of voters is odd) but
> need not be transitive. (Arrow, 1973, p. 127)

In Condorcet's "paradox of voting"—only the simplest of many—
a 'sum' of rational decisions *cannot* be made rational due to specific
interactions between the rational choices (that we try but fail to
'sum'). Condorcet offered a voting method that eliminates some
effects of intransitivity, but fails to define a winner of an election if
candidate *A* beats *B*, *B* beats *C*, and *C* beats *A*—despite absolute
transitivity of individual voters' choices. Methods are still being
designed in our times to cope with voting paradoxes (Eppley, 2003).

The logical structure of Condorcet's paradox can be applied
to many situations, including those in which not many, but the
only decision maker makes rational transitive choices paradoxically
resulting in intransitivity.[3] For sciences, Baumann's paradox of "a
better theory" is of special interest.

> Let us now assume that the following is true about
> theories A, B, and C. With respect to explanatory power,
> A is better than B, and B is better than C. With respect to
> support by data, B is better than C, and C is better than
> A. With respect to simplicity, C is better than A, and A
> is better than B. [...] Given our assumptions, we would
> have to say that all things considered, C is a better theory

---

[3]One can see that some of the deterministic models of intransitive relations
described above have the same, Condorcet-like composition, applied there to
composition of tools, sensitive zones, and non-sensitive zones. Perhaps some
generalization, including Condorcet-like compositions among others determining
intransitivity, is possible, but Condorcet-like compositions are more evident.

than A (it "beats" A 2:1 on the relevant criteria). Also, A
is better than B, and B is better than C. (Baumann, 2005,
pp. 234–235)

In contrast with a situation of elections, in which the only candidate must be elected, Baumann underlines that intransitivity of
the relation "is a better theory than" does not contain something
"wrong or irrational about the way the scientists evaluate the different theories" (*ibid.*, p. 238). "If pluralism is true, then there is rational
acceptability of intransitive rankings of theories" (*ibid.*, p. 236), and
the same concerns many situations of everyday life.

Another kind of paradox of transitivity resulting in intransitivity
is the "mere addition" paradox.[4] Temkin (1987, 1996) has designed a
complex of proofs to show how (nearly) indistinguishable, continual
transitions between situations—compared paradoxically but necessarily—result in final intransitive choices; examples include increase
of money, decrease of pain from torture to hangnail, population
increase with simultaneous decrease of life quality, etc.

Intransitive preferences and illusory correlations

One should note a puzzling feature of intransitive relations between
the voters' preferences described above in Condorcet's paradox
(see Poddiakov, 2006). Although the majority of voters agree—
not disagree—that A is preferable to B, B to C, and C to A, all three
coefficients of rank correlations between the voters' preferences
are negative (namely, $-.5$), as calculated with Spearman's formula,
$r_s = 1 - 6 \sum d^2/(n^3 - n)$, where d is difference of ranks, and n is
the number of ranks. Yet a negative correlation between different
persons' preferences is usually considered as characteristic of their
disagreement with one another, not of their agreement.

A ternary chain of negative correlations between pairs of three
characteristics is in opposition to the beliefs of many people, who
think that if X negatively correlates with Y, and Y negatively correlates with Z, then X must *positively* correlate with Z. This is
one of many ways in which correlation coefficients acquire surplus
meaning in the interpretations of psychologists (Valsiner, 1986).
Here the rationality of assumed transitivity of relations ends—

---

[4][This transposition into the domain of ethics of the Paradox of the Heap
(analyzed in Chap. 4) was named and studied by Parfit (1984). (Ed.)]

quite counter-intuitively; the 'negativeness' or 'positiveness' of coefficients of correlations is free to take on any version of transitive or intransitive forms. Though X positively and significantly correlates with Y, and Y positively and significantly correlates with Z, X can negatively and significantly correlate with Z.

This intransitivity of correlational relations is rarely emphasized. As a result, a scholar who reads in one article that characteristic X (*e.g.*, a personal trait, frequency of heart rate, etc.) positively correlates with Y, and, in another article, that characteristic Y positively correlates with Z, can without testing come to the ill-grounded conclusion that X positively correlates with Z. S/he will be almost sure of that conclusion, and special instructive and demonstrative examples are necessary to overcome such a belief.

Interplay: intransitivity resulting in transitivity

Can intransitive interactions lead to a transitive result? To answer that question, Poddiakov (2000/2006) designed a special *cellular automaton*.

A cellular automaton is a mathematical model of a space, consisting of many *cells*. Each cell can be in one of a given set of *states*, from which it can transit into another state under the influence of adjacent cells, according to given *transition rules*. Even cellular automata with very simple transition rules can display unexpected structural phenomena, including 'self-organization' of initial structures, emergence of complex structures out of chaos, and ordering, destruction, and 'death' of structures (Dooley, 1997; Prigogine & Stengers, 1984; Weisstein, 2000/2010).

Poddiakov (2000/2006) devised a cellular automaton with 3,600 small square cells in the shape of a big square (60×60), displayed for convenience on a computer screen. Each cell can be in one of nine states, depicted by one of nine colors (or, as in Fig. 12.2, nine gray-scale patterns). A computer, by a random process, puts the automaton into an *initial state* in which each of 400 medium (3×3) subsquares of the large square is assigned a color. It is important to note that, after this random initialization, *the transformations of the automaton from state to state are entirely deterministic*. Specifically, the transition rules are as follows. If at time t a given cell has color 1, and there is a cell of color 2 that *touches* the given cell (*i.e.*, the two cells have an edge or a corner in common), then at time $t + 1$

the given cell will have color 2. Similarly, a cell that has color 2 at time t, and touches a cell of color 3, will have color 3 at time $t + 1$; and so on. Finally, a cell that has color 9 (the last color) at time t, and touches a cell of color 1, will have color 1 at time $t + 1$.

This and similar automata can be used to simulate the dynamics of n combative species replacing one another (described above, with $n = 3$ species), or the dynamics of preferences in the model can-vassers for canvassers, where groups of canvassers with intransitive "persuasion power" relations like those illustrated in Fig. 12.1 try to persuade each other.

Interesting results can be obtained by modifying the automaton to allow some elements to have no interactions with any others (neither influencing, nor being influence, by them). If there are relatively few non-interacting cells (from none to approximately 30% of the total), one can see dynamical effects of "ordering out of chaos" and "emergence": small wholes turn into a large, whole, dynamical pattern of well-differentiated and well-ordered structures containing several "centers of crystallization", out of which several long, thin spirals unfold, each of them cycling perpetually through the colors $1, \ldots, 9$ (see Fig. 12.2). No one color dominates, and at any time the numbers of cells of each color are approximately equal.

On the other hand, if there are relatively many non-interacting cells (more than about 30%), then one color can achieve relative or

Figure 12.2: The cellular automaton simulating multiple intransitive interactions with no non-interacting cells. The initial state ($t = 0$) is to the left; the state at $t = 300$ is to the right.

absolute dominance. In *relative dominance* (Fig. 12.3, top) no color is eliminated, but a large number of cells become—and remain—of the dominant color. In *absolute dominance* (Fig. 12.3, bottom) all but one color is eliminated: the final picture is absolutely monochrome. That is, an initial pattern containing both elements that interact with each other *intransitively*, with no dominating preferences, and elements that have no interactions at all ('neutrals' or 'bystanders'), can result in a pattern *identical* to that which would be produced

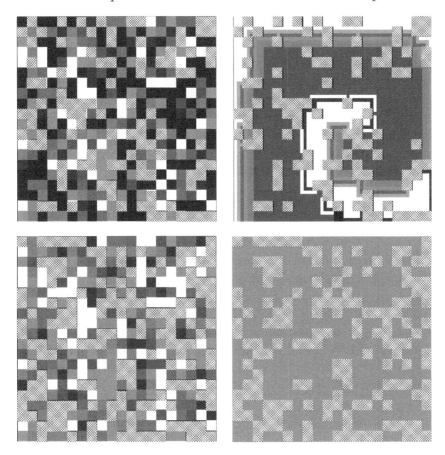

Figure 12.3: The modified cellular automaton with non-interacting cells (cross-hatched): at the top, with about 25% of the cells non-interacting, the result is relative domination of one color; on the bottom, with about 37%, the result is absolute domination of one color. The initial state is to the left; the state at $t = 300$ is to the right.

by a purely transitive subordination of colors; and the greater the proportion of non-interacting elements, the greater the probability of converging to such a monochrome stable state. This serves as an additional confirmation of the general idea that transitivity is fundamentally a property of non-interacting objects.[5]

Real complex and developing systems: interplay between transitivity and intransitivity

In the next sections we examine biological, psychological, and social examples of systems that are *open* (*i.e.*, dependent on exchange relations with their environments), *self-reproducing* (autopoietic), and *highly variable in their forms of expression*. We show that in a system of that type we can expect such a complex interplay between transitivity and intransitivity that the application of either pure transitivity or pure intransitivity rules in formal models is inadequate to the nature of the phenomena.

Cyclic processes and 'resource pump' reasoning

Any cyclic process of transitions from one state to a second, from the second to a third, etc., and finally back to the first again, can be interpreted in terms of intransitivity of 'preferences' in a system. Thus, a system in state *A* finds state *B* to be preferable (*e.g.*, from the point of view of energy), so the system transits towards *B*— operationally, the 'preference' is the propensity to make this transition; but then state *C* is preferable to *B*, so the system moves towards *C*; and so on. Real systems very often realize such cyclic, intransitive processes.[6]

---

[5][An alternative interpretation of the role of the non-interacting cells is that they create—within the original space of 3,600 cells laid out with the straightforward geometry and connectivity of a large square—a new space (occupied by the other, interacting cells) whose intricate (though stable) geometry and connectivity can profoundly affect the automaton's dynamics. This interpretation is complementary, not contradictory, to that of the authors; each can shed light on the other. (Ed.)]

[6][If all the 'preferences' could be measured, simultaneously, on a single monotonically ordered numerical scale—*e.g.*, 'free energy' measured in calories —then the preferences would, indeed, be globally transitive; of course, as shown clearly by the examples that follow, the 'preferences' in a natural system need not be commensurable on a single monotonically-ordered scale. (Ed.)]

In chemistry, perhaps the best known example is the *Belousov–Zhabotinsky* cyclic reaction (Belousov, 1959; Zhabotinsky, 1964), in which different colors of a liquid solution sequentially replace one another again and again. (Many chemists thought such a reaction absolutely impossible, and for many years publication of works on this cyclic reaction encountered great difficulties.)

In biochemical systems, the citric acid cycle, discovered by Hans Krebs in the 1930s-40s (Krebs, 1964) is an example of an *autogenerative* biological energy production system. It takes place in the mitochondria, and consists of eight transformations. While producing the necessary energy resources for the organism—hydrogen (H_2), at four loci within the cycle—the cycle reproduces itself in order to maintain stability in the life of the organism (see Fig. 12.4). This autogeneration is made possible by the central role given to a catalyst —Acetyl Coenzyme A (Lipmann, 1964)—that binds the chemical substances at different transformation "bridges" within the cycle, thus allowing their transformation into another substance.

The Krebs Cycle is the common pathway for all biochemical energy processes. It breaks down all food substances that the cell receives—glucose, fatty acids, etc. It involves the release of energy (hydrogen electrons) into binding compounds $NADH_2$ and $FADH_2$. The process extracts extra hydrogen from water molecules present

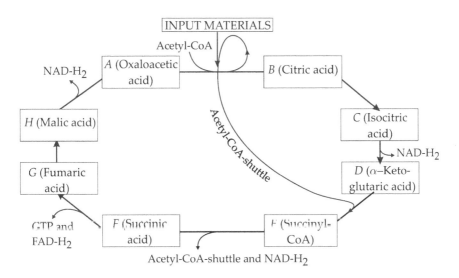

Figure 12.4: The citric acid (Krebs) cycle.

in the cell. It is a stable, evolutionarily ancient mechanism of energy production, common to all known life forms. What is depicted in Fig. 12.4 is a formal and simplified cyclical basic biochemical process in which intransitivity rules. It is important to emphasize that no single part $A, \ldots, H$ in the cycle

$$\cdots \to A \to B \to C \to D \to E \to F \to G \to H \to A \to \cdots$$

dominates the rest more than any other. All are equally important to the dynamically functioning whole. Yet, simultaneously, each is antecedent to the next, and hence we can see transitive formal relations forming part of an intransitive cyclical energetic chain. This aspect of the cyclical causal system—that *all parts are relevant*—was taken into consideration when a name had to be given to the cycle. The difficulty of naming the cycle parallels that of the fluidity of the names of Hindu gods and goddesses who are constantly transforming themselves into other forms. Yet to name a dynamic complex requires a moment of stability. Krebs himself described this in his 1953 Nobel Lecture:

> It is convenient to use a brief term for the [cycle]. Its essential feature is the periodic formation of a number of di- and tricarboxylic acids. As there is no term which would serve as a common denominator for all the various acids, it seemed reasonable to name the cycle after one, or some, of its characteristic and specific acids. It was from such considerations that the term «citric acid cycle» was proposed in 1937. (Krebs, 1964, p. 403)

As things go in the history of names, the shortest name for the systemic cycle became that of its discoverer—the label "Krebs Cycle" covers the example of the basic energy producing intransitive relation in the biological world.

All biological systems are organized in one or another kind of complex cyclical (= self-regenerating) relations, and hence intransitivity rules the biological, social, and psychological domains. All reasonable causal models for use in these domains are non-linear (Weiss, 1978), and may take the form of *systemic-direct* or *systemic-catalyzed* causality (Valsiner, 2000, p. 74).

That cyclic processes exist in the real world means intransitivity of such preferences is a regular feature of the world. 'Resource pump' reasoning in these situations acquires a different and novel sense. Support of functioning and development of ordered structures (including cyclic structures) requires energy and other resources (*e.g.*, reactants) as Ilya Prigogine has proved (Prigogine, Allen, & Herman, 1977; Prigogine & Stengers, 1984). So an "energy pump" providing cyclic transitions is not a fallacy of nature, but a way to support the existence and development of ordered structure, whether it is the Belousov–Zhabotinsky reaction, or the Krebs Cycle, or the development of an embryo (gametogenesis; Jablonka, 2001, p. 109). Yet there is a difference: biological systems, in contrast to (say) economic 'resource pump' systems, produce new resources rather than spending existing resources in fixed amounts at regular intervals, creating a de-escalatory downward cycle where each step entails proportionately more expenditure than the one before. A "money pump" is situated in the realm of consumption, not that of production. Production is achieved by combining the autoregenerative (cyclic) nature of the basic system with the functioning of specific catalyzed transformations within the cycle (Fig. 12.4).

An example of circularity and, respectively, intransitivity of neurophysiological processes is the brain dynamics of *change of dominants* discovered by A. A. Ukhtomskii (1927), who showed that concentration of arousal (*i.e.*, emergence of a dominant) in an area of the brain induces inhibition in adjacent areas, but after that arousal, regulation is replaced by inhibition, and vice versa. At a level of aural impression, Wilhelm Wundt (1876/1976) had earlier shown that perception of the same monotonous rhythm can change by oscillating between two rhythms with different points of perceived dominant. At a level of visual perception, some illusions are the sequential, involuntary change of perception of a picture X as, first, a picture of object A, then as a picture of object B, and again as a picture of object A, and so on (Block, 2002).

Variable social dominance hierarchies

Processes analogous to intransitive combative relations between species are found in human social dynamics. In a public-goods game, cooperators (players who contribute their money to a public pool), "loners" (players who do not join the group), "free riders" (or "defectors")—players who join the group and receive part of the profits, but do not contribute to the public pool—"coexist through rock–paper–scissors dynamics" (Semmann et al., 2003, p. 390). If "defectors have the highest frequency, loners soon become most frequent, as do cooperators after loners and defectors after cooperators. On average, cooperation is perpetuated at a substantial level" (*ibid.*). (Concerning 'resource pump' reasoning, we note that the functioning of this intransitive cycle of social dynamics is supported by money from experimenters with a grant to pay participants.)

In experiments on the organization of public-goods games participants do not need to invent anything new: they make choices between options offered by experimenters. Situations not of simple cyclic functioning, but of invention and development, are more intriguing. A most interesting turn to intransitive relations is observable in individual and group development, *e.g.*, the case of Japanese macaques developing new techniques of food extraction (sweet-potato washing or candy-eating) that entailed a break in the regular dominance hierarchy (adult males > adult females > juveniles). It was precisely the actions by the most sub-dominant monkeys (juveniles) that led the invention and proliferation of the new technique:

> The "sweet-potato-washing sub-culture" observed by me at Kosima was started by a one-and-a-half-year youngster (111♀). Here again, the sub-culture spread to its playmates and to some of their mothers also, and when the mother once learned the sub-culture, it was always transmitted to her baby, and the family to which 111♀ belonged was the first in which all the members washed sweet potatos [*sic*]. In this troop, no obvious case of successive caring the young by adult males has been noticed and none of them learned the sweet-potato-washing. (Kawamura, 1959, p. 46; Kawamura, 1963, p. 83)

The observations of innovation in behavioral techniques here indicate a practice of transitivity-breaking.[7] It may be, precisely, that if a transitive relation (such as a dominance hierarchy) is established, it leads to its own demise by constructive efforts to undermine it on behalf of the lower levels of the hierarchy (*cf.* also examples like "reflexive management", p. 376 below). In the biological world,

---

[7][A different kind of subversion of a primate dominance hierarchy in a social learning context—in this case, among African vervet monkeys—has recently been reported by van de Waal, Renevey, Favre, and Bshary (2010). (Ed.)]

both transitive and intransitive forms of relations exist in mutual interdependence that is guaranteed by the organismic nature of biological systems. Transitivity leads to intransitivity (as an act of overcoming the former), which in its turn may lead to new episodic formations of transitive relations.

In general, primates' social behavior is of special interest in comparison with social behavior of other animals. In groups of many animal species (not necessarily primates), there can occur intransitive chains of dominance in which an individual A dominates B, B dominates C, and C dominates A (Chauvin, 1969); but in primates dominance hierarchies are not fixed—rather, they are usually re-negotiated as the troops of animals encounter new demands (Goodall, 1986, pp. 319–334; Strum, 1987, p. 77). The primate pattern of flexibility of dominance hierarchies is of course very familiar to human psychologists from the classic "Robbers' Cave" experiment in the social psychology of group conflict creation and resolution (Sherif, Harvey, White, Hood, & Sherif, 1961). In the behavior of primates (including humans), variability prevails, in both interindividual and intra-individual forms (Molenaar, 2004; Molenaar, Huizinga, & Nesselroade, 2002; Valsiner, 1984).

The task of establishing transitivity relations becomes complicated if there is high variability within limited time for all comparable objects. Such variability may occur if the phenomenon is context-dependent: in context X one transitive order may apply to relations (*e.g.*, parents > children > pets, and parents > pets, in 'freedom of speech' inside a family), while the same phenomenon in context Y can show a different (intransitive) order—the angry parent hits the child, the child hits the dog, and the dog bites the parent! Family relations are not transitive: a father can dominate over his child, the child can dominate over her or his mother, and the mother can dominate over the father (Druzhinin, 2000, p. 40).

Variability has not been easily accepted in the behavioral sciences (see Maruyama, 1963, 1999, also Siegler, 2007). Statements like *A is X* are sought after—rather than conditions under which seemingly opposite claims like *A is X and/although/because A is non-X* may be the case. Psychology continues on the road of looking for ontological essences (*competences*) that are assumed to underlie the manifest phenomena (*performances*), rather than looking for the cyclical systemic organizational forms that generate a variety of outcomes— at times seemingly mutually opposite ones.

In this respect, consider a frequent question in the minds of sociologists and psychologists: dominance relations between genders. Ordinary human beings—certainly in 'Western' societies—have devoted much passion to the 'struggles' and 'fights' for 'liberation'— usually of the 'women's worlds' from the 'dictatorship' of males. What is implied in these 'fights' is the notion of transitive hierarchy and its desired reversal (male dominance over females becomes 'equality', which in all ways means women's dominance over men, given the centrality of women in reproduction of the species; Rogers, 1975). However, anthropologists in their work in non-'Western' societies may find the gender relations far more complex than can be covered by a simple label of the form *dominant over....*

For example, Meigs (1990) found among Hua males in the New Guinea Highlands three parallel gender ideologies—one male-chauvinistic, the second subdominant and envious of female reproductive power, and the third egalitarian. As the carriers of these three ideologies are the same persons, movement between the ideologies may be an example of shifts between different forms of hierarchy. These shifts are continuous in time—so in our real-life gender relations we may observe a chain of events where males may be dominant over females in domain X who in their turn are dominant over males in domain Y, and so on *ad infinitum*.

Other evidence of intransitivity in social role relations can be found in mythology (*e.g.*, the Kali myth, Menon & Shweder, 1994) or in symbolic power (the *devadasi* case, Valsiner, 1996). The latter is of wide relevance in Indian family structure where the sub-dominant young wife, who enters into a marriage and is governed by mother-in-law and husband, is still in a dominant position over both of the others by way of the power of reproduction (Menon, 2002). Such flexibility of power roles indicates the functional role of intransitivity relations in systems where the production of novelty is central.

In general, in human psychological functioning—regulated by semiotic mediators (Peirce, 1893, 1896; Valsiner, 1999, 2001, 2007) —transitivity of relations may be exceptional, and intransitivity of relations may play an important role. Petrovsky (1996, p. 417–425) considers intransitivity of human preferences to be a law reflecting cognitive and emotional complexity of a personality. *Intransitivity* is related to a person's *transitions to novel systems of preference criteria*. Petrovsky has designed formalisms to estimate a number of different criteria used by a person, based on the number and features of observed violations of transitivity (*cf.* Lages et al., 1999). A special kind of dynamics of preferences has been shown in the framework of *regret theory*: "an individual compares the outcomes within a given prospect giving rise to the possibility of disappointment when the outcome of a gamble compares unfavourably with what they might have had" (Starmer, 2004, p. 132), resulting in a cycle of preferences that violates transitivity.

Transitivity as a special case of intransitivity

By its nature, any cycle is intransitive ($A \rightarrow B \rightarrow C \rightarrow A \cdots$)—yet transitivity can occur in triplet sub-structures of such cycles (three in this example: $A > B > C$ entails $A > C$; $B > C > A$ entails $B > A$; $C > A > B$ entails $C > B$). A full cycle of at least three parts is intransitive, whereas all three-unit relations extracted from it are transitive. This being so, all cyclical processes are intransitive but include sub-parts that are transitive when considered separately.

Thus, transitivity is an artifact lying between cycles (where it is an analytic sub-part) on the one hand, and mutuality in dyadic relations ($A <> Z$) on the other. For an illustration of the latter, assume we have the cycle $A \rightarrow B \rightarrow C \rightarrow A \cdots$. Let us group $B \rightarrow C$ into one unit Z; now the cycle becomes $A \rightarrow Z \rightarrow A$, in which the redundant appearance of A is due purely to the imposition of a privileged 'initial moment' in the flow of time. Freed from this flow, the reformulated cycle becomes labeled as a *mutuality relation* $A \leftarrow \rightarrow Z$ in which the underlying cyclical processes are out of sight, underneath the new label: *i.e.*, mutuality of relations between two parts of a cyclical chain can be translated into a binary relation $A <>$ non-A *precisely because* the focal part (A) maintains a constant relation with the rest of the cycle (non-A, *i.e.*, Z). Of course such translation eliminates the systemic order from consideration.[8]

Fig. 12.5 is particularly relevant for figuring out where systemic self-maintenance becomes open for innovation of the system itself,

---

[8] An area which has been moving in the opposite direction to this translation is the current interest in the dialogical nature of the self (Hermans, 2001, 2002). Starting from static (and essentialistic) models of the self, *dialogical self* (DS) theory returns to the opposition of united parts (self $<>$ non-self), trying to restore the process mechanisms of that relation. DS theory thus moves from an ontology of the self (A) to that of a dialogic relation of self with non-self ($A <> Z$), and reaches the need to study the cyclic process involved: $A \rightarrow B \rightarrow C \rightarrow A$.

and when it merely fluctuates between momentary dominance of any of the three components over the other two. In fact, in case of an intransitive hierarchy, it is not possible to answer the question "which of the parts is dominant over the others?" They all are— over all the others!

In fact, the systemic cycle in Fig. 12.5 leads us to further elaboration of the meaning of the posited relationship between parts of the cycle. Thus, if we say that *P is dominant over Q* or that *Q is dominant over S* (etc.), the notion *dominant* refers to the initial condition of the to-be-performed transformation in the cycle. It is equivalent to *leads to* in terms of systemic pathways (see Fig. 12.4 on the transformations within the Krebs Cycle).

The autoreproductive system in Fig. 12.5 is open to change. Under certain conditions (trajectory Y) it continues as a cycle along the lines of the intransitive relation. Yet under other conditions (trajectory X) it can become closed—and transitive (in case of trajectory X, $P > Q$ and $Q > S$ lead to $P > S$). This move to transitivity equals elimination of the autoreproductivity of the system —and means the extinction of the system. Hence it is clear that all organisms that maintain their systemic nature operate under conditions of intransitivity cycles.

The "locus of rupture" in Fig. 12.5 is interesting also as to its potential for the innovation of the system—its development. Aside

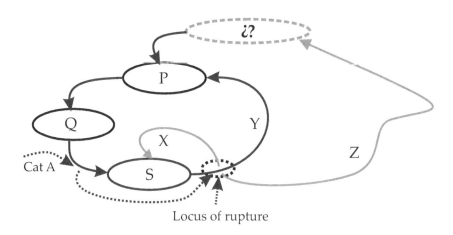

Locus of rupture

Figure 12.5: The intransitive hierarchy and its innovating rupture.

from the system-detrimental redirection of the *leads to* relation from the regular (Y) to the self-destructive (X) trajectory, the rupture can lead to a trajectory where a new component of the system (¿?) is created and integrated into the system. This *trifurcation* (maintenance < > extinction < > innovation) is regulated by catalytic systems (Cat A in Fig. 12.5; see also autocatalysis in Fig. 12.1). The following excerpt from earlier work (Valsiner, 2006) situates our present analysis philosophically and historically.[9]

> We reach here a major epistemological generaliza-
> tion about what the theories of *causality* or *relationships*
> mean in psychological and biological systems. Instead
> of the usual acceptance of *linear direct* causality models
> (A causes B [. . .]), an appropriate model of causality is a
> systemic catalyzed approach (system A–B–C results in X
> if catalytic condition Z is present). Such understanding
> has productively been put into place in biochemistry
> (Krebs, 1964). From that viewpoint, mere discovery of
> statistical "relationships" between variable X and vari-
> able Y in a correlational analysis reveals little about the
> actual functioning of the system in which X and Y are
> systemically linked. Correlational data do not explain—
> they need explanation themselves!
>
> The move to the use of systemic catalyzed causality
> models replaces our focus of analysis from the structure
> of the system as it is (once the structure is described)
> to that of under what conditions that system might be
> modified. This focus is nothing new in science: In 1927,
> Kurt Lewin emphasized the conditional-genetic nature
> of unitary complex phenomena (*konditional-genetische Zu-
> sammenhänge*—p. 403) where, through the study of var-
> ied conditions of functioning (*Bedingungsstruktur*) of the
> system, its potentials for transformation into a new state
> —as well as conditions of its breakdown—could be re-
> vealed. Vygotsky's use of the same epistemological mind-
> set led him to elaborate the "method of double stim-
> ulation" as the methodological tool for developmental
> psychology [. . .].

---

[9]See Valsiner (2000, particularly p. 74 and pp. 78–81) for more on linear direct causality models and the "method of double stimulation".

It is interesting that psychology by the 21st century has lost the insights stemming from the conditional-genetic syntheses that were there in the 1920s. The reasons for such forgetting are mostly due to the social dynamics within that science, rather than to systematic analysis of ideas (Toomela, 2007, 2009; Toomela & Valsiner, 2010). The development of a scientific discipline is itself an example of intransitive relations, where PRESENT may be viewed as "better than" RECENT PAST—yet ANCIENT PAST (if we were to consider the 1920s as such) may be "better than" PRESENT!

Hierarchical levels in organized complex systems

Complex systems are hierarchically organized and dynamically self-regulated. Any hierarchical organization entails a transitive structure, like that in Fig. 12.6(a). Yet dynamic self-regulation counters this transitivity with intransitivity—through cyclical loops of regulation that feed from the lowest to the higher levels, as in Fig. 12.6(b). The whole dynamic hierarchy entails coordination of 'top-down' and 'bottom up' regulatory processes where new levels of the hierarchy can emerge (see Fig. 12.5). Yet each level is qualitatively unique and irreducible to any other levels. In contemporary developmental science it is the theoretical scheme of probabilistic epigenesis (Gottlieb, 1992, 1997, 1999, 2003). The perspective of probabilistic epigenesis entails a cyclical view of relations between genes and environment (Gottlieb, 1998, Fig. 6, p. 798).

Fig. 12.6 depicts a modified model of the probabilistic epigenesist system. It is obvious that the scheme here leads to an emphasis on the specific kinds of relationships between levels—first of all, between adjacent levels (*e.g.*, Johnston & Edwards, 2002). Whereas most of the interest in the elaborations of the probabilistic epigenesist idea concentrates on the three lowest levels (genetic, neural activ-

(a)   (b)

Figure 12.6: Hierarchical organization: transitive or intransitive?

ity, behavior), for our purposes we modify the traditional scheme (by adding the level of higher mental functions, instead of environment —the latter is assumed to function at all levels).

The scheme in Fig. 12.7 shows the unity of intransitive loops $(A - B - D - E)$ and transitive parts $(A - B - C)$ in the structure. It is precisely the open intransitive cycles that make development of the organism possible. Higher mental functions both **result from the ruptures of intransitivity cycles** and **feed forward into further construction of these cycles**. The same intermediate outcome (D) at the behavioral level is guaranteed both through the mechanisms of the lower (neural) level (B) and its adjacent higher (mental functions) level (E), and it feeds further into modification of the neural and mental organizational forms. The emergence of novelty at the higher level allows for the emergence of new quality—reflexivity—in the functioning of the other levels. Through such proliferation of hierarchy new recursive regulatory mechanisms emerge.

Intransitivity and reflexivity

Any kind of multi-level hierarchical organization of systemic phenomena to the issue of possible reflexivity of one level in relation to others. In what sense can parts of the system at a higher level

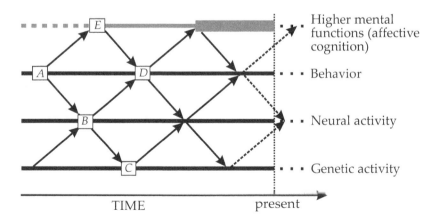

Figure 12.7: Multi-level system of probabilistic epigenesis (after Gottlieb, with modifications).

of organization 'know'[10] the properties of functioning of the lower levels? One can show that reflexivity can serve a base for a special domain of intransitivity in human relations.

Understanding of the others and social relations between humans requires reflexive thinking, in Lefebvre's (1977) terms, or recursive thinking, in Flavell, Miller, and Miller's (1977/2002) terms. Both kinds of terms describe reasoning like 'I think that you think that she thinks, ...,' etc. In Lefebvre's (1977) theory of reflexive control, the *depth of recursive reasoning* (*i.e.*, the length of the chain 'I think that you think that...') is operationalized through the concept of *rank of reflexion*:

> The *rank of reflexion* of a persona can be introduced to characterize the "depth of simulation" represented by the number of successive imbeddings of the other personæ by the persona in question. (Lefebvre, 1977, p. 75)

A gambler has zero rank of reflexion, if s/he knows the declared rules of action in a game together with a payoff matrix, but does not take into account other gamblers' points of view. A gambler has the first rank of reflexion, if s/he knows the rules of action, knows the payoff matrix, and takes into account the other gamblers' points of view *while assuming that that they have zero rank of reflexion*. In general, a gambler who has rank of reflexion R assumes that the other gamblers have rank of reflexion at most $R - 1$. Using this assumption as a hypothesis, s/he reconstructs the others' reasoning, then chooses an optimal strategy for further action (Pospelov, 1974).

Flavell has shown that the understanding of recursive thought can be an important part of the development of theory of mind. A brilliant example of a child's recursive reasoning in a specially designed game is the following (Flavell et al., 1977/2002, p. 225):

> ...he might feel that we, that we know that he thinks that we're going to pick this cup so therefore I think we should pick up the dime cup, because I think he thinks, he thinks that we're going to pick up the nickel cup, but

---

[10]Here we do not imply conscious knowledge, but merely a functional regulatory role of one level over another. Thus, regulatory genes "know" how and when to unblock the functioning of protein-producing genes—without any conscious knowledge. Of course, in the human case, at some levels of organization such knowing becomes conscious and communicable.

then he knows that we, that we'll assume that he knows that we, that we'll assume that he knows that, so we should pick up the opposite cup.

Lefebvre (1977, p. 47, p. 74) has introduced some formalisms to describe such reasoning: for example,

$$T + \{T + [(T + Ty)]x\}y.$$

Here x, y, z represent different persons, and T an arena of their joint attention, thinking, actions, etc.; Tx, Ty, Tz (read T *from x's position*, etc.) represent images of the arena presented by x, y, z, respectively; Txy, Txz, Tyz, etc. (read Tx *from y's position*, etc.) represent images of images of the arena presented by x, y, z, respectively; and so on.

It is essential for the issue of intransitivity that reflexivity provide an opportunity to transform objectively transitive relations of superiority into their opposites. A participant of an interaction must be able, based on her or his knowledge of a rival's beliefs, attitudes, knowledge, etc., to create and design novel conditions in which variables considered very significant by the rival lose their significance —while variables seemingly unimportant to the rival become, by contrast, of great importance. In consequence, things that seemed to have positive valence to the rival (*e.g.*, his or her merits) will now seem to have negative valence (*e.g.*, turning into his or her demerits) even though objective transitive relations of superiority previously seemed unshakable (Poddiakov, 2000/2006).

An antagonistic game: Fox and Dragon

Let us consider a confirming example from the theory of reflexive games, used to teach students reflexive reasoning (Alekseeva, Kopylov, & Maracha, 2003). Students in an institute are presented with the following problem. In a fairytale forest, among ordinary brooks, lakes, etc., are ten numbered wells containing poisoned water. Whoever drinks from any of the wells dies in a hour. The only antidote is to drink (within the hour) water from any other well with a larger number. For example, if one drinks from the fifth well, it is necessary to drink from the sixth well, or the seventh, etc.: then the bane from the well with the smaller number will be neutralized, and the drinker will stay alive

and healthy. Note that water from the tenth well cannot be neutralized by anything. Now, a terrible dragon owns all ten wells
and allows citizens of the forest to use the first nine wells only—
although he can use all ten. The dragon does not like a certain
fox, and makes him accept the following duel: each of them should
bring a glass of water for the other, and the other should drink it.
It is known that after the duel the dragon has died and the fox has
stayed alive and healthy. How was it possible?

Alekseeva et al.'s (2003) article does not contain any answers.
We offer the following. The dragon decides to give the fox the
water from the tenth well (because its water contains the most
powerful bane, which has no antidote), and to drink water from
the tenth well after drinking water brought by the fox (because
that water is also the most powerful antidote). Understanding the
dragon's reasoning, the fox brings him ordinary water. The dragon
—following his decision—drinks it, drinks water from the tenth well,
and dies: the thing that seemed to him to be the most powerful antidote acted on him as the most powerful poison. Meanwhile, the fox
stays alive and healthy because he drank water from (any) one of the
first, second, . . . , ninth wells before the duel. What seemed to the
dragon a certain bane for the fox, water from the tenth well, was really an antidote. In our terms, this means that a transitive hierarchy
of superiority relations, which seemed absolutely stable and unshakable, became shaky and turned into an 'upside-down' hierarchy—
by the interaction of the 'absolute' hierarchy with conditions that
were unexpected by one participant.[11]

A cooperation game: Wife and Husband

One can see that, in general, reflexion can promote success—like the
fox's success in the problem above. Another example of success is
from a different kind of game, *cooperation games*, in which "the in-

---

[11]Alekseeva et al. (2003) also asked college students to state what principle can
be learned by solving Fox and Dragon, and in which teaching situations it would
be appropriate to present the problem. This question, like the problem, is not
answered in the article. We suggest that Fox and Dragon can be a teaching metaphor
for artful reflexive behavior in any organization that seems, so to speak, to be a
'terrarium of like-minded persons' ruled by an authoritarian 'dragon' prepared to
apply hard sanctions to opponents. The necessity of 'aikido' metaphors (*e.g.*, 'do
not hinder a rival's action aimed against you; rather, use his own action against
him') can seem a reasonable possible behavior in such an extreme situation.

terests of the players coincide, but can be realized only if the players can coordinate their choices" (Rapoport, 1977, p. 20). Rapaport (*ibid.*, p. 22, following Schelling, 1960) provides the following amusing example of successful reflexive coordination,

> a man and his wife who lost each other in a crowded department store both cheerfully head for the lost and found [...]. [T]his spot has no "objective" advantages. The choice depends on a shared sense of humour and on the knowledge that the sense of humour is shared.

Inconsistency of ranks of reflexion with practical success

Although reflexion can promote success and only good gamblers are able to increase ranks of reflexion, it has been shown in reflexive game theory that this increase may lead to a danger of getting into troubles of 'surplus wisdom' and sequential loss. A good gambler with a higher rank of reflexion overestimates a rival and thinks that the rival's rank of reflexion is high as well; if the rival really has a lower rank of reflexion, this overestimate may lead the better gambler to lose (Pospelov, 1989). In other words, reflexion is capable of turning strict and objectively stable transitive hierarchies of superiority relations into 'upside-down' ones.

Hide-and-Seek. To show in more detail this inconsistency of ranks of reflexion with success in practical activity, let us consider a version of the game Hide-and-Seek.[12] Table 12.1 shows degrees of illumination in several rooms, known to both participants, as well as the Hider's and Seeker's ranks of reflection and room choices (known only to themselves).

Table 12.1: Participants' ranks of reflexion and choices of rooms.

	Rank of reflexion				
	0	1	2	3	4
Hider's room choice	the darkest	any but the lightest	any but the darkest	the lightest	the darkest
Seeker's room choice	the lightest	the darkest	any but the lightest	any but darkest	the lightest

---

[12]We could use Dragon and Fox here, but prefer to present a new situation.

Hider hides in some room. Seeker chooses a room in which to seek Hider. They have the following strategies: other things being equal, Seeker prefers to look for Hider in the lightest room, because it is easier, and Hider prefers to hide in the darkest room. Increases in ranks of reflexion mean that one participant catches on to the other's catching on, and so forth. One can see that after the second rank of reflexion, strategies begin to repeat themselves.

This is an illustration, not an exact proof, of the following statement of game theory: for two-player games, increasing ranks of reflexion above the second rank does not provide anything new in strategies (though psychological difficulties continue to increase). In fact, a higher rank of reflexion can lead to loss. If Hider has zero rank of reflexion, s/he always chooses the darkest room. In this case, if Seeker has the first rank of reflexion, s/he wins because chooses the same room; however, if Seeker has the third rank of reflexion and so chooses among all rooms **except** the darkest one, s/he loses. In other words, the worse gambler, who does not think about the rival's thoughts at all, can win against the better gambler, who makes a long sequence of recursive reasoning.[13]

However, a clear conclusion about intransitivity of superiority cannot be inferred immediately from this fact. So let us introduce another version of the game that does show intransitivity (Poddiakov, 2000/2006, 2006). Our version is that each of two participants

---

[13][Psychology has been anticipated by literature—specifically, mystery stories. "It's a very useful device in fiction, because you can prove very nearly anything by it. But, deep down inside me, I've never really believed in it. You remember the famous passage in which Dupin shows how it is possible to anticipate the way a person's mind will work, and uses as an example the schoolboy's game of evens or odds. You have a marble concealed in your left or your right hand, and the other fellow gets the marble if he picks the correct hand: so on as long as your marbles last. After estimating the intelligence or stupidity of your opponent, you put yourself mentally in his place, think what he would do, and win all the marbles. Well, it won't work. I've tried it. It won't work because, even if you have two minds exactly adjusted, the one thing they will differ over is what constitutes strategy. And, if you try any such games when the other fellow is probably only leaving it to chance, you'll build up such an elaborate edifice that you can't remember where you started . . . ." (Carr, 1944, p. 220)

The speaker is standing in for his author (who delighted in paradoxes). "Dupin" is C. Auguste Dupin, Edgar Allan Poe's detective *avant la lettre* (Poe, 1845). (Ed.)]

should make a combined choice. Namely, s/he has to choose both her or his role in the game (to be Seeker or Hider) and a rank of reflexion which s/he is obliged to realize in the process of the play. For example, one participant may choose one of two cards. One reads "I hide myself. I have the second rank of reflexion", and the other reads "I seek. I have the third rank of reflexion". (This is a case of meta-reflexivity.)

Now one can see the following. A Seeker with rank of reflexion 1 wins when playing against a Hider with rank of reflexion 0; a Hider with rank of reflexion 2 wins when playing against a Seeker with rank of reflexion 1; and a Seeker of rank of reflexion 3 wins when playing against a Hider with rank of reflexion 2. Yet a Hider with rank of reflexion 0 wins when playing against a Seeker with rank of reflexion 3! Thus, one cannot state that a higher rank of reflexion is better than a lower one. The profitability of a given rank is determined by its interaction with the rival's rank—it does not constitute an absolute value in itself.

This intransitivity of levels of reflexivity works in real life as well. Inconsistency of ranks of reflexion with success of practical activity can be expressed, for example, if you miss a rendezvous with your friend because the meeting place appears (to you) not sufficiently exact, and you cannot imagine either where s/he thinks you are or what s/he thinks you are thinking about where s/he is. In more extreme cases, this inconsistency may lead to victims and deaths in military operations, in which some preliminary rendezvous may be hard or impossible to keep (Poddiakov, 2002, 2006).

Gifting. More examples of inconsistency between reflexion and success of practical activity can be found even in such a pleasant and exciting activity as *gifting*. Mutual gifting is a kind of cooperation game, with the shared aim of causing pleasure. It necessitates reflexion to imagine which gift can make another person happy and, in many cases, to keep some preliminary secrecy from that person. A romantic and, at the same time, amusing example of inconsistency of ranks of reflexion with success in practical activity can be found in O. Henry's story *The Gift of the Magi* (Henry, 1919), in which a loving young husband sells his golden watches to buy very nice and expensive combs for his wife, who has long beautiful hair—while at the same time she cuts and sells her hair to buy an excellent golden chain for his watches. In the short term, it would seem better if they had not imagined each other's assumed preferences so vividly and, respectively, had chosen other gifts (or other ways to raise money with which to buy them). Yet, as the story gives a clue, this romantic behavior can be marvelous from the point of view of longer-term life perspective.

Reflexive management.   A most artful way to work with supe-riority/inferiority relations, based on reflexion, is not just to take others' reflexion into account, but to engage in *reflexive manage-ment*, which has the aim of shaping in the others definite images of a situation and of the others themselves. Efforts in that direc-tion can be found in humanistic psychotherapies, which provide a client with a 'push for' emergence and development of a novel and positive self-image and images of others, and novel behavioral strategies extending the client's human potential. Negative exam-ples arise when rivals' doctrines are (re)constructed using manipu-lative reflexive control (Lefebvre, 1977). For example, an advanced strategic behavior in knowledge-based economics is *Trojan horse teaching*, in which a 'teacher', ostensibly helping his or her rival to learn something, really teaches the rival useless or disadvantageous things. This can be a way to control others' development, stimulat-ing emergence of structures desired by the 'teacher' and hindering emergence of undesirable ones (Poddiakov, 2001, 2004). Yet—as entailed by the open-systemic nature of all developmental processes —this process is in principle not controllable. Management efforts in intransitive systems become acts of negotiation of the relationships between components of the system. In contrast, management efforts in transitive systems can be complete, requiring no negotiation.

The problem of measurement in complex systems

All that was said above referred to complex systems themselves, and did not address the question of ways to get information from them. How can one obtain data on superiority relations in an intransitive system? Actually, in simple situations —those where transitivity applies—comparison of two (or more objects) does not cause any problems. For example, if it is necessary to compare two persons' heights, one can place the persons near each other, using the first as a measure for the second (or vice versa); or if the persons are in different places, one can use some mediating tools (*e.g.*, a measurement stick with the metric system on it). In cases of work with complex systems, measurement of their characteristics becomes much more difficult.

Interactions with tools

A first problem is that an instrument (a tool) of measurement can interact with the object measured, and this interaction should be taken into account. For example, using a thermometer to measure the temperature of water in a glass, one should take into account that s/he gets data on temperature of not the water itself, but on the temperature of the system *glass and water and thermometer*. If the thermometer itself is very cold or very hot, or made from materials that react with water and glass exothermically or endothermically, the result of measurement can be very different. Likewise, the measurement of blood pressure uses the relationship of the act of exclusion of blood from the arm to that of the measurer's return to the hearing of the heartbeats—documenting those on a conventional scale of the height of a column of mercury in a tube. This measurement procedure is vulnerable to all kinds of situational conditions— placement of the arm, placement of the cuff on the arm, placement of the stethoscope on the arm, acoustic perception capacity of the measurer, etc. Yet the resulting measure—the numerical recording of "blood pressure"—becomes accepted as a valid "vital sign" in routine practice, without any questioning of whether all these conditions were in place.

Any theory of measurement needs to include a sub-theory of tools to measure its characteristics, and an analysis of the interaction of the measuring tools with the objects measured (Pyatnitsyn & Vovk, 1987). For example, for a complex system with many factors of potential impact, the complete design of experiments requires the complete (exhaustive) combinatorial testing of all interactions of all elements of the system with all tools of measurement. Physical sciences spend about 90% of their time proving that their instruments were not flawed by apparatus errors, and that the discovered facts are not erroneous (Knorr-Cetina, 1999).

Taking account of measurement tools is necessary not only in natural sciences, but in social sciences as well—or even more so. Social conventions have led to the uncritical acceptance of the validity of measurement techniques that are labeled "standardized". That label cannot substitute for a careful analysis of how these instruments actually work. Very often, the use of different instruments (surveys, experimenters, etc.) to measure the (purportedly) 'same' characteristic leads to different (or even opposite) results, because of specific interactions of participants with the instruments. If a tool of measurement is included into a theory of a system explored, the problem of conclusions about superiority relations gets more difficult. One should work not with statements like

A is superior to B by relation R,

B is superior to C by relation R,

etc., but with statements like

A is superior to B by relation R measured by tool T,

B is superior to C by relation R measured by tool T,

etc., where tool T may interact with A differently than with B or C, etc. The result is that the qualities 'measured' by these tools are different—all measurements are results of tool-to-object interaction. It may well be that it is the *discrepancy* between results of different tool uses, rather than any *concordance* between them, that characterizes the 'measured' object best. A way to make such conclusions more robust is to use different tools—but this is not absolute, due to the openness of complex systems.

Algorithmic undecidability of some comparisons

In complex systems, equivalent algorithms of measurement applicable to different cases are often impossible. This is a problem of *algorithmic undecidability* (as it is called in mathematical logic and computer science). Many well-defined general problems, towards the solution of which one might hope to apply algorithmic methods, can be proved to have no general solution algorithm. This does not mean that one or another individual problem of this type cannot be solved. An individual problem may have an algorithmic solution, which in some cases may be very simple and transparent; yet there cannot be any general algorithm applicable to the whole class of all problems of the type, nor any algorithm to separate the class into subclasses and subsequently apply more specific algorithms appropriate to features of the subclasses. It can even happen that *every* problem of the given type is solvable by *some* algorithm— yet there is no algorithmic way to discover that algorithm.

This seems a challenge for the intellect, since common sense appears to say that a well-defined general problem, of which every individual instance is solvable, surely must have a whole, general algorithm, and that only a lack of clever people has kept such a general algorithm from being discovered. Yet, based on studies of Gödel, Turing, Church, and many others (see Penrose, 1990), common sense must revise its belief: it has been proved that algorithmically undecidable problems do exist.

Interestingly, many of us and even our school-aged children can and do successfully solve different *individual* problems, belonging to algorithmically undecidable problem classes. This situation arises in various contexts: finding proofs of equivalence of two or more mathematical expressions (*e.g.*, 'trigonometric identities'), a type of problem often posed in secondary school; finding proofs of equivalence of two computer programs (*i.e.*, do the programs, presented with equivalent initial data, always come to equivalent results?); constructing complex technical devices (automata) from simpler devices, etc. The metamathematical proofs of Gödel et al. apply to show that these and many other classes of problems *do not and cannot* have general algorithms to produce a solution —yet despite this fact, human thinkers *can and do* solve problems in these classes successfully, in individual ways, by invention and application of special and even unique methods and techniques.

Equivalence

A key point of measurement is the problem of equivalence. Establishing a standard (a unit) of measurement is the core of measurement. To the extent that determining equivalence is impossible, measurement is also impossible. Algorithmic undecidability signifies that one cannot use 'equivalent' tools (instruments and algorithms of their application) to compare systems A, B, C, etc.: it is impossible that there be a uniform, invariably successful algorithm. Thus even here, in the simplest case of making a conclusion about superiority relations, one should work with statements like

A is superior to B by relation R measured by tool $T(1)$,
B is superior to C by relation R measured by tool $T(2)$, etc.

—or even

A is superior to B by relation R measured by tool $T(1)$
in world-context $K(1)$

and the like. It can be impossible to make any definite conclusion whether A is superior to C or not. This uncertainty is multiplied by the possibility of interactions between the system explored and the tools of its exploration.

Let us consider an example. Which of three programs A, B, and C (whether they are in a computer, in a brain, etc.) is better in a definite relation: speed of running, or defense from running forever, etc.? (The problem of determining whether a program halts or will keep running forever is algorithmically undecidable, but some palliative solutions are possible.) We can compare these three programs only by using one or more "measuring" programs, which can occur in different, cooperative and combative relations with the programs explored and between one another. Based on all said above about interactions in complex systems, one cannot give any definite answer to the question whether program A is superior to C or not.

Solving a problem, belonging to a class of algorithmically undecidable problems, necessarily involves creativity and heuristics. Heuristics manage the process of search by transforming, narrowing and widening, creating and cutting off directions of the search.

Developmental studies of (in)transitivity mastery: two research and teaching orientations

Concerning age-related and individual development of understanding of transitivity and intransitivity, two kinds of research orientations can be found unequally represented. There are many studies of the development of transitive inference in children (Flavell et al., 1977/2002), but few on the understanding of regular intransitivity. *Yet understanding of intransitivity is no less important for cognitive development than understanding and mastery of transitivity.* Both forms of relations—and transitions from one to the other— are part of our multi-faceted life.

The same applies to teaching the understanding of intransitivity. Based on our everyday life experience, we know that children play Bear–Housewife–Cowboy; Bulldog–Mongoose–Cobra, etc., in all of which competitive relations are intransitive: character A beats character B, character B beats character C, and character C beats character A (as in Rock–Paper–Scissors). Similarly, we know that adults tell children tales in which the first character (*e.g.*, a mouse) is afraid of the second (*e.g.*, a cat) and controlled by it, the second character is afraid of and controlled by the third, and so on—but the last, and seemingly the strongest, character in the chain is afraid of and controlled by the first.[14] It seems not only a tendency of folklore to play with paradoxes, but also a way to express real intransitivity of some superiority relations, a cultural tool to make a human ready for encounters with intransitivity in the real world.

Science-based teaching to stimulate intransitivity understanding is possible as well. Roberts (2004) has designed a complex of learning exercises on intransitivity for university students. Poddiakov has designed a complex of problem-solving toys to stimulate understanding of multi-variable interactions in preschoolers and secondary school students. Some of the toys realize intransitive relations, either supportive or combative ones. Let us consider some "intransitive" toys having Condorcet-like compositions of their parts.

- Intransitive Hungry Monkeys A, B, and C are so posed that A can feed B, B can feed C, and C can feed A, but not vice versa (as in Fig. 12.8).
- Plastic models of intransitive Mobile Assault Towers are such

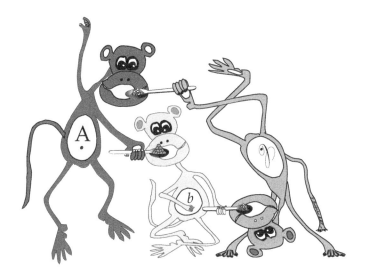

Figure 12.8: An intransitive family of hungry monkeys.

---

[14][See also Heraclitus, Fragment XXV: Ζῆ πῦρ τὸν γῆς θάνατον, καὶ ἀὴρ ζῆ τὸν πυρὸς θάνατον· ὕδωρ ζῆ τὸν ἀῆρος θάνατον, γῆ τὸν ὕδατος; that is, "Fire lives in the death of earth, air lives in the death of fire, water lives in the death of air, and earth in the death of water" (Patrick, 1888, pp. 684, 649). (Ed.)]

that tower *A* has a 'gun' on the top that can attack other towers' tops, a middle level that is proof against all attacks, and a vulnerable bottom level. Tower *B* is vulnerable at the top, has a 'gun' at the middle level, and is invulnerable at the bottom. Tower *C* is invulnerable at the top, vulnerable at the middle level, and has a 'gun' at the bottom. (Compare with Fig. 12.1.) In combat, *A* beats *B*, *B* beats *C*, and *C* beats *A*.

- Three Intransitive Double-Gears (see Fig. 12.9) are such that, when assembled into pairs, double-gear *A* rotates faster than *B*, *B* rotates faster than *C*, but *C* rotates faster than *A*.

Yet in general there seem to be many fewer teaching methods for mastery of intransitivity than for mastery of transitivity.

General Conclusions

The objective impossibility of universal, exact prescriptions for solving problems like 'which is better?' has profound implications. It means that a complex system should solve such a problem either by modifying a previous method of solution (because its exact reproduction cannot result in success) or by inventing an essentially new method. The objective impossibility of universal, exact prescriptions under conditions of uncertainty of the dynamically changing environment makes the system necessarily open to creating novelty. Such openness may be limited to specific moments of encounter with the environment—the stability of established intransitivity cycles may be broken by episodic ruptures. Creative search for novelty becomes more and more necessary in systems of increasing hierarchical integration, as the adaptation at each level of the hierarchy requires openness to innovation that is not detrimental to the survival of the whole system. Hence the emergence of novelty is enhanced at higher levels of organization—

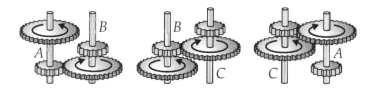

Figure 12.9: An intransitive set of three double-gears.

yet remains systemically conservative, *i.e.*, does not let the complex system dissipate under occasional failures of some part of it. The creation of novelty preserves the non-novel maintained status of the system.

We have demonstrated that intransitivity is one of the main relations that makes the functioning of complex systems and their development possible. Transitivity of relations is a sub-component of basic intransitivity cycles, and cannot serve as the criterion of rationality. Intransitivity of relations in multi-level functional systems enables the emergence of reflexivity which could be considered as the actual basis for rationality. In other terms—rationality is a strategic, rather than a logical, invention of the human mind.

Understanding of intransitivity develops throughout historical, cultural and individual development, and is not less important for human cognitive and cultural development than development of transitivity understanding. Studies of the development of intransitivity understanding should be added to, and related to, developmental studies of transitive inferences, creating a whole picture in this area. Making sense of intransitivity is an integral part of the image of a world that necessarily involves interacting complex dynamical systems.

References

Ainley, S. (1978). *Mathematical Puzzles*. Englewood Cliffs, NJ: Prentice Hall.

Alekseeva, L. N., Kopylov, G. G., & Maracha, V. G. (2003). Issledovatel'skaia deiatel'nost' uchashtikhsia: formirovanie norm i razvitie sposobnostei [Exploratory activity of students: Formation of norms and development of abilities]. *Issledovatel'skaya rabota shkol'nikov, 4*, 25–28.

Anand, P. (1993). The philosophy of intransitive preference. *The Economic Journal, 103*, 337–346.

Arrow, K. (1963). *Social Choice and Individual Values*. New Haven, CT: Yale University Press.

Arrow, K. (1973). Social choice and justice. In P. P. Wiener (Ed.), *Dictionary of the History of Ideas: Studies of Selected Pivotal Ideas, Vol. 4* (pp. 276–284). New York, NY: Scribner.

Bar-Hillel, M., & Margalit, A. (1988). How vicious are cycles of intransitive choice? *Theory and Decision*, *24*, 119–145.

Baumann, P. (2005). Theory choice and the intransitivity of "Is a better theory than". *Philosophy of Science*, *72*, 231–240.

Belousov, B. L. (1959). Periodicheski deistvuiushchaya reaktsiya i ee mekhaniem [A periodic reaction and its mechanism]. *Sbornik referatov po radiatsionnoi meditsine*, *147*, 145.

Block, J. R. (2002). *Seeing Double: Over 200 Mind-Bending Illusions*. London, GB: Routledge.

Boddy, L. (2000). Interspecific combative interactions between wood-decaying basidiomycetes. *FEMS Microbiology Ecology*, *31*, 185–94.

Borda, J.-C. de. (1781). Mémoire sur les élections au scrutin [Memoir on elections by vote]. In *Mémoires de l'Académie Royale des Sciences* (pp. 657–665). Paris, FR: Imprimerie Royale. (Translation and commentary by A. de Grazia published as Mathematical Derivation of an Election System, *Isis 44*(1/2), 1953, pp. 42–51)

Carr, J. D. (1944). *To Wake the Dead*. New York, NY: Books, Inc.

Chan, K. M. A. (2003). Intransitivity and future generations: Debunking Parfit's mere addition paradox. *Journal of applied philosophy*, *20*(2), 187–200.

Chauvin, R. (1969). *Le Comportement Animal [Animal Behavior]*. Paris, FR: Masson et Cie.

Condorcet, N. (2009). *Essai Sur L'Application de L'Analyse à la Probabilité Des Decisions Rendues à la Pluralité Des Voix [Essay on the Application of Analysis to the Probability of Majority Decisions]*. Whitefish, MT: Kessinger Publishing. (Original work published 1785)

Dawes, R. M. (1998). Behavioral decision making and judgment. In D. Gilbert, S. Fiske, & G. Lindzey (Eds.), *The Handbook of Social Psychology, Vol. 1* (pp. 497–548). Boston, MA: McGraw-Hill.

Deshpande, M. N. (2000). Intransitive dice revisited. *Teaching Statistics*, *22*(3), 80.

Dooley, K. J. (1997). A complex adaptive systems model of organization change. *Nonlinear Dynamics, Psychology and Life Sciences*, *1*(1), 69–97.

Druzhinin, V. N. (2000). *Psikhologiya Semi [Psychology of Family]*. Ekaterinburg, RU: Delovaya Kniga.

Eppley, S. (2003). *Benevolent Strategic Indifference and Group Strategy Equilibria: Minimal Defense and Truncation Re-*

sistance as Criteria for Voting Rules.   Retrieved January 26, 2010, from http://alumnus.caltech.edu/~seppley/ Strategic%20Indifference.htm

Fishburn, P. C. (1991). Nontransitive preferences in decision theory. *Journal of Risk and Uncertainty, 4*, 113–134.

Flavell, J. H., Miller, P. H., & Miller, S. A. (2002). *Cognitive Development.* Upper Saddle River, NJ: Prentice Hall. (Original work published 1977)

Gigerenzer, G. (2002). *Adaptive Thinking: Rationality in the Real World.* New York, NY: Oxford University Press.

Goodall, J. (1986). *The Chimpanzees of Gombe.* Cambridge, MA: Harvard University Press.

Gottlieb, G. (1992). *Individual Development & Evolution: The Genesis of Novel Behavior.* New York, NY: Oxford University Press.

Gottlieb, G. (1997). *Synthesizing Nature/Nurture.* Mahwah, NJ: Lawrence Erlbaum Associates, Inc.

Gottlieb, G. (1998). Normally occurring environmental and behavioral influences on gene activity: From central dogma to probabilistic epigenesis. *Psychological Review, 105*(4), 792–802.

Gottlieb, G. (1999). *Probabilistic Epigenesis and Evolution.* Worcester, MA: Clark University Press. (23rd Heinz Werner Lectures)

Gottlieb, G. (2003). Probabilistic epigenesis of development. In J. Valsiner & K. J. Connolly (Eds.), *Handbook of Developmental Psychology* (pp. 3–17). London, GB: Sage.

Henry, O. (1919). The gift of the Magi. In *The Four Million* (pp. 16–25). New York, NY: Doubleday, Page & Company.

Hermans, H. (2001). The dialogical self: Toward a theory of personal and cultural Positioning. *Culture & Psychology, 7*(3), 243–281.

Hermans, H. (Ed.). (2002). *The Dialogical Self* (Vol. 12, No. 2 of *Theory & Psychology*). (Special issue)

Ivin, A. A. (1998). *Logika [Logic]* (2nd ed.). Moscow, RU: Knowledge.

Jablonka, E. (2001). The systems of inheritance. In S. Oyama, P. E. Griffiths, & R. D. Gray (Eds.), *Cycles of Contingency* (pp. 99–116). Cambridge, MA: MIT Press.

Johnston, T. D., & Edwards, L. (2002). Genes, interactions, and the development of behavior. *Psychological Review, 109*, 26–34.

Kawamura, S. (1959). The process of sub-culture propagation among Japanese macaques. *Primates, 2*, 43–60.

Kawamura, S. (1963). The process of sub-culture propagation among Japanese macaques. In C. H. Southwick (Ed.), *Primate Social*

Behavior: An Enduring Problem. Selected Readings (pp. 83–90). Princeton, NJ: D. Van Nostrand Company, Inc. (reprinted from Kawamura, 1959)

Kerr, B., Riley, M. A., Feldman, M. W., & Bohannan, B. J. M. (2002). Local dispersal promotes biodiversity in a real-life game of rock-paper-scissors. *Nature, 428*, 171–174.

Kirkup, B. C., & Riley, M. A. (2004). Antibiotic-mediated antagonism leads to a bacterial game of rock-paper-scissors *in vivo. Nature, 428*, 412–414.

Knorr-Cetina, K. (1999). *Epistemic Cultures*. New York, NY: Cambridge University Press.

Kozielecki, J. (1979). *Psychological Decision Making (volume 24, theory and decision library)* (G. Eberlein, Ed.). Boston, MA: D. Reidel Publishing Company. (revised edition based on the original Polish *Psycholgiczna Teoria Decyzji*, 1975, Warszawa, PL: Państowowe Wydawnictwo Nauowe. Theory and Decision Library, Vol. 24)

Krebs, H. A. (1964). The citric acid cycle. In *Nobel Lectures, Physiology or Medicine 1942–1962* (pp. 399–410). Amsterdam, NL: Elsevier.

Kuhn, T. S. (1962). *The Structure of Scientific Revolutions*. Chicago, IL: University of Chicago Press.

Lages, M., Hoffrage, U., & Gigerenzer, G. (1999). *Intransitivity of Fast and Frugal Heuristics.* Max Plank Institute for Human Development. Retrieved June 3, 2011, from http://ideas .repec.org/p/xrs/sfbmaa/99-49.html

Lefebvre, V. A. (1977). *The Structure of Awareness: Toward a Symbolic Language of Human Reflexion* (A. Rapoport, Trans.). Beverly Hills, CA: Sage. (foreword by A. Rapoport)

Lewin, K. (1927). Gesetz und Experiment in der Psychologie [Law and experiment in psychology]. *Symposion, 1*(4), 375–421.

Lipmann, F. (1964). Development of the acetylation problem: A personal account. In *Nobel Lectures, Physiology or Medicine 1942–1962* (pp. 413–438). Amsterdam, NL: Elsevier.

Maruyama, M. (1963). The second cybernetics: Deviation amplifying mutual causal processes. *American Scientist, 51*, 164–179.

Maruyama, M. (1999). Heterogram analysis: Where the assumption of normal distribution is illogical. *Human Systems Management, 18*, 53–60.

Meigs, A. (1990). Multiple gender ideologies and statuses. In

P. R. Sanday & R. G. Goodenough (Eds.), *Beyond the Second Sex: New Directions in the Anthropology of Gender* (pp. 101–112). Philadelphia, PA: University of Pennsylvania Press.

Menon, U. (2002). Neither victim nor rebel: Feminism and the morality of gender and family life in a Hindu temple town. In R. A. Shweder, M. Minow, & H. Markus (Eds.), *Engaging Cultural Differences* (pp. 288–308). New York, NY: Russell Sage Foundation.

Menon, U., & Shweder, R. A. (1994). Kali's tongue: Cultural psychology and the power of 'shame' in Orissa. In S. Kitayama & H. Markus (Eds.), *Emotion and Culture* (pp. 237–280). Washington, DC: American Psychological Association.

Molenaar, P. C. M. (2004). A manifesto on psychology as idiographic science: Bringing the person back into scientific psychology, this time forever. *Measurement: Interdisciplinary Research and Perspectives*, 2, 201–218.

Molenaar, P. C. M., Huizinga, H. M., & Nesselroade, J. R. (2002). The relationship between the structure of inter-individual and intra-individual variability. In U. Staudinger & U. Lindenberger (Eds.), *Understanding Human Development* (pp. 339–360). Dordrecht, NL: Kluwer.

Nowak, M. A., & Sigmund, K. (2007). Biodiversity: Bacterial game dynamics. *Nature*, *418*, 138–139.

Parfit, D. (1984). *Reasons and Persons*. Oxford, GB: Clarendon Press.

Patrick, G. T. W. (1888). A further study of Heraclitus. *The American Journal of Psychology*, *1*(4), 557–690.

Peirce, C. S. (1893). Evolutionary love. *The Monist*, *3*, 176–200.

Peirce, C. S. (1896). The regenerated logic. *The Monist*, *7*(1), 19–40.

Penrose, R. (1990). *The Emperor's New Mind*. New York, NY: Oxford University Press.

Petrovsky, V. A. (1996). *Lichnost' V Psikhologii [Personality in Psychology]*. Rostov-na-Donu, RU: Feniks.

Piotrowski, E. W., & Makowski, M. (2005). Cat's dilemma — transitivity vs. intransitivity. *Fluctuation and Noise Letters*, *5*(1), L85–L95.

Poddiakov, A. N. (2000). Otnosheniya prevoshodstva v strukture refleksivnogo upravleniya [Relations of superiority in structure of reflexive control]. In A. V. Brushlinskii & V. E. Lepskiy (Eds.), *Refleksivnoe Upravlenie: Tezisy Mezhdunarodnogo Simpoziuma* (pp. 37–38). Moscow, RU: Institute of Psychology of

Russian Academy of Sciences.

Poddiakov, A. N. (2000/2006). *Issledovatel'skoe povedenie: strategii poznaniia, pomosht', protivodeĭstvie, konflikt [Exploratory Behavior: Cognitive Strategies, Help, Counteraction, and Conflict].* Moscow, RU: PER SE. Retrieved July 15, 2011, from http://hse.ru/data/866/913/1235/ip.rar

Poddiakov, A. N. (2001). Counteraction as a crucial factor of learning, education and development: Opposition to help. *Forum Qualitative Sozialforschung/Forum: Qualitative Social Research*, 2(3). Retrieved June 15, 2011, from http://www.qualitative-research.net/fqs-texte/3-01/3-01poddiakov-e.htm (online journal, not paginated)

Poddiakov, A. N. (2002). Myshlenie i resenie zadats [Thinking and problem solving]. In V. A. Volodin, L. V. Petranovskaya, & T. Kashirina (Eds.), *Entsiklopediya dlya deteĭ: Arkhitextura dushi. Psikologiya litsnosti. Mir vzanmootioshenij. Psikhoterapiya [Encyclopedia for Children: Architecture of the Soul. Psychology of Personality. World of Relationships. Psychotherapy]* (pp. 124–152). Moscow, RU: Avanta+.

Poddiakov, A. N. (2004). *'Trojan Horse' Teaching in Economic Behavior* (Working Paper). Moscow, RU: Moscow State University—Higher School of Economics. Retrieved June 15, 2011, from http://ssrn.com/abstract=627432 (Internet version, via Social Science Research Network, of the paper presented at the conference "Cross-Fertilization Between Economics and Psychology", Drexel University, Philadelphia, PA)

Poddiakov, A. N. (2006). Neperekhodnost' (Netranzitivnost's) noshenii Prevoskhodstva I Priniatie Reshenii [Intransitivity of superiority relations and decision making]. *Psychologia. Journal Visshey shkoly ekonomiki*, 3, 88–111. Retrieved June 15, 2011, from http://creativity.ipras.ru/texts/poddyakov_3-03pp88-111.pdf

Poe, E. A. (1845). The purloined letter. In E. Leslie (Ed.), *The Gift: A Christmas, New Year, and Birthday Present* (pp. 41–61). Philadelphia, PA: Carey and Hart.

Pospelov, D. A. (1974). Igry Reflexivnye [Reflexive games]. In *Entsiklopediya Kibernetiki, T. 1* (p. 343). Kiev, UA: Glavnaya Redaktsia USE.

Pospelov, D. A. (1989). *Modelirovanie rassuzhdenij. Opyt analiza myslitel'nyh aktov [Modeling Reasoning. Experience in the Analysis of Mental Acts]*. Moscow, RU: Radio i svyaz'.

Prigogine, I. R., Allen, P. A., & Herman, R. (1977). Long term trends and the evolution of complexity. In E. Laszlo & J. Bierman (Eds.), *Goals in a Global Community* (pp. 1–63). New York, NY: Pergamon.

Prigogine, I. R., & Stengers, I. (1984). *Order Out of Chaos*. London, GB: Heineman.

Pyatnitsyn, B. N., & Vovk, S. N. (1987). Induction and multifactor experimentation. In *Inductive Logic and the Shaping of Scientific Knowledge. (Induktivnaya logika i formirovanie nauchnogo znaniya.)* (pp. 144–172). Moscow, RU: "Nauka".

Rapoport, A. (1977). Foreword. In Lefebvre, V. (1977), *The Structure of Awareness: Toward a Symbolic Language of Human Reflexion* (pp. 9–37). Beverly Hills, CA: Sage.

Reichenbach, T., Mobilia, M., & Frey, E. (2007). Mobility promotes and jeopardizes biodiversity in rock–paper–scissors games. *Nature, 448*, 1046–1049.

Roberts, S. C. (2004). A ham sandwich is better than nothing: Some thoughts about transitivity. *Australian Senior Mathematics Journal, 18*(2), 60–64.

Rogers, S. C. (1975). Female forms of power and the myth of male dominance: A model of female/male interaction in a peasant society. *American Ethnologist, 2*, 727–756.

Schelling, T. (1960). *The Strategy of Conflict*. Cambridge, MA: Harvard University Press.

Semmann, D., Krambeck, H.-J., & Milinski, M. (2003). Volunteering leads to rock-paper-scissors dynamics in a public goods game. *Nature, 425*, 390–393.

Shafir, S. (1994). Intransitivity of preferences in honey bees: Support for comparative evaluation of foraging options. *Animal Behaviour, 48*, 55–67.

Sherif, M., Harvey, O. J., White, B. J., Hood, W. R., & Sherif, C. W. (1961). *Intergroup Conflict and Cooperation: The Robbers' Cave Experiment*. Norman, OK: University of Oklahoma Book Exchange.

Siegler, R. S. (2007). Cognitive variability. *Developmental Science, 10*(1), 104–109.

Starmer, C. (2004). Development in nonexpected-utility theory: The hunt for a descriptive theory of choice under risk. In C. Camerer, G. Loewenstein, & M. Rabin (Eds.), *Advances in*

Behavioral Economics (pp. 104–147). Princeton, NJ: Princeton University Press.

Strum, C. C. (1987). *Almost Human*. New York, NY: W. W. Norton.

Temkin, L. (1987). Intransitivity and the mere addition paradox. *Philosophy & Public Affairs, 16*, 138–187.

Temkin, L. (1996). A continuuum argument for intransitivity. *Philosophy & Public Affairs, 25*, 175–210.

Temkin, L. (1999). Intransitivity and the person-affecting principle: A response. *Philosophy and Phenomenological Research, LIX*(3), 777–784.

Toomela, A. (2007). Culture of science: Strange history of the methodological thinking in psychology. *IPBS: Integrative Psychological & Behavioral Science, 41*(1), 6–20.

Toomela, A. (2009). How methodology became a toolbox—And how it escapes from that box. In J. Valsiner, P. Molenaar, M. Lyra, & N. Chaudhary (Eds.), *Dynamic Process Methodology in the Social and Developmental Sciences* (pp. 45–66). New York, NY: Springer.

Toomela, A., & Valsiner, J. (2010). *Methodological Thinking in Psychology: 60 Years Gone Astray?* Charlotte, NC: Information Age Publishing.

Tversky, A. (1969). Intransitivity of preferences. *Psychological Review, 76*, 31–48.

Tversky, A., & Kahneman, D. (1986). Rational choice and framing of decisions. *Journal of Business, 59*, 251–278.

Ukhtomskii, A. A. (1927). Dominanta kak faktor povedeniya [Dominant as the behavior factor]. *Vestnik Kommunisticheskoi Akademii, 22*, 215–241.

Valsiner, J. (1984). Two alternative epistemological frameworks in psychology: The typological and variational modes of thinking. *Journal of Mind and Behavior, 5*, 449–470.

Valsiner, J. (1986). Between groups and individuals: psychologists' and laypersons' interpretations of correlational findings. In J. Valsiner (Ed.), *The Individual Subject and Scientific Psychology* (pp. 113–152). New York, NY: Plenum.

Valsiner, J. (1996). Devadasi temple dancers and cultural construction of persons-in-society. In M. K. Raha (Ed.), *Dimensions of Human Society and Culture* (pp. 443–476). New Delhi, IN: Gyan Publishing House.

Valsiner, J. (1999). I create you to control me: A glimpse into basic processes of semiotic mediation. *Human Development, 42,* 26–30.

Valsiner, J. (2000). *Culture and Human Development.* London, GB: Sage.

Valsiner, J. (2001). Process structure of semiotic mediation in human development. *Human Development, 44,* 84–97.

Valsiner, J. (2006). Developmental epistemology and implications for methodology. In W. Damon & R. M. Lerner (Eds.), *Theoretical Models of Human Development, Vol. 1 of Handbook of Child Psychology* (6th ed., pp. 166–209). New York, NY: Wiley.

Valsiner, J. (2007). *Locating the Self: Looking for the Impossible? Or Maybe the Impossible is the Only Possibility.* Retrieved from http://www.tu-chemnitz.de/phil/ifgk/ikk/cs/secure/Valsiner.pdf (Paper presented at the conference "Culturalization of the Self", Chemnitz, DE, December 1, 2007)

van de Waal, E., Renevey, N., Favre, C. M., & Bshary, R. (2010). Selective attention to philopatric models causes directed social learning in wild vervet monkeys. *Proceedings of the Royal Society: Biological Sciences, 277*(1691), 2105–2111. Retrieved June 15, 2010, from http://rspb.royalsocietypublishing.org/content/277/1691/2105

Weiss, P. (1978). Causality: Linear or systemic? In G. Miller & E. Lenneberg (Eds.), *Psychology and Biology of Language and Thought* (pp. 13–36). New York, NY: Academic Press.

Weisstein, E. (2010). *Cellular Automaton.* Retrieved June 3, 2010, from http://mathworld.wolfram.com/CellularAutomaton.html

Wundt, W. (1976). *Untersuchungen zur Mechanik der Nerven und Nervencentren [Investigations of the Mechanics of the Nerves and Nerve Centers].* Erlangen & Stuttgart, DE: Ferdinand Enke Verlag. (Original work published 1876)

Zhabotinsky, A. M. (1964). Periodicheskii protsess okisleniya malonovoi kisloty rastvore (issledovanie kinetiki reakcii Belousova) [Periodic processes of malonic acid oxidation in a liquid phase (study of kinetics of the Belusov reaction)]. *Biofizika, 9,* 306–311.

Zinoviev, A. A. (1972). *Logicheskaja fizika [Logical Physics].* Moscow, RU: Nauka.

Addendum

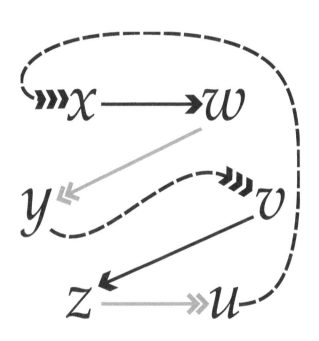

13

Mathematical Modeling and Diagrammatic Reasoning

Lee Rudolph

> Diagrammatic reasoning is the only really fertile reasoning. (Peirce, 1905, p. 542)

> I predict that it will become increasingly useful to formulate hypotheses about[...]behavior in terms of a logic which can handle such hypotheses. Such a logic will be immeasurably superior to the strange little schematic diagrams to which psychological theorists have been addicted. (Carroll, 1959, p. 45)

The epigraphs suggest some natural questions about the use(s) of diagrams—broadly understood to include 'figures' of all sorts—in mathematics and, especially, mathematical modeling.

- Can such diagrams (sometimes) be part of "reasoning"—the dialectical part (to use the classical term recalled in Chap. 2) of a "process of persuasion"—or must they (always) contribute only to the rhetorical part of such a process?
- Whatever the apportionment between dialectic and rhetoric of the purposes such diagrams serve (or the effects they have), is use of diagrams in fact "fertile"—and, if so, of *what*, for *whom*, in *which circumstances*? For example, a reader who refers to "strange little schematic diagrams" need not have found those

Keywords: arrows, conceptualization, diagrammatic reasoning, exhibits, flowcharts, figures and diagrams, illustrations, visualization.

diagrams to be sterile: he may have found them to be fertile ground for sowing and reaping misapprehension or worse.

- Supposing there are some (classes of) such diagrams that both function dialectically and are fertile of desirable consequences, can we say anything useful about *how* they function (or what they have in common), and *how* (and in what contexts) modelers—particularly modelers in the human sciences —might most successfully use (or refrain from using) them?

I hope to return to these questions systematically elsewhere.

For now, in this somewhat miscellaneous chapter I analyze diagrams from earlier chapters semi-systematically; I discuss arrows in mathematics; and I give two brief case studies—one from this volume (suggesting that sometimes no diagram can serve either a dialectic or a rhetorical purpose), one from the literature of set theory and logic (illustrating diagrammatic group self-persuasion).

The ineluctable modality of the visible

The first thing to note about mathematical diagrams, as they appear in printed media, is that they are essentially *visual* and *static*, and usually accompanied by *text* (in which I include mathematical formulæ, etc.). In mathematics, other "visual aids"—ranging from diagrams drawn (by Self or Other) in sand (*cf.* Archimedes) or on paper, blackboards, whiteboards, etc., through orreries and planeteriums for general edification and specialized devices for investigation of unsolved problems in celestial mechanics,[1] string or plaster "mathematical models" of interesting or important mathematically-defined surfaces and solids,[2] magic-lantern and stereopticon shows,

---

[1]"I am still at Saturn's Rings. At present two rings of satellites are disturbing each other. I have devised a machine to exhibit the motions of the satellites in a disturbed ring; and Ramage is making it, for the edification of sensible image worshippers." (Maxwell, 1857/1882, p. 295; written to a friend taking holy orders.)

[2]Addressing the Kansas Academy of Science, Emch (1896, p. 92) reported that in Zürich he had seen "the first model of a surface of the third order with its 27 straight lines", constructed by Fiedler, who "always used to have it suspended at the ceiling of his study or parlor". He predicted (p. 93) that "Collections of mathematical models will become, for the mathematical departments of the universities, polytechnics, and colleges, what the museums of natural history are for the biological departments", and he was right: Kidwell (1996) has given a fascinating history of a century of use of such models in American mathematical pedagogy, until the Great Depression put the last major model-maker out of business.

and animated cartoons starring Donald Duck (Disney et al., 1959), to PowerPoint presentations and their likely future extensions— are (of course) mostly visual,[3] are often dynamic, and may be accompanied by spoken as well as printed text; but here I concentrate on static, annotated diagrams on paper, like those in this volume.

Conceptualization *vs.* depiction

At the beginning of his essay "Redrafting the Tree of Life", Stephen J. Gould observes of scholars (and humans in general) that

> we are all primates, first and foremost, and primates are preeminently visual creatures among mammals [...]. Astute observers have always recognized this "ineluctable modality of the visible" (to quote James Joyce in *Ulysses*), and many mottoes and clichés of our culture recognize the predominance of sight [...]. (Gould, 1997, p. 30)

Gould then specializes to biologists.

> Hardly any major subject in biology can be more intrinsically spatial, and therefore more eminently pictorial, than our views on relationships among organisms. [...] [P]ictures embody theories, while theories both inform and constrain our notion of empirical reality [...]. We study pictures as summaries of the theories we have lived by; but we must also acknowledge pictures as powerful impediments to the development of more fruitful visions—for we cannot conceptualize what we cannot depict [...]. (*ibid.*)

Gould's essay is an overview and discussion of the development (evolution?) of "the iconography of evolutionary trees" (*ibid.*, p. 48). Given his context, he presumably has rhetorical (and perhaps ideological) reasons to give such a narrow, reductive description of the uses to which the community of biologists puts "pictures".

---

[3]Of course "visual aids" can have haptic components: "proof by hand-waving" is not *only* a common turn of phrase! Indeed, "proof by hand-holding" is not *even* a common turn of phrase, but years ago, tutoring in topology, I found it useful to draw figures on a blind student's palm. Somewhat conversely, the linguist Susanna Cumming tells me (personal communication) that, working with a blind graduate advisee, she found she had to draw diagrams (on paper or a whiteboard) to help *herself* think about fine points of transformational grammar.

For all I know of biologists, Gould's description may be quite accurate (or might have been in the 1990s). I am certain it would be wrong if transposed to the community of mathematicians, and I hope it would be wrong if transposed to the (overlapping) community of mathematical modelers.[4] Mathematicians both "conceptualize" and "depict" (with wide variations among sub-disciplines, individuals, and situations), but each of the two activities most certainly can and does occur—fruitfully—in the absence of the other; nor is it even true that every conceptualization can be somehow 'improved' by depiction (see pp. 414–415 for an example).

A menagerie of diagram-types

The following list is a collection, not a classification. Diagrams of all the types listed appear somewhere in this volume (see Table 13.1). The words and phrases used all have common meanings, and it is handy to keep those available, so when using them as diagram types I distinguish them typographically, thus:

- exhibits and illustrations;
- aids to visualization and aids to conceptualization;
- flowcharts;
- definitions and examples;
- calculations and proofs.

Note that graphs, surely the most common form of diagram or figure in the literature of human (and other) science, do not appear in this list. Only in Chap. 11 did any graphs appear, and there their function was not to further any reasoning about the models being presented, but simply to illustrate the software implementations of those models (of course the graphs being illustrated do have both dialectical and rhetorical functions *when the software is actually being used*, rather than—as in Chap. 11—only being *mentioned*).. For discussions of that familiar diagram type (and much more of relevance to this chapter), see the three classics by Huff (1954) and Tufte (1983, 1984).

---

[4]The evidence from the intersection of the sets of biologists and mathematicians, *i.e.*, mathematical modelers in the biological sciences, is mixed; *cf.* the commentaries by Gould (1996) and Root-Bernstein (1996) in an issue of *Art Journal* devoted to "Contemporary Art and the Genetic Code".

Table 13.1: Diagram types that appear in this volume.

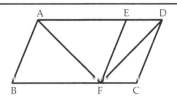

An exhibit: Fig. 7.3.

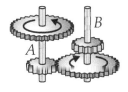

An illustration: Fig. 12.9 (detail).

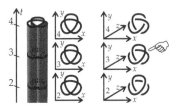

A type 1 aid to visualization:
Fig. 8.10 (detail).

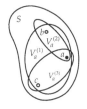

A type 2 aid to visualization:
Fig. 4.1 (detail).

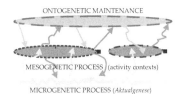

An aid to conceptualization:
Fig. 1.2.

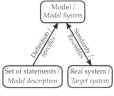

A flowchart: Fig. 1.4 (detail).

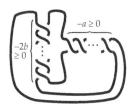

A definition: Fig. 8.9 (detail).

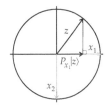

An example: Fig. 3.3.

A calculation: Fig. 8.8 (detail).

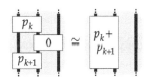

A proof: Fig. 8.6 (detail).

Exhibits

An *exhibit* is a diagram or figure that is in its own right—as an object in the world—a subject of inquiry and discussion, not only (like every other diagram or figure that anyone actually looks at) a subject of inspection or contemplation.

So-called "optical illusions"[5] (like Sander's "Parallelogram Illusion", Fig. 7.3, thumbnailed in Table 13.1) and instances of raw material for visual Gestalt formation (like the sub-figures of Fig. 7.4 on p. 204) are typical exhibits.[6] To help fix ideas, here is an account of how Fig. 7.3 works—not how it works as an "optical illusion" (a question in the psychology of perception), but how it works as an exhibit in a publication (or perhaps during the private process of modeling).

(1) The figure's *structure*, as an object in the world, is that of a depiction of a purely mathematical object—meaning in this case a depiction that maps an abstract construction in Euclidean geometry to a more or less carefully crafted physical representation of an instance of the construction (*e.g.*, abstract points are mapped to small, roundish blobs of ink, abstract line segments are mapped to narrow, visually "straight" smears of ink, etc.).

(2) The figure's *function*, as a subject of inquiry and discussion, is to illustrate that in some circumstances the mapping in (1) can carry a proposition P about an abstract Euclidean construction that is true of the construction onto a proposition $\widetilde{P}$ about the physical representation (no matter how well crafted) that is (a) *objectively true*—meaning in this case that measurements with a trusted physical instrument (*e.g.*, a ruler or a pair of dividers) confirm $\widetilde{P}$ (according to some accepted rules for using the instrument, some accepted meanings of "true" for propositions about necessarily approximate measurements, etc.)—

---

[5]See pp. 8–9, 199–200, and 204 (note 4) for discussions.

[6]To the extent that the thumbnails and details in Table 13.1, or (many of) the diagrams and figures in the referenced books by Huff and Tufte, are understood to be (possibly idealized or edited) *copies* of other diagrams or figures, copied and presented precisely for the purpose of analyzing the functions and structures of their originals in their original contexts, they too are exhibits. By analogy with mathematicians' (and philosophers') employment of "philosophers' quotes" (p. xi) as a signal to the reader that a word or phrase is being *mentioned*, not *used*, one might employ some device—to be called, perhaps, a "mathematicians' frame"— as a signal to the viewer that a diagram or figure is being *exhibited*, not used.

and yet (b) *subjectively false*—meaning in this case that a viewer's perception of the physical figure disconfirms $\widetilde{P}$.

Note that, in (1), I used the phrase "abstract construction in Euclidean geometry". What I take Euclidean geometry to be abstracted *from* (*pace* Kant and other apriorists) is the physical geometry of the natural and constructed worlds; this latter importantly includes "points" that are small roundish blobs, "lines" that are thinnish, straightish smears, and so on. When I assert in (1) that the depiction "maps an abstract construction" to a "physical representation of an instance of that construction", I am disclosing my presupposition that it cannot be possible that *the figure functions as an exhibit* (in whatever manner—here, in the manner described in (2)), unless *the mapping both **respects** (in some manner) and is respected by* the abstraction processes that lie in the cultural/historical evolutionary and individual/cognitive/perceptual background.

This particular kind of highly structured reciprocal relation between the *depiction* and the *depicted* is characteristic of an exhibit, so a suitably modified version of (1) should apply to any exhibit.[7] Again to help fix ideas, here is an account of how the sub-figures of Fig. 7.4 —a "circle consisting of dots", an "oval consisting of dots", and a "circle consisting of squares"—work, as exhibits.

(1′)   Each sub-figure's structure, as an object in the world, is that of a depiction of a purely mathematical object—meaning in this case a depiction that maps an abstract object in the mathematical theory of *finite topological spaces* (namely, a "finite circumplex",[8] as described by Rudolph, 2006b, 2012, and Rudolph, Han, & Charles, 2009) to a more or less carefully crafted physical representation of an instance of that construction: specifically, in each sub-figure, each of 45 *closed points* (belonging to a finite circumplex that contains 90 points altogether) is mapped to one of 45 small printed glyphs ("dots" in the first two sub-figures, "squares" in the third), and each of 45 *open points*— any one of which contains exactly two closed points in its topological *closure* in the finite circumplex—is mapped to the region on the page containing both exactly two "adjacent" printed

---

[7]It wouldn't be difficult, and might be enlightening, to formalize this in the category-theoretic language of morphisms, functors, and inverses (Chap. 9).

[8]Motivation to use finite topological spaces (in particular, finite circumplexes) as models in psychology came from Shepard (1978) and Russell (1980).

glyphs *and the print-free part of the page "between" them.*[9]

(2′) Each sub-figure's function, as a subject of inquiry and discussion, is to illustrate that in some circumstances a substantially accurate physical representation of an abstract mathematical object that *is not* Euclidean, and *cannot* even be "embedded" (in a technical topological sense) in a Euclidean space, will nonetheless *be perceived* as representing substantially the same object (*i.e.*, the same in those respects currently in attention) as a substantially accurate physical representation of an abstract mathematical object that *is* Euclidean (namely, in the three sub-figures, a circle, an oval, and a circle again).

Note that though the phrasing of (1′) is very similar to that of (1), the phrasings of (2′) and (2) aren't much alike. If I were to give an account of the working of some exhibit drawn from a wider class of figures and diagrams than those having somehow to do with mathematics, I could expect to have to generalize (1) somewhat further, but the generalization would remain recognizable *as* a generalization of (1): in short, exhibits (as exhibits) all have similar *structures*. In contrast, even an exhibit drawn from the class of figures and diagrams intended for use only in the exposition (or exploration) of mathematics, can (as an exhibit) perform *functions* like those in (2), those in (2′), or neither.

Note also that it is entirely possible to present a diagram that

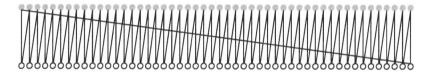

Figure 13.1: A Hasse diagram for the 90-point circumplex: each open point of the circumplex is mapped onto a shaded disk and each closed point onto a white disk; the relation between an open point and a closed point that belongs to its *closure* maps onto to the relation between a shaded disk and a white disk that is directly joined to it by a line segment.

---

[9]*Mutatis mutandis*, the same description applies to versions of the figure on computer monitors, etc. I think it likely—but have not tried to find out—that effectively the "same" description could be applied to a version of the figure haptically impressed upon the skin of a blindfolded or non-sighted collaborator.

depicts an abstract mathematical structure and that is not an exhibit in my sense of the word because the diagram is not itself, as an object in the world, the intended subject of inquiry and discussion. Fig. 13.1 illustrates this situation: it is an example (as defined below), not an exhibit, although it depicts (using a standard mathematical construction known as a *Hasse diagram*) precisely the same 90-point circumplex featured in each sub-figure of Fig. 7.4.

Note, finally, that despite having one and the same abstract mathematical structure underlying each of them, the sub-figures of Fig. 7.4 look quite different. This is nearly inevitable, given that they, as physical objects, occupy part of the physical world (a page of a book or a computer screen) in which much more of abstracted Euclidean plane geometry than the small fragment represented in the abstract theory of finite topological spaces *can* be seen—and therefore *must* be seen, so long at least as its viewers are functioning as "primates, first and foremost", and not *also* as mathematically trained primates somewhat capable of temporarily and partially subduing "the ineluctable modality of the visible".

Unlike abstract finite circumplexes, abstract Euclidean circles and triangles—and *a fortiori* 'circular' and 'triangular' physical objects—have *shapes* that distinguish them from each other. Or, rather, these objects have such shapes to those persons (forming a large majority) whose eyes, optic nerves, visual cortices, and so on are able—and have learned how, after enough 'normal' training time in visual perception for their bodies-in-the-environment—to function 'normally', whether that training has taken place (a) in the usual manner, starting directly after birth (or perhaps somewhat earlier; see Del Giudice, 2011), or (b) in exceptional circumstances, *e.g.*, after surgery performed in adulthood to reverse congenital blindness, as in 66 cases described by Senden (1932/1960).

In his account of Senden's research, Zeeman (1962, p. 244) wrote that "The striking feature common to all [those] patients seems to have been an immediate perception of topological features, but an inability for some months even to detect the difference between a triangle and a circle." Though I would be delighted to be proved wrong, I suspect Zeeman of some hyperbole in implicitly suggesting that there is no level of 'ability to detect' between 'ability to detect straight edges, angles, and similar accoutrements of Euclidean geometry' and 'ability to detect topological features'.

In particular, I find it hard to imagine that any of Senden's subjects would, had they been tested for it, have displayed much (or any) "inability [...] even to detect the difference between" (1) the depiction of a topological circle in the usual, geometrically "round" manner of Fig. 8.1(a) on p. 230, and (2) the depiction of a topological circle in the unusual geometrically complicated manner of Fig. 8.1(b), reproduced in Fig. 13.2(a). However, for my purposes—and, I think, for Zeeman's—"an inability to [...] to detect the difference between a triangle and a circle" is remarkable enough, and all that is needed.

On the other hand, I also find it hard to imagine that anyone—no matter with how much mathematical training—could learn either to *look at* the oddly shaped depiction of a 1,050-point circumplex in Fig. 13.2(b) and *see* just a depiction of a "finite circumplex". The ease with which the physical world can overload proposed exhibits with extraneous structure makes their correct use (by my standards) challenging. Taking due account of the fact that my "structures" are mathematized, I think it fair to equate my "extraneous structure" to what has been called *excess* (or *surplus*) *meaning*" in the context of metaphor by literary critics, cognitive scientists, education theorists, and others.

Illustrations

An *illustration* is a diagram or figure that represents, more or less directly (allowing the use of sub-culturally appropriate conventions, *e.g.*, presence or absence of perspective and shading, possibly the indication of "hidden lines", etc.), an object in the world in a particular state; textual labels, arrows, etc., are optional components of an illustration. Thus, for instance, the detail from Fig. 12.9 reproduced in Table 13.1 is meant to portray two actual, physical double-gears;

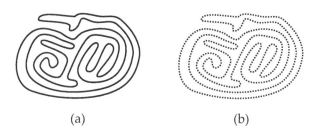

(a)   (b)

Figure 13.2: (a) A "simple closed curve" (Fig. 8.1(b)). (b) The 1,050-point circumplex maps onto a 'dotted-line' depiction of (a).

even if a photograph had been used instead of a drawing, for the full intended effect to be achieved, the curved arrows (or equivalent extra text) would have had to be added. It may be thought that, if a *moving* picture (of some sort) had been used, then neither icons (arrows) nor text would have to be added; whether or not that is true (and, if it is, to what extent) is an empirical question that has probably been thoroughly studied, but I am not aware of the answer.[10] Other illustrations in this volume are the "intransitive family of hungry monkeys" (Fig. 12.8) and the screenshots in Chap. 11.

In mathematical modeling generally, illustrations are more common than exhibits; with the (increasingly frequent) exception of screenshots, they tend—understandably—to appear most often where what is being modeled is (some aspect of) a macroscopically visible physical *thing*. Their function, as I understand it, is to engage a reader's attention by encouraging a personal, *imagined* (or remembered) *physical encounter* with the thing illustrated (or its remembered simulacrum) in the state illustrated; labels, arrows, etc., serve as instructions for imagined (or remembered) interaction with the object. On this account, providing the reader with the actual object could perform more or less the same function (as could, no doubt, an appropriate "virtual reality" show). I have noticed that some people classify figures with such a function under the head of "visualization", but this is inconsistent with my use of that term.[11]

Illustrations can be, and are, used in mathematical modeling of any domain in which physical objects (of the appropriate nature) are being modeled; in such cases they may even be of primary importance in the model. In human sciences, for an illustration to play a primarily *dialectic* role in a mathematical model seems unlikely to me, though certainly one can play a *rhetorical* role; *e.g.*, Figs. 12.8 and 12.9 may help to 'personalize' the abstract notion of

---

[10]My guess is that it is false: I have the impression that I, for one, am often very bad at *seeing* what someone means to *show* me (whether in a moving picture or a live demonstration) without extensive verbal assistance, particularly if—like a schoolchild confronted by Poddiakov's intransitive set of gears (p. 382)—what I am meant to see is strongly counter to what I am predisposed to see.

[11]Recurring to my own experience, I find it common for some "ineluctable modality" *other* than that "of the visible" to be the one with which I feel most saliently engaged in such imagined (or remembered) encounters—*e.g.*, if I imagine myself playing with the double-gears in Fig. 12.9, I am immediately overwhelmed by the ineluctable modality of the haptic. That is why, for me, it feels odd to speak of "visualization" in such cases; and this oddness has influenced my definitions.

intransitivity for a reader of Chap. 12, just as the toy objects they illustrate are designed to do for schoolchildren.

Aids to visualization and conceptualization

To give a proper description of the essential qualities of a diagram or figure that make it what I am calling an *aid*, I find it helpful to use several technical terms, some from mathematical *category theory* (introduced and discussed by Yair Neuman in Chap. 9). To avoid, not needless but possibly overwhelming, formalization, I have not properly set up the (mathematical) categories to which they are meant to apply. I do *not* guarantee that there is some fully satisfactory, fully formalized set-up along the lines I am sketching (though I believe there is), but I do disclaim that I am writing *only* metaphorically (see Rudolph, 2006a, for thoughts and opinions on the virtue of using metaphors as a companion to mathematics in mathematical modeling, and the vice of using them as a substitute for it).

An aid has *structural members* of different types, that are subject to *coherence conditions*. The structural members are:

(i) a *physical ambient* (*e.g.*, a sheet of paper, a blackboard or whiteboard, or the screen of a computer monitor),

(ii) a *physical diagram* 'in' or 'on' that ambient (*e.g.*, ink and ink-free areas on a printed page, chalk and the absence of chalk on a blackboard, or variously illuminated phosphor dots or liquid crystals on a computer monitor);

(iii) a spatialized mathematical structure[12] that is used in the mathematical model under discussion; and

(iv) a *mapping* (or *morphism*) *from* the spatialized mathematical structure that is used in the model *to* the spatialized mathematical structure that is a conventional mathematization of a pre-mathematical spatial structure of the physical ambient of the physical diagram.

The coherence conditions that these members must satisfy if they are to function together as an aid are that:

(v) (some of) the physical elements of the diagram, (some of) the relations among those, and so on, when taken (whether at first or second sight, or only after substantial reflective cogitation)

---

[12]See Chap. 6, pp. 175–178, for a discussion of "mathematical spaces" and the spatialization of mathematical structures in general.

to stand in for mathematical elements, mathematical relations among those, and so on, appertaining to the mathematical structure of the Euclidean plane, can be assembled into a mathematical structure upon which there is defined

(vi) a mapping (or morphism) that is a (one-sided) *inverse* of the morphism in (iv).

There is enough slop in (i)–(vi) as they stand, and there would remain enough slop even if they were fully formalized, to account for the unfortunate fact that a proffered aid can often be misleading, and can occasionally lead to outright *mis*understanding.

I am not sure that I can find any formal or semi-formal justification for dividing aids into the three sub-types illustrated in Table 13.1. Informally, the sub-type seems to derive from the aid's location on a certain scale that feels meaningful to me but which I can't quite pin down with a name—"concreteness" is close, but doesn't entirely capture the meaning. This being the case, I only sketch some possible distinctions among the sub-types.

***Aids to visualization* (Type 1).** This subtype certainly includes cases in which the object of the aid is a mathematical space that is used in the model, and actually occurs *ab initio* in a Euclidean space $\mathbb{E}_n$ of some dimension n. When $n = 1$ or $n = 2$, the usual conventional identification of the Euclidean line or plane (more precisely, of an arbitrarily large, but bounded, part thereof) with the page (etc.) prescribes a way to depict the structure, more or less accurately, on the page (etc.). Of course some detail is inevitably lost (or elided), no matter how simple the structure; and for very complicated structures, it may be literally impossible to capture all the complications (within any prescribed degree of 'accuracy') in any such depiction.[13] When $n = 3$, there are many standard

---

[13] An example of this impossibility is any *actual* "fractal" in any technical sense —say, to be precise, the sense of being "approximately self-similar at each of an infinite sequence of scales diverging to infinity". Of course, in that (or any other generally accepted) technical sense, *nothing* in the physical (or social) world is or possibly could be a fractal: most purported examples (*e.g.*, shorelines) are in fact approximately self-similar at no more than a handful of different scales. The abuse of this term is counterproductive and a prime example of "excess meaning"(p. 404) gone wild—the excess meaning in this case lying in the imputation of infinitely many levels of complexity where only a finite (and usually small) number of levels can even be shown to exist in any reasonable sense, and the counterproductivity lying in the cessation of further attempts at understanding those actual 'levels' when (purportedly) confronted with the specter of infinity.

ways (multiple views, exploded views, etc.) to portray the object of visualization. Fig. 8.10 shows one (equally standard, though less well known outside mathematics) way to portray an object of visualization that originates in 4-dimensional space, called *slicing*.

Aids to visualization (Type 2). This subtype certainly includes cases where the object of the aid is a mathematical space that is used in the model, but does not occur *ab initio* in any Euclidean space and may not even be capable of occurring as a subspace of any Euclidean space.[14] For this type of aid to work properly, the viewer should ideally be led only to consider those features of the (Euclidean) diagram depend on properties of $\mathbb{E}_2$ that it *does* share with the mathematical space to be visualized; in any case, the viewer should *not* be led to (mis)apply properties of $\mathbb{E}_2$ that it does *not* share with the mathematical space to be visualized. Fig. 4.1 works as an aid to visualization of "vicinities" in a V-space because the formal properties of the "vicinities" in the example are captured by the (physical) properties of the oval regions in the diagram; it would not (I think) have worked as an aid to visualization of the "closeness" relation in a V-space, even though the notions of "vicinities" and "closeness" are *logically interchangeable*, because the "closeness" relation derived from the depicted oval regions *is not much like* the (pre-mathematical) relation of "closeness" in the diagram.

Aids to conceptualization. This sub-type certainly includes cases where the object of the aid is not *ab initio* a mathematical space at all, but only acquires a spatial structure through the figure.

Flowcharts

The editors of the Oxford English Dictionary have given two definitions of "flow chart" (flow chart, 2011, definition **C2** under the heading "Compounds" for the main entry "flow, n."), separated by a semi-colon but not otherwise labeled. Here they are, split apart, with my own labels added in square brackets.

---

[14]For instance, a finite circumplex cannot be a subspace of any Euclidean space: every point of a Euclidean space is closed (in the technical topological sense), but half the points in a finite circumplex are not closed. So-called *infinite dimensional spaces* are an entirely different (and very important) class of counterexamples: every subspace of $\mathbb{E}_n$ has dimension at most n (this may or may not be 'obvious', but it is non-trivial to *prove*).

[C2-1] a diagram showing the movement of goods, materials, or personnel in any complex system of activities (as an industrial plant) and the sequence of operations they perform or processes they undergo;

[C2-2] also, a diagram in which conventional symbols show the sequence of actual or possible operations and decisions in a data processing system or computer program [...].

Nowhere in the OED is the variant spelling "flowchart" attested, and the most recent usage example for "flow chart" is from 1968.[15] More precisely, now much more than in 1968 there exist two distinct usage communities of academic writers—call them (1) "flow chart"ists and (2) "flowchart"ists —which when intersected with the community of writers of mathematical English become of nearly *equal* size.

My guess (which I have not checked) is that, presently, the "flow chart"ist and "flowchartist" *usage communities* have come to coincide more or less with the two *functional communities* implicitly defined by [C2-1] and [C2-2], respectively.

(1′) The first of these communities is distinguished by its members' concern with the (literal) "movement of goods, materials, or personnel" and other *physical* entities through *physical* structures (like "an industrial plant"), typically under *human direction and control*; the (relevant) work that such persons do may be called "operations research" (when more 'theoretical') or "logistics" (when more 'applied').

(2′) The second of these communities is distinguished by its members' concern with the (figurative) 'movement' of *abstract* entities such as "data" (and other information, including *information*

---

[15]These facts are not unrelated: the *Mathematical Reviews* database attests only two instances of "flowchart" before 1969 (in 1964 and 1968), compared to 67 of "flow chart" in the same period, whereas without restrictions on the date of publication, the two spellings occur 290 and 287 times, respectively; similar results can be gleaned from the JSTOR corpus, though instead of increasing its share of the usage of the two spellings combined from less than 3% to a bit more than 50% in *Mathematical Reviews*, in JSTOR "flowchart" increased its share only from approximately 4.8% to approximately 28.4%. This distinction between *Mathematical Reviews* and JSTOR is obviously due to the broader range of subject areas represented in the JSTOR corpus which, in particular, contains journals in Management and Economics. (For information on *Mathematical Reviews* and JSTOR, see p. 280, note 1, and p. 54, note 16, respectively.)

about information, e.g., which "operations and decisions" are "actual or possible" and which are not) 'through' more or less *abstract* (though possibly spatialized) structures (like "a data-processing system or computer program"), often only very indirectly under *human direction and control* (this indirection being a direct consequence of the fact that "direction" and "control" are themselves, or can be treated as, just some more "information about information"); the (relevant) work that such persons do may be called "systems analysis" (when 'high level') or "computer programming" (when 'low level').

Whether or not that guess is correct, I will stick with flowchart as the name of a diagram type, both because I have always been a "flowchart"ist and because—certainly in this volume, and (it seems to me) throughout current literature in the human sciences—all or almost all diagrams of this type, whether prepared by "flow chart"ists or "flowchart"ists, are used in the manners and for the purposes sketched in (2′), rather than those sketched in (1′).

So: what is a flowchart? Given my identification of (2) with (2′), it is natural for me to start from [C2-2], changing it in two ways. First, although there are in fact several somewhat different standard collections of "conventional symbols" that have been established for various somewhat different purposes (for further information, see *Flowchart*, n.d.), for my purposes it is sufficient (and necessary, given that my purposes include describing the flowcharts that actually appear in this volume) to consider just *two* "conventional symbols", each of which can (contrary to practice in many standard collections) be instantiated graphically in widely different ways: *containers* and *arrows*. Second, I do not wish to maintain the restriction to "a data-processing system or computer program", I find it satisfactory to cut that phrase down to "system" (which I leave undefined). However, I *do* want to retain the word "sequence" (while allowing its meaning to expand somewhat).

In short, then, a flowchart is a diagram consisting of one or more *containers* connected (including self-connections) by one or more *arrows* indicating *flow* of information (data, control, etc.). A given flowchart is understood to be a representation of the totality of all possible sequences of containers alternating with arrows as allowed by the flowchart; each such sequence is understood to represent a *flow* (temporal or extra-temporal, causal or acausal) of *information* (data, control, etc.) from one container to the next, definitely or indefinitely, all depending on the particular flowchart.

Definitions

The most common kind of definition is a figure that either depicts a unique configuration (in some mathematical space, representable in the Euclidean plane) that is its *definiendum* (*e.g.*, the Borromean rings defined by Fig. 8.1(h)) or a generic configuration (of the same sort, but with parts left to be 'filled in' according to some ancillary scheme) the collection of specializations of which are its *definienda* (*e.g.*, the "general mind-knot" in Fig. 8.9(a), copied in Table 13.1, becomes when appropriately filled in the particular mind-knots in Fig. 8.9(b) and (c)). There are also definitions that depict a *process*, typically in before-and-after style (but occasionally with some intermediate steps as well); examples in this volume are the definitions of Reidemeister moves and crossing changes (Figs. 8.2 and 8.4).

Examples

The most common type of example is a figure that depicts a unique configuration (in some mathematical space, representable in the Euclidean plane) that is purportedly representative of a larger class of similar figures.

Calculations

A calculation is a figure depicting a *particular* mathematical object, in which either implicit (Euclidean) properties or explicit properties (usually instances of a definition, either of a configuration or configuration-type or of a process that can be applied to configurations), lead (when the figure is viewed either in conventional, lineal 'reading order' or as directed by arrows or the like) to a (mathematical) conclusion *about that particular object*.

Proofs

A proof is a figure depicting a *general* mathematical object of some type, in which either implicit (Euclidean) properties or explicit properties (usually instances of a definition, either of a configuration or configuration-type or of a process that can be applied to configurations), lead (when the figure is viewed either in conventional, lineal 'reading order' or as directed by arrows or the like) to a (mathematical) conclusion *about all objects of that type*. So-called *diagram-chasing* in category theory (*e.g.*, Fig. 9.3(b)) is a very pure sort of proof.

Of course, just as every calculation in ordinary arithmetic is a proof (albeit of a banal theorem), so every calculation is a proof.

Psyche and arrows

Among the many types of possible visual components of a diagram, arrows of different sorts may be the most pervasive. They can be used as elements of every type of diagram discussed in this chapter (and of graphs as well); in Table 13.1, only the examples of a type 2 aid to visualization, a definition, and a proof incorporate no arrows, and for each of these types there is an instance elsewhere in this volume in which one or more arrows play a role (*e.g.*, Figs. 4.2, 9.3, and 9.2, respectively). Table 13.2 gives examples of some functions of arrows in diagrams. It is interesting that one of these—arrows used to indicate causal sequence—seems not to have appeared (at least, in paramathematical diagrams) until 1940; certainly Wold (1949) attributes it to Tinbergen (1940) without suggesting any earlier use, nor have I found one.[16]

Arrows also appear in typography (see Table 13.3 for examples from mathematical typography in this volume), where they perform many of the same functions that they do in (non-textual portions of) diagrams, often (though not invariably) subject to one or another standardized relationship matching their several (typographical) forms with their several (textual) functions. Again, it is interesting how recently arrows first came to be used to indicate mapping in mathematics.

> Topologists even used the arrow notation a few years before Eilenberg and Mac Lane [in 1945]. Mac Lane says 'the arrow $f: X \to Y$ rapidly displaced the occasional notation $f(X) \subset Y$ for a function. It expressed well a central interest of topology. Thus a notation (the arrow)

---

[16]Cajori (1928–1929) records no such use *by mathematicians* before 1929. My guess is that the much earlier, not particularly mathematical use by chemists (as in Fig. 12.4) is probably unrelated; but I can't prove it.

Table 13.2: Some functions of arrows in diagrams.

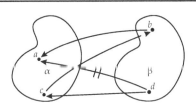

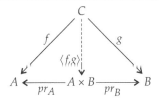

Concrete   mapping   (correspon-   Abstract   mapping   (morphism):
dence): Fig. 4.2.   Fig. 9.3.

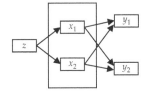

Spatial sequence (geometric vec-   Temporal sequence: Fig. 3.1.
tors): Fig. 3.4.

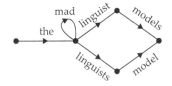

Extra-temporal sequence (genera-   Extra-temporal sequence (transfor-
tion): state diagram for a "finite   mation): Fig. 8.3 (detail).
state grammar" of a "subpart of En-
glish" (after Chomsky, 1957, p. 19).

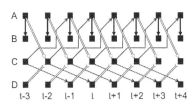

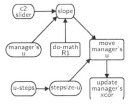

Causal sequence: after Wold's mod-   Flow of information (data, control,
ification (1949, p. 13, Fig. 5) of Tin-   etc.): Fig. 11.5(a).
bergen's "arrow scheme".

Table 13.3: Some standard arrow types in mathematical texts, with their usages and meanings/functions, and exemplary uses in this volume.

ARROW	USAGE	MEANING OR FUNCTION (IN CONTEXT); EXAMPLE(S).
$\rightarrow$	$A \rightarrow Z$	Mapping, morphism, or correspondence (from a 'source object' A to a 'target object' Z); p. 115, p. 261 n.
	$x \rightarrow y$	Limiting behavior (a 'variable' x 'approaches the limit' y); p. 145 n.
$\mapsto$	$a \mapsto f(a)$	Mapping, morphism, or correspondence (from a 'source element' a to its 'image element' $f(a)$); p. 124, p. 285.
$\leftrightarrow$	$A \leftrightarrow Z$	One-to-one correspondence or 'isomorphism' (between 'objects' A and Z); not used in this volume (a non-standard nonce use of this arrow appears in Chap. 8, pp. 244–250).
$\Rightarrow$	$P \Rightarrow Q$	Logical implication (proposition P implies proposition Q; if P, then Q); p. 92, p. 343.
$\Longleftrightarrow$	$P \Longleftrightarrow Q$	Logical equivalence (propositions P and Q imply each other; P if and only if Q); p. 130, p. 174.
$\circlearrowleft$	$A\circlearrowleft$	Self-map or 'endomorphism' (mapping from an object A to itself); p. 261.

led to a concept (category)' (Mac Lane [1971], p. 29). But Eilenberg and Mac Lane were the first to declare arrows were as important as spaces. (McLarty, 1990, p. 335)

Two case studies

Against "diagrams everywhere"

In this volume, Chap. 5 is unique among the contributed chapters in that it has no diagrams. As a thought experiment, I asked myself whether it would have been improved by one.

In particular, I found the Thomsen Condition puzzling. In my attempts to understand it, I first used a common heuristic ("consider your puzzle in a special case, where it may become obvious"); the result appears on p. 141, note 1. I kept trying, and came up with the graphic representation of the condition presented in Fig. 13.3: given six monaural presentations (say, x, y, z on the left and u, v, w on the right), join them into a loop by six arrows, alternating left and right, and using three types of arrows in the order 1, 2, 3, 1, 2, 3. Now the Thomsen Condition is the requirement that, if for two of the three types of arrows, the two binaural presentations defined by that type of arrow sound equally loud, then so does the binaural presentation defined by the third type of arrow.

Has anything been gained by this? More precisely, who (if anyone) would gain (what, and how) if that diagram, with its accompanying interpretation, had been included in Ng's chapter? I can barely convince myself that, by the acts of *conceiving* and *drawing* the figure, *I* gained (a very little) extra insight into the Thomsen condition;[17] I am quite confident that hardly anyone would gain anything worthwhile from the acts of *viewing* or *contemplating* it (and many readers would understandably, and unnecessarily, be put off by it).

Brute facts?

As promised, here is an example of group self-persuasion by a diagram. The group in question consisted of a number of learnèd and influential set-theorists and mathematical logicians.

> What reasons could be given for or against the alternating pattern? [...] [T]hose favoring the alternating picture were without a new assumption, but they were supported by the brute fact that almost any human being will judge WWW to be a "more natural" continuation of V than V‾‾‾‾.

Figure 13.3: A graphic presentation of the Thomsen Condition.

---

[17]Months later, correcting page proofs, I find that any such extra insight has evaporated entirely.

It is not obvious to me that the proposition "that almost any human being will" make the judgment described is true, much less that it is a "brute fact"; there are many propositions about the behavior of "almost any human being" of which the truth or falsehood can only be decided empirically, and I see little reason to believe that this is not one of them. Still, the continuation labeled QPD turned out to be, provably, the correct one.

References

Cajori, F. (1928–1929). *A History of Mathematical Notations, Vols. I & II*. Chicago, IL: Open Court.

Carroll, J. B. (1959). An operational model for language behavior. *Anthropological Linguistics*, 1, 37–54.

Chomsky, N. (1957). *Syntactic Structures*. The Hague, NL: Mouton.

Del Giudice, M. (2011). Alone in the dark? Modeling the conditions for visual experience in human fetuses. *Developmental Psychobiology*, 53(2), 214–219.

Disney, W. (Producer), Banta, M., Berg, B., Haber, H. (Writers), Clark, L., Luske, H., Meador, J., & Ritherman, W. (Directors). (1959). *Donald in Mathmagic Land* [Animated cartoon]. US: Walt Disney Productions.

Emch, A. (1896). Mathematical models. In B. B. Smyth (Ed.), *Transactions of the Twenty-sixth and Twenty-seventh Annual Meetings of*

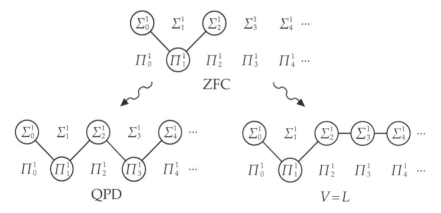

Figure 13.4: Which continuation is "more natural"? (After Maddy, 1988, pp. 739–740 and p. 742).

the Kansas Academy of Science (1893–1894) (pp. 90–93). Topeka, KS: Kansas State Printing Co. Retrieved July 23, 2011, from `http://www.jstor.org/stable/3623917`

flow chart. (2011). *OED Online.* New York, NY: Oxford University Press. Retrieved December 29, 2011, from `http://www.oed.com/view/Entry/71998?#eid4157456`

Flowchart. (n.d.). Wikipedia, The Free Encyclopedia. Retrieved December 29, 2011, from `http://http://en.wikipedia.org/wiki/Flowchart#Symbols`

Gould, S. J. (1996). The shape of life. *Art Journal, 55*(1), 44–46.

Gould, S. J. (1997). Redrafting the tree of life. *Proceedings of the American Philosophical Society, 141*(1), 30–54.

Huff, D. (1954). *How to Lie with Statistics.* New York, NY: W. W. Norton.

Kidwell, P. A. (1996). American mathematics viewed objectively: The case of geometric models. In R. Calinger (Ed.), *Vita Mathematica: Historical Research and Integration With Teaching* (pp. 197–208). New York, NY: Cambridge University Press.

Mac Lane, S. (1971). *Categories for the Working Mathematician.* New York, NY: Springer.

Maddy, P. (1988). Believing the axioms. II. *Journal of Symbolic Logic, 53*(3), 736–764.

Maxwell, J. C. (1857/1882). To the Rev. L. Campbell (on taking priest's orders). In L. Campbell & W. Garnett (Eds.), *The Life of James Clerk Maxwell: With Selections From His Correspondence and Occasional Writings* (pp. 293–297). London: Macmillan. (Letter written December 22, 1857)

McLarty, C. (1990). The uses and abuses of the history of topos theory. *The British Journal for the Philosophy of Science, 41*(3), 351–375.

Peirce, C. S. (1905). Prolegomena to an apology for pragmaticism. *The Monist, 16*, 492–546.

Root-Bernstein, R. (1996). Do we have the structure of DNA right? — Aesthetic assumptions, visual conventions, and unsolved problems. *Art Journal, 55*(1), 47–55.

Rudolph, L. (2006a). Mathematics, models and metaphors. *Culture & Psychology, 12*, 245–265.

Rudolph, L. (2006b). Spaces of ambivalence: Qualitative mathematics in the modeling of complex, fluid phenomena. *Estudios Psicologías, 27*, 67–83.

Rudolph, L. (2012). *The Hole in Emotion Space: Topological Consequences for Circumplex Models of Affect.* (In preparation)

Rudolph, L., Han, L., & Charles, E. P. (2009). *Modeling Emotional Development via Finite Topological Spaces and Stratified Manifolds.* (Draft manuscript, Clark University)

Russell, J. A. (1980). A circumplex model of affect. *Journal of Personality & Social Psychology, 39*, 1161–1178.

Senden, M. von. (1960). *Space and Sight; The Perception of Space and Shape in the Congenitally Blind Before and After Operation* (P. Heath, Trans.). Glencoe, IL: Free Press. (Original work published 1932)

Shepard, R. N. (1978). The circumplex and related topological manifolds in the study of perception. In S. Shye (Ed.), *Theory Construction and Data Analysis in the Behavioral Sciences* (pp. 29–80). San Francisco, CA: Jossey-Bass.

Tinbergen, J. (1940). Econometric business cycle research. *The Review of Economic Studies, 7*, 73–90.

Tufte, E. R. (1983). *The Visual Display of Quantitative Information.* Cheshire, CT: Graphics Press.

Tufte, E. R. (1984). *The Visual Display of Qualitative Information.* Cheshire, CT: Graphics Press.

Wold, H. O. A. (1949). Statistical estimation of economic relationships. *Econometrica, 17* (Supplement: Report of the Washington Meeting, July, 1949), 1–22.

Zeeman, E. C. (1962). The topology of the brain and visual perception. In M. K. Fort (Ed.), *Topology of 3-Manifolds and Related Topics* (pp. 240–256). Englewood Cliffs, NJ: Prentice Hall.

Contributor Biographies and Contact Information

Ralph Abraham (Chap. 11) has been a mathematician and chaos theorist since 1958, and Professor of Mathematics at the University of California at Santa Cruz since 1968. He founded the Visual Math Institute in Santa Cruz, and is the author of mathematics textbooks including *Foundations of Mechanics* (with Jerrold Marsden) and *Dynamics, the Geometry of Behavior* (with Christopher Shaw) and philosophical books including *Chaos, Gaia, Eros* and *Chaos, Cosmos, and Creativity* (with Rupert Sheldrake and Terence McKenna).

abraham@vismath.org
http://www.vismath.org

Jerome R. Busemeyer (Chap. 3) is Professor of Psychological and Brain Science, and Professor of Cognitive Science, at Indiana University. His research interests include dynamic, emotional, cognitive, and neural models of decision making, quantum psychodynamics, neural network models of function learning, and methodology for comparing complex dynamic models of behavior. His most recent book, *Cognitive Modeling* (with Adele Diederich) was published by Sage in 2010. jbusemey@indiana.edu
http://mypage.iu.edu/~jbusemey/home.html

Rainer Diriwächter (Chap. 7) is Associate Professor of Psychology at California Lutheran University. He is co-editor of the books *Striving for the Whole: Creating Theoretical Syntheses* (with Jaan Valsiner) and *Innovating Genesis: Microgenesis and the Constructive Mind in Action* (with Emily Abbey), and author of numerous articles on holistic psychology, cultural psychology, and the history of psychology. His current interests emphasize emotional experiences and psychological synthesis. rdiriwae@callutheran.edu

Damir D. Dzhafarov (Chap. 4) is a National Science Foundation Post-Doctoral Fellow in the Logic Group at the University of Notre Dame. His research interests are in mathematical logic, specifically in computability theory and reverse mathematics; among the journals in which he has published papers are *Order*, *Notre Dame Journal of Formal Logic*, *Theoria*, and *The Journal of Symbolic Logic*.

damir@math.uchicago.edu
http://www.math.uchicago.edu/~damir

Ehtibar N. Dzhafarov (Chap. 4) is Professor of Psychological Sciences at Purdue University. His current interests include subjective topology and geometry of stimulus spaces derived from discrimination and categorization functions, general theory of subjective dissimilarity and subjective distance, and theory of numeric and non-numeric mathematical representations of behavioral relations. His most recent book, *Measurement and Representation of Sensations* (with Hans Colonius), was published by Psychology Press in 2006.

ehtibar@purdue.edu
http://www1.psych.purdue.edu/~ehtibar

Dan Friedman (Chap. 11) is Distinguished Professor at the University of California, Santa Cruz. He has broad research interests in applied economic theory, with emphasis on learning and evolution, laboratory experiments, and financial markets. His most recent book, *Morals and Markets: An Evolutionary Perspective on the Modern World*, was published by Palgrave–MacMillan in 2008.

dan@ucsc.edu
http://leeps.ucsc.edu/people/Friedman

Akio Kawauchi (Chap. 8) is Professor of Mathematics at Osaka City University. His interests include knot theory and low-dimensional combinatorial topology. An active promoter of scientific applications of knot theory, he convened the 2006 International Workshop on Knot Theory for Scientific Objects, held in Osaka. He is the author of many papers and the book *A Survey of Knot Theory*, published by Birkhäuser in 1996. kawauchi@sci.osaka-cu.ac.jp
http://www.sci.osaka-cu.ac.jp/~kawauchi

Yair Neuman (Chap. 9) is Associate Professor at the Department of Education, Ben-Gurion University of the Negev. His expertise is in

interdisciplinary research where he draws on diverse disciplines to address problems from an unusual perspective. He has published extensively in disciplines including Psychoanalysis, Theoretical Biology, Mathematical Modeling, Semiotics, and Information Sciences. His most recent book, *Reviving the Living: Meaning Making in Living Systems*, was published by Elsevier in 2008.   `yneuman@bgu.ac.il`
`http://bgu.academia.edu/YairNeuman`

Che Tat Ng (Chap. 5) is Professor of Mathematics at the University of Waterloo, Canada. He has published many papers on topics including general topology, inequalities arising in geometry, and information theory. In the field of functional equations and their applications, he has contributed to the characterization of information measures, and has recently become more involved with numerical representations of behavioral relations.   `ctng@uwaterloo.ca`
`http://www.math.uwaterloo.ca/PM_Dept/Homepages/ng.shtml`

Alexander Poddiakov (Chap. 12) is Professor of General and Experimental Psychology at the Higher School of Economics, Moscow. His interests include economic psychology, development of exploratory behavior, creativity and strategies of social interactions, help and counteraction in human development, "Trojan horse" teaching, and psychology of complex problem solving. His most recent book, *Competition and Conflicts of Participants in the Educational Process* was published in 2010 by Fiero, Moscow.   `alpod@gol.ru`
`http://epee.hse.ru/Poddiakov`

Lee Rudolph (Chaps. 1, 2, 6, 10, and 13) is Professor of Mathematics at Clark University, an affiliate of the Social, Evolutionary, and Cultural Psychology program at Clark, and a participant in the Kitchen Seminar hosted by the SEC program. He is a low-dimensional topologist who has recently been working on applications of topology to robotics as well as to social and cultural psychology. His most recent book is *A Woman and a Man, Ice-Fishing* (Texas Review Press, 2006); its title poem first appeared in the *New Yorker*.`lrudolph@clarku.edu`
`http://black.clarku.edu/~lrudolph`

Jaan Valsiner (Chaps. 1 and 12) is Professor of Psychology at Clark University. He is an author or editor of over 30 books and more than 300 book chapters and journal articles, among them *Culture and the Development of Children's Action* (Wiley, 1987; second edition, 1997), *The Guided Mind* (Harvard University Press, 1997), *Culture*

In Minds and Societies (Sage, 2007), and *A Guided Science: History of Psychology in the Mirror of its Making* (Transaction Publishers, in press). He is the original convener of the Kitchen Seminar.

jvalsiner@clarku.edu
http://kitchenseminar.org/About.html

Paul Viotti (Chap. 11) is Assistant Professor of Political Science at California State University, Chico. He is currently investigating Americans' attitudes toward economic inequality. He is also a visual artist, and author (with Scott Draves, Ralph Abraham, Frederick David Abraham, and Julian Clinton Sprott) of "The aesthetics and fractal dimension of electric sheep".   pviotti@csuchico.edu
http://www.viotti.com

Index of Cited Authors

Index of Notations and Concepts

432

454